7th International Symposium on the Science and Technology of Light Sources

Kyoto, Japan
27-31 August 1995

ISBN: 978-1-5108-4111-6

Printed from e-media with permission by:

Curran Associates, Inc.
57 Morehouse Lane
Red Hook, NY 12571

Some format issues inherent in the e-media version may also appear in this print version.

Printed by Curran Associates, Inc. (2017)

For permission requests, please contact Foundation for the Advancement of the Science & Technology of Light Sources (FAST-LS) at the address below.

Foundation for the Advancement of the Science & Technology of Light Sources
FAST-LS
Belmayne House
99 Clarkehouse Road
Sheffield, United Kingdom
S10 2LN

www.fast-ls.org

Additional copies of this publication are available from:

Curran Associates, Inc.
57 Morehouse Lane
Red Hook, NY 12571 USA
Phone: 845-758-0400
Fax: 845-758-2633
Email: curran@proceedings.com
Web: www.proceedings.com

Symposium Proceedings

The Illuminating Engineering Institute of Japan

The 7th International Symposium on the Science & Technology of

Light Sources

Kyoto–Japan
27th–31st August 1995

Proceedings of the Seventh International Symposium on the Science & Technology of Light Sources (LS:7)

Kyoto – Japan

27 – 31 August 1995

Invited Papers
and
abstracts of Landmark
and Poster Papers

Edited by
R. Itatani
Niihama National College
of Technology
and
S. Kamiya
Matsushita Electronics Corp.

Published by
The Illuminating Engineering
Institute of Japan

Published by: The Illuminating Engineering Institute of Japan
3rd Floor, Suitaya Biru
2-8-4 Kanda Tsukasacho, Chiyodaku
Tokyo, 101 Japan

Distributed by: LS:7 Organizing Committee
c/o Dr. T. Urayama
Dept. of Electrical & Electronics Engineering,
Ritsumeikan University
1916 Nojicho, Kusatsushi,
Siga Pref. 525-77 Japan

Catalogued Title: Proceedings of the Seventh International
Symposium on the Science and Technology of
Light Sources

PREFACE

This volume contains both the full texts of the invited papers and the abstracts of the accepted contributed papers that are to be presented at the Seventh International Symposium on the Science and Technology of Light Sources (LS:7) being held in Kyoto, Japan, from the 27th to the 31st August, 1995.

The LS symposia are the only international symposia which focus on the science and technology of light sources. Previous symposia were held in Loughborough (1975; 90 delegates), Enschede (1979; 126), Toulouse (1983; 150), Karlsruhe (1986; 198), York (1989; 250) and most recently in Budapest (1992; 150). Kyoto symposium is the first in the series of symposia to be held outside Europe. 105 contributed papers will be presented and number of the participants is expected to be over 200.

The format of the Kyoto symposium has followed the tradition of the series of the symposia. Namely seven main sessions contained in the scientific programme of LS:7 comprise a long poster session bracketed by two Invited papers and two Landmark papers presented to the full Symposium. The Landmark papers included in the programme were selected from the Contributed paper by the Organizing Committee. Selection has been made regarding the novelty, timelines, quality and significance to the general programme of the Symposium. In addition special consideration was given to the papers of younger scientists.

Each paper in the volume has been assigned a number according to its position in the running order as planed in the programme and further letter designation according to which of the three possible categories it belongs, i.e. I (Invited), L (Landmark) or P (Poster).

Ryohei Itatani
Shigeru Kamiya

July 1995

Editors

ORGANIZING COMMITTEE

LS:7 Organizing Committee

Prof.Dr.R.Itatani(Chairman)	Niihama National College of Technology, Japan
Dr.S.Kamiya(Secretary)	Ex-Matsushita Electronics Corp., Japan
Dr.T.Urayama(Treasurer)	Ritsumeikan University, Japan
Prof.Dr.L.Bartha	Academy of Science, Hungary
Dr.R.S.Bergman	GE Lighting, USA
Dr.B.Ditchek	Osram-Sylvania Inc., USA
Dr.R.Devonshire	University of Sheffield, UK
Dr.A.G.Jack	Philips Lighting B.V., The Netherlands

LS:7 Local Committee

Prof.Dr.R.Itatani(Chairman)	Niihama National College of Technology
Dr.S.Kamiya	Ex-Matsushita Electronics Corp.
Dr.T.Urayama	Ritsumeikan University
Dr.M.Nanjo	Electrotechnical Laboratory
Mr.M.Ohmori	Illuminating Eng. Institute of Japan
Mr.Y.Ohtagaki	Iwasaki Electric Corp.
Mr.Y.Yuge	Toshiba Lighting & Technology Corp.
Dr.K.Watanabe	Mitsubishi Electric Corp.
Dr.Y.Watanabe	Kanagawa University

LS:7 Corresponding Members

Prof.Z.Cai	Fudan University, China
Prof.D.Karabourniotis	University of Crete, Greece
Dr.J.Maya	Matsushita Electric Works, USA
Prof.H.P.Popp	University of Karlsuruhe, Germany
Prof.P.Ramakrishnan	Indian Institute of Technology, India
Dr.E.Schnedler	Philips Laboratory, Germany
Dr.D.O.Wharmby	GE Lighting, UK
Dr.J.F.Waymouth	Consultant, USA
Dr.G.Zissis	Paul Sabatier University, France

LS:7 Financial Supporters

The LS:7 Organizing Committee gratefully acknowledge the following organizations and business firms for their financial support to organize and execute the Symposium.

Illuminating Engineering Institute of Japan
Lamp Manufacturers Association of Japan

Asahi National Lighting Co., Ltd.
Azuma Kogyo Co., Ltd.
CKD Corp.
Elevam Corp.
Fuso Electric Industrial Co., Ltd.
Harison Electric Co., Ltd.
Hitachi Ltd.
Ichiko Industries Ltd.
Ikeda Electric Co., Ltd.
Iwasaki Electric Co., Ltd.
Japan Storage Battery Co., Ltd.
Kansai Electric Power Co., Ltd.
Kasei Optonix Ltd.
Koito Manufacturing Co., Ltd.
Kondo Philips Lighting Co., Ltd.
Matsushita Electric Works Ltd.
Matsushita Electronics Corp.
NEC Home Electronics Ltd.
NGK Insulators Ltd.
Nichia Chemical Industries Ltd.
Nippo Electric Corp.
Nippon Electric Glass Co., Ltd.
Osram-Merco Ltd.
Sakata Manufacturing Co., Ltd.
Sankyo Denki Co., Ltd.
Soei Tsusho Co., Ltd.
Stanley Electric Co., Ltd.
Tokyo Kagaku Kenkyusho Co., Ltd.
Toshiba Glass Co., Ltd.
Toshiba Lighting and Technology Corp.
Ushio Inc.

LS:7 Exhibitors

The LS:7 Organizing Committee acknowledge the following companies for preparing the interesting exhibition.

JASCO Corp.
Kasei Optonix Ltd.
Minolta Camera Sales Co., Ltd
NGK Insulators Ltd.
NIchia Chemical Industries Ltd.
Nippon Electric Glass Co., Ltd.
Ohtsuka Electronics Corp.
Opto Research Corp.
Saes Getters Japan Co., Ltd.
Shimadzu Corp.
Soei Tsusho Co., Ltd.
Topcon Corp.
Toshiba Ceramics Co., Ltd.
Toshiba Glass Co., Ltd.

Scientific Programme

August 28, Morning
9:00 – 12:30

Session 1
(Fluorescent lamps)
Chairman : R. Itatani

August 28, Afternoon
13:30 - 17:00

Session 2
(Electronics and Related)
Chairman : John F. Waymouth

August 29, Morning
9:00 - 12:30

Session 3
(HID Lamps)
Chairman : Brian M. Ditchek

August 29, Afternoon
13:30 - 17:00

Session 4
(Excimer, Dielectric Barrier Discharge and Xe Discharge)
Chairman : A.G.Jack

August 30, Morning
9:00 - 12:30

Session 5
(Modelling)
Chairman : R. Devonshire

August 30, Afternoon
13:30 - 17:00

Session 6
(Diagnostics, Measurements and Miscellaneous)
Chairman : Rolf S. Bergman

August 31, Morning
9:00 – 12:00

Session 7
(Materials and Environment)
Chairman : S. Kamiya

Scientific Programme

Texts of Invited, Landmark and Poster Papers

LIGHTING INDUSTRIES IN JAPAN

T. HANADA

TOSHIBA LIGHTING & TECHNOLOGY CORPRATION

2-2-13 MINAMISHINAGAWA SHINAGAWA-KU TOKYO 140 JAPAN

1. Introduction

Since the first incandescent lamp was introduced by T.Edison in 1879, the efficacy of light sources have continuously improved through many kinds of light emitting systems.

The energy problems which came out in 1973, however, accelarated the speed for improving the luminous efficacy to save electric energy. In Japan, the production of light sources has shown steady increase by 1991. After that, however, it shows slight decrease depending upon the economic recession. Although total quantity seems saturating, higher efficacy new lamps are rapidly increasing in quantity, and replacing the conventional lamps.

2. Lamp market in Japan

In the past 10 years the total production of lamps has increased by 3-8% per annum, however, the growth of production of lamps has declined for the last few years as Fig.1 shows. On the other hand, the value of sales of lamps amounted to US$4,770 million in 1991. The increase of total sales amount showed 9.7% per annum in the past 5 years. (1)

2-1 Incandescent lamps

The production of incandescent lamps for general lighting service is stabilizing. The production of tungsten halogen lamps grew remarkably. (The production in 1990 increased by about 10 times that of 1980.) However, it decreased gradually in these years affected by the economic recession, and the volume in 1994 has shrinked to the level in 1987.

2-2 Fluorescent lamps

The fluorescent lamps are predominating light sources for general lighting in Japan. The tube diameter has modified from T12 to T9 to improve efficacy. After T8 lamp was introduced as the standard type lamp in Europe. In 1991, the lamp was commercialized in Japan too, combined with high-frequency operation ballast(control gear). We are expecting great increase of T8 lamp market meeting the the energy saving policy.

2-3 Compact fluorescent lamps

The production of compact fluorescent lamps both of with ballast and without ballast is only 5% of the total fluorescent lamps. Many types of compact fluorescent lamps have been commercialized and the production has increased by 30% in the last 3 years.

2-4 High intensity discharge lamps (HID lamps)

The production of HID lamps has increased considerably up to 1991. However, it has decreased slightly in the last 3 years. Low wattage HID lamps are now widely used for indoor lighting such as offices or shops.

2-5 Electronic ballasts(Electronic control gears)

Electronic ballasts for general lighting service were also actively studied for energy saving since the first oil crisis in 1970's. In Japan, practical electronic ballast was introduced in 1978. The ballast operates usual fluorescent lamps supplied by commercial A.C.power.

Circuit topology for both of inverter and rectifying circuit of power supply have been actively investigated and improved for high efficient lightings. In 1991, high frequency electronic ballast system was developed for fluorescent lighting combined with T8 lamps. The new system is effective for higher energy saving and is growing very quickly.

Elecronic ballasts for HID lamps were also developed. Now the inverter technology would be indispensable for energy saving.

3.Trends of lamp development

In this section, I will focus on the recent technological trends of lamp industries in Japan. Energy crisis and the requirement of the better visual environment have been the driving forces to develop compact lamps with higher efficacy, better color rendition during the last decades. Progress of new materials and process technology have also contributed to the development of new lamps. Recently, another important social issue for the R&D of lamps is the requirement of environmental conservation. Lamp industries are currently trying to eliminate harmful materials and to reduce the exhaustion of such materials throughout the lamp life from its manufacturing to its disposal. For examples, the reduction of mercury quantity in discharge lamps, the replacement with energy efficient and long life lamps to reduce the emission of carbon dioxides into the atmosphere.

To exploit a new market is another important subject to activate the lighting industries. Recent example is the back lighting of the liquid crystal displays. In the followings, I will describe the technological trends of each lamp and consider the several issues of the future.

3-1 Tungsten halogen lamps

The important issue of this lamp is to improve the luminous efficacy. The appearance of tungsten halogen lamps with infrared reflective films, which is one of the solutions to this issue, has opened the way to the new generation of incandescent lamps. The use of these lamps are diverging today from general lighting to automobile head lamps, photo copiers, etc. At present, it is remarkable that the efficacy of this lamp has been improved by 20-30% compared with the conventional tungsten halogen lamps. We can expect such lamps will expand more searching new application fields.

The combination of this lamp with dichroic mirror can greatly reduce the infrared radiation to an object as much as 90%. Therefore, this lamp can prevent the object from the heat damage even under the higher illumination levels. Since multilayer films of TiO_2-SiO_2 absorb UV light also, this lamp is effective to prevent fading. Thus, these lamps are especially suitable for the lighting of shops or fine art museums. The development for both of heat resistant, high quality infrared reflective films and coating process technology will be the key subjects to achieve the further improvement of the efficacy. This type of tungsten halogen lamp was developed for the tubular bulb at an early stage. Today, progress of the coating technology has made it possible to form multi-layer thin films over the more complex bulb geometry such as spherical or oval shapes, resulting in more luminous efficacy.

Tungsten halogen lamps with dichroic mirror with front glass cover have been introduced into the market to prevent from the breakage of the bulb. (2)(3)

3-2 Fluorescent lamps

The inventions of three band rare earth phosphors and the progress of electronic ballasts have largely changed the lighting environment of fluorescent lamps in quality, application area and maintenance.

The steady research of the materials and manufacturing processes of the three band phosphors has improved the color rendering up to 88 keeping high luminous efficacy. Today, these lamps are widely accepted in home lighting as well as shop and office lighting because of their excellent properties. High color temperature of 5,000K and 6,700K is more preferred to low color temperature of 3,000K and 3,500K in Japan. The advantages of high frequency operation : high efficacy, dimming at any level, quick start, reduction of flicker, reduced weight, etc., have been recognized by the customers. The electronic ballasts are rapidly replacing the conventional choke coil ballasts these days.

The specially designed fluorescent lamp systems for high frequency operation have been launched into market. The system characteristics are shown in Table 1. The lamp of the system uses T8 bulb and the luminous efficacy is 100 lm/W. These lamps meet the IEC standard for high frequency operated fluorescent lamps. They have binominal wattages, and customers can choose the light depending on their needs. Operating two of these lamps in high power mode gives the equivalent luminous output of three conventional T10 rapid-start cool white fluorescent lamps operated with choke coil ballast, and the power consumption is 23% less. (4)

3-3 Compact fluorescent lamps

Compact fluorescent lamps both of with and without ballasts are widely marketed and are growing rapidly. Their operating circuits are quickly changing from choke coil ballast to electronic ballast. As some of their lamp sizes are close to those of the incandescent lamps, they achieved more capability to fit the existing fixtures of incandescent lamps. Many kinds of the compact fluorescent lamps are introduced into market to meet the customer's requirements. Today, we easily find the compact fluorescent lamps applications in many places where incandescent lamps were used before. Since compact fluorescent lamps are widely used in small fixtures, optimum luminous characteristics should be kept even in higher ambient temperature. The use of amalgam gives the wider tolerance range of temperature.

Electrodeless fluorescent lamps have been introduced into market 5 years ago. The radio frequency power of 13.6MHz is inductively coupled to the discharge plasma that generates UV. This UV is converted to visible light via phosphors, which is the same mechanism as the usual fluorescent lamps. The advantage of the electrodeless lamp is the extremely long lifetime as long as 40,000-60,000 hours. At present, these lamps are used only to the specific places due to the high cost and limited shape. Improvement of compactness and its cost would be required to penetrate into the general lighting market. (5)

Harmonic distortion is one of the important issues to reduce the electro-magnetic interference(EMI). Several techniques have been developed, and are in practical use now.

3-4 Small fluorescent lamps

The liquid crystal displays(LCDs) are rapidly increasing in use all over the world because of their thinness, less energy consumption and reduced weight. The LCDs need back lighting, and the fluorescent lamps have many suitable features, such as better color rendition, efficacy, and life performance for the purpose.

Today, the tube diameter has been reduced to 2.6mm, and the surface brightness of 32,000cd/m^2 is obtained at 200mm lamp length with 2.25W. This lamp is applicable to 9"LCD. Further size reduction is attempted to improve the efficacy of the back lighting system. The cold cathode, although its efficacy is low, is generally used because of the reliability throughout the lifetime. It is strongly required to reduce the electrode loss for the back lighting..

In some application area of fluorescent lamps, temperature free is required, thus low pressure Xe discharge lamps have been developed. The lamps are used for such as the indicator of the automobile or office automation equipments. The most desirable points are the development of higher efficiency phosphors which convert the vacuum ultra-violet to visible light and the fundamental analysis of this discharge transferring the electric power into UV radiation. (6)

3-5 HID lamps

The trends of HID lamps would be smaller size with higher color rendition. Especially, compact metal halide lamps with high color rendition are now expanding the application into the area such as indoor lighting, stadium lighting, and the optical instruments like over head projectors and liquid crystal projectors, etc.

Currently, HID lamps with unique technology have been developed and marketed. (a) The HID lamp with built-in starter of non-linear ceramic capacitor gives the advantage of quick start. (b) HID lamps with twin tubes have been introduced into market. One of them comprises a mercury vapor arc tube and a HPS one, and allow the users to select the color temperature they prefer. Another one has two HPS arc tubes, and the each arc tube is operated alternatively every time the lamp is turned on. (c) The HID lamps of which the outer bulb is coated with fluoroplastics (common trade name : Teflon) have been introduced into market to prevent from the tube explosion.

Inductively coupled metal halide lamps as a means of electrodeless HID lamps have been studied. Fig.2 shows the structure of this lamp system. The advantage of this light source is the higher possibility of selecting the filling materials that can lead to both the better color rendition and higher luminous efficacy. This lamp will need further research to emerge in the practical use. As another example of electrodeless HID lamp, microwave operated metal halide lamps have also been developed. These light sources are used as strong UV generators. (7)

3-6 Other applications

The high UV output low pressure mercury discharge lamps have been developed to treat the water or to react materials optically. These lamps are operated with square wave power to prevent mercury depletion that greatly reduces the UV output.

For the purpose of the photo-chemical application, 50kW HPS

lamps have been developed, which are used to synthesize the materials of artificial fibers.

Exima lamps have been studied as narrow band UV sources.

Large scale color display systems have been developed and are creating new markets these days. As for the picture cells, fluorescent lamps, incandescent lamps or CRTs are used. Among them, fluorescent lamps give the advantage of higher efficiency, higher brightness, and longer life. Several tens thousands of multi-channel fluorescent color tubes are arranged to form a large display, which can be clearly seen from several ten meters to a hundred meter away.

High brightness blue LED has been invented in 1994. Its brightness is about one hundred times higher than the conventional ones. The three color LEDs of red, green and blue would be used as the new light sources applicable to traffic signals or elements of some large scale displays. (8)

4. Summary

Today, customers require the electric light sources with high quality, maintenance free, and environmental conservation. The new light sources will open the new market. Research on fundamental phenomena, materials, process and computer simulation will accelerate the development of new light sources. I have a dream that new mechanism of light generation based on the interaction between photon and matter would give rise to the practical light sources for the future lighting industries. The good cooperation of scientific research in universities and technological development in industries will open the new era of lighting for the next century.

References

1.Machinery Statistics,Ministry of International Trade & Industry"

2.Yuge,Y.,LS5,1989,#66I

3.Bergman,R.S.,et al,LS6,1992,#43I

4.Kakitani,T.,et al,IESNA,1993,#48P

5.Sinomiya,M.,et al,J.of IES,20-1 1991 44

6.Sakurai,T.,et al,LS6,1992,#26P

7.Stewart,C.N.,et al,LS6,1992,#78L

8.Nakamura,S.,et al,Appl.Phys.Lett.,64-13 1994 1687

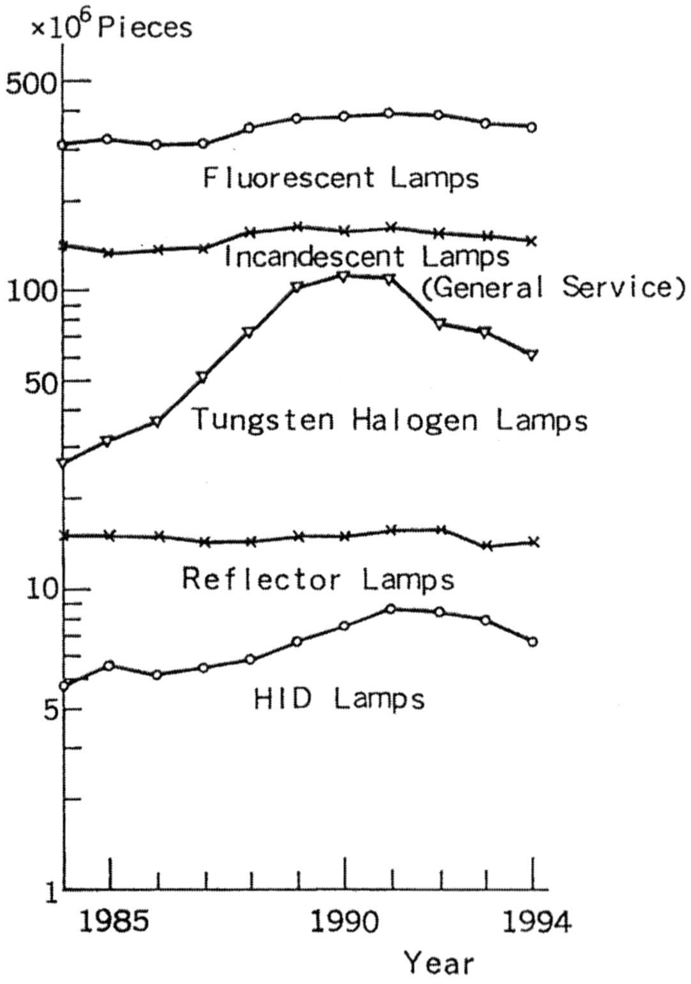

Fig.1 Production of lamps in Japan (Domestic)

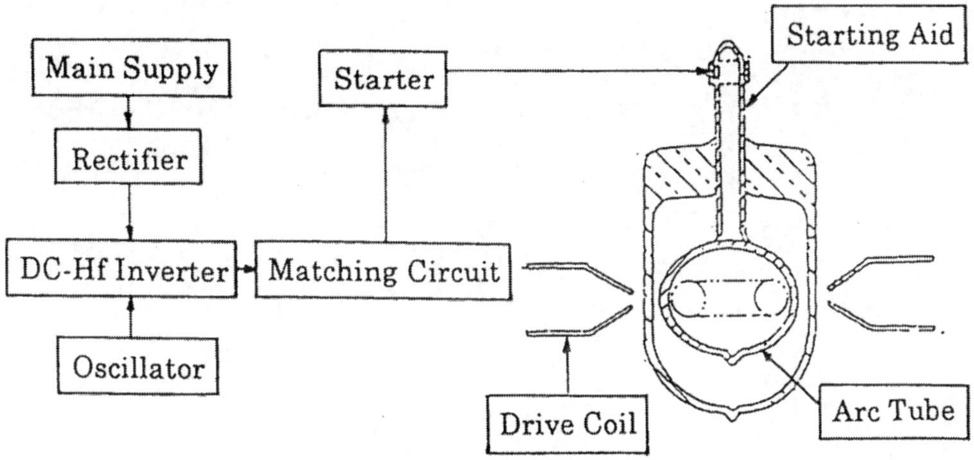

Fig.2 Inductively coupled HID lamp and circuit

Table 1 Characteristics of fluorescent lamps for high frequency operation

lamp type	dia. (mm)	length (mm)	base	wattage (W)	current (A)	lumen (lm)	color temp.* (K)	Ra	life (hrs)
FHF16	25.5	590	G13	16	0.255	1,400	5,000	88	8,500
				23	0.425	2,000			
FHF32	25.5	1,200	G13	32	0.255	3,200	5,000	88	12,000
				45	0.425	4,500			
FHF50	25.5	1,500	G13	50	0.355	5,200	5,000	88	12,000
				60	0.550	6,400			

* Color temperatures of 3000,3500,4200,5000,6700 K are available. Ra=88

Low power compact electrodeless lamps

D O Wharmby
GE Lighting, Melton Road, Leicester LE4 7PD, UK
S-A El-Hamamsy
GE Corporate Research and Development, Schenectady, NY 12301, USA

In 1994 GE Lighting introduced 'Genura', the first compact electrodeless lamp. This is a self-contained lamp with integral circuit, for operation on 230/240V mains supplies. The lamp is intended as a high efficacy replacement for the incandescent reflector lamp in the many applications for which a broad light distribution is either desirable or acceptable. It incorporates a low pressure mercury rare-gas discharge which generates UV, and this is converted to visible light by a phosphor. The lamp operates by magnetic induction, and is sometimes known as a solenoidal electric field lamp or H-discharge. H-discharges have been extensively investigated in the scientific literature: references are contained in a recent review (1).

1. Performance

Figure 1 shows a diagram of the Genura lamp indicating the drive circuit, coil and discharge vessel. The similarity between the outlines of the 100W R80 incandescent reflector lamp and that of Genura means that it will fit satisfactorily in the majority of fixtures intended for incandescent reflector lamps, and in many other fixtures as well. Genura has a relatively wide beam, making it suitable for a variety of downlight applications: it is not intended for spot applications where tight beam control is required. The lamp can be operated in any orientation.

The Genura lamp operates at a nominal 23W and gives a light output of 1100 lumens at 100 hours. The efficacy at 100 hours is 48 lm/W. Life is 10,000 hours. The comparable European incandescent R80 reflector lamp in contrast give about 1000 lumens at 100W with a life of 1000 hours. The colour temperature is 3000K and the colour rendering index R_a is 82.

Typical applications are for hotels and offices and public places where cost of light and high labour cost for replacing lamps, coupled with pleasing aesthetic appearance are important to users. Examples of applications will be shown. At present, Genura is most suitable for applications in which the lamp is switched on for long periods - typical of many commercial applications: time to reach 80% of full light output is 8 minutes.

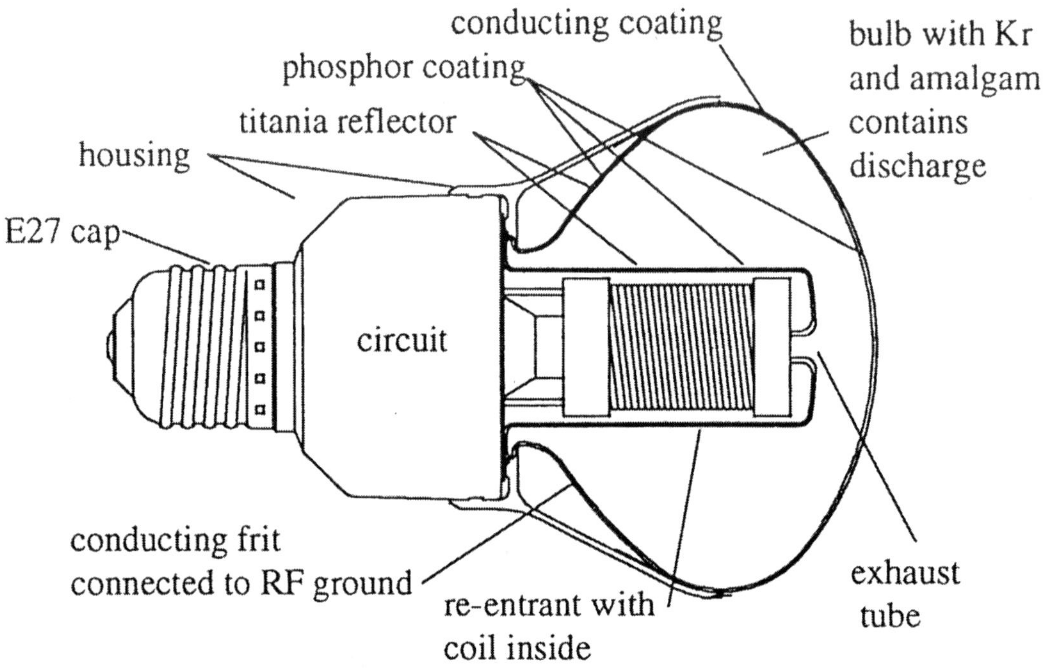

Figure 1 - Showing the bulb, coatings and exciter coil for the Genura reflector lamp.

2. Construction

The Genura lamp comprises a soda-lime glass bulb shown in figure 1. This is filled with krypton and contains an amalgam. A re-entrant runs up the centre of the bulb to contain the lamp coil. The exhaust tube runs through the centre of the coil and has dimples in it to confine the amalgam to a well-defined point in the ballast, thus ensuring that the mercury vapour pressure remains near optimum for a lamps run in free air and in a wide range of fixtures.

The bulb has several coatings. Before sealing to the re-entrant, the bulb is coated internally with a transparent electrically conductive coating, needed to reduce electromagnetic interference (EMI) to the required levels [2] (see section 8). Next a coat of Alon provides a glass protection coat on both bulb and re-entrant. The re-entrant and the neck region of the bulb are then coated with a titania reflector. Finally the glassware receives a coat of triphosphor.

After the bulb is sealed and pumped, it is coated externally in the reflector region with a silver frit, with gaps to prevent induced current flow. When the lamp is assembled the frit is inside the housing and is connected to the RF ground to provide a capacitative connection to the internal conducting film (see section 8).

The exciter coil comprises about 10 turns of insulated wire on a ferrite core and having an inductance of about 10μH. Ferrite permeability is relatively unimportant, but the ferrite must have low loss at the temperature and magnetic flux density levels reached in lamp operation.

3. Operation

Initial breakdown is caused by the longitudinal component of the electric field E_z from the coil. Part of this field extends into the discharge space, the breakdown current flowing through parasitic capacitance. The electric field which drives the H-discharge is the azimuthal electric component E_ϕ provided by the changing magnetic flux in the coil. At any time during the transition from the breakdown phase to the stable H-discharge, E_ϕ must be greater than the maintenance field for the H-discharge at every value of the electron density which prevails during the transition. An analogous requirement applies to electroded discharges. The H-discharge forms as single turn secondary around the coil. The time taken is usually a few milliseconds: thus the user is unaware of any delay in the formation of the H-discharge after switch-on.

The simple model for the stable H-discharge is that of a transformer with a single turn secondary of inductance L_a (the inductance of the arc or plasma) and resistance R_a (the arc or plasma resistance). The primary is the excitation coil with unloaded inductance L_u and unloaded resistance R_u. The standard expression for the loaded impedance of the exciter coil (the impedance when the plasma is turned on) is

$$R_l + jX_l = R_u + k^2\omega L_u Q_a/(1+Q_a^2) + j(\omega L_u - k^2\omega L_u Q_a^2/(1+Q_a^2)) \qquad (1)$$

$$k = \sqrt{k_1 k_2} = \sqrt{\frac{N_1}{N_2}\frac{L_{12}}{L_{11}} \cdot \frac{N_2}{N_1}\frac{L_{12}}{L_{22}}} = \frac{L_{12}}{\sqrt{L_{11} L_{22}}}$$

The angular frequency is ω and the coupling constant between the coil and single turn plasma loop is k. It is important to recognise that the loaded impedance only depends on the plasma $Q_a = \omega L_a/R_a$ and not just the plasma resistance. One way to view the action of the primary coil is that it transforms the plasma resistance (usually a few ohms) into the correct value to ensure the desired power dissipation for the chosen circuit configuration.

Many fixtures are metal and can form a shorted turn around the lamp. Since fixtures can be highly conducting, their Q is large. If this is the case equation 1 shows that the resistance reflected into the primary is small, and so there is relatively little power dissipation in a high Q secondary. The main effect of the fixture is to generate a magnetic field opposing the coil field, which reduces the effective primary inductance L_u. The effect of this on lamp operation is that it reduces the loaded resistance of the coil (through equation 1). Depending on the load characteristics of the circuit, this changes the power in the discharge: for example on a constant voltage circuit the power is increased by a reduction in loaded resistance. This is undesirable because the light output changes and because it may alter the switching characteristics of the FETs (see section 6), causing them to fail. The solution to this difficulty is to include in the housing for each lamp a shorted turn which shields the fixture from the coil. Proper choice of the size and disposition of this shorted turn means that the lamp can be operated reliably in all fixtures.

4. 1-dimensional plasma/coil model

A plasma/coil model restricted to the radial dimension gives some insight into the operation of these lamps. The plasma/coil model[3] is based on the fluorescent lamp model of Dakin [4], except that the electric field E_ϕ and the electron temperature T_e are allowed to vary radially, and the radiation transport calculation must take into account the obstructing effect of the re-entrant. Because electron density n_e can be much higher than in normal fluorescent lamps, rare gas excitation and ionisation can be significant and are included. Starting with an assumed E_ϕ variation, the plasma calculation yields the radial variation of plasma conductivity. Representing the coil by a current sheet at the edge of the reentrant, Maxwell's equations are then solved for the fields with these conductivity values. Iteration to convergence at an assumed input power per unit length then yields plasma properties, including UV production and plasma current.

Measurements of the variation of plasma current over a wide range of powers and at different frequencies show excellent agreement with calculations. We made the measurements on a lamp with cylindrical geometry, having a re-entrant which passes right through the lamp. This allowed us encircle the plasma current with Rogowski coil(5) in order to measure that current. The plasma current is sensitive to the calculated values of n_e and T_e which in turn determine the radial variation of the excited states and the UV output. The satisfactory agreement between calculated and measured plasma current over a wide range of power gives us confidence that this simple plasma/coil model is essentially correct.

5. Choice of Operating Frequency

The operating frequency of the electrodeless system is determined by considering both the efficiency of coupling the energy from the coil into the plasma, and the EMI regulations at the frequencies of interest(2). The equation for the loaded impedance of the coil also gives us the coupling efficiency of the coil/lamp system. The ratio of the power dissipated in the reflected load resistance to the total power dissipated in the coil is the coupling efficiency η. As the same current flows through the unloaded resistance of the coil and the reflected arc resistance the ratio of powers is given by ($Q_u = \omega L_u / R_u$)

$$\eta = \frac{k^2 Q_u \dfrac{Q_a}{1+Q_a^2}}{1 + k^2 Q_u \dfrac{Q_a}{1+Q_a^2}}$$

The efficiency is increased by increasing the coupling coefficient, k, the unloaded coil Q_u, and the arc factor, $Q_a/(1+Q_a^2)$. The coupling coefficient is determined by the geometry of the system and the coil design. Q_u is also determined by the geometry and the coil design but it increases with frequency as well. The arc factor reaches a maximum value when the arc Q_a is equal to unity. Figure 2 shows the variation of coupling efficiency and the arc factor with frequency for an electrodeless fluorescent lamp with an open ferrite core. It is clear that the efficiency is very low at low frequencies and rises above the 90% level at about 1.5 MHz. Thus from an efficiency consideration the operation of the lamp should be at a frequency above 1.5 MHz.

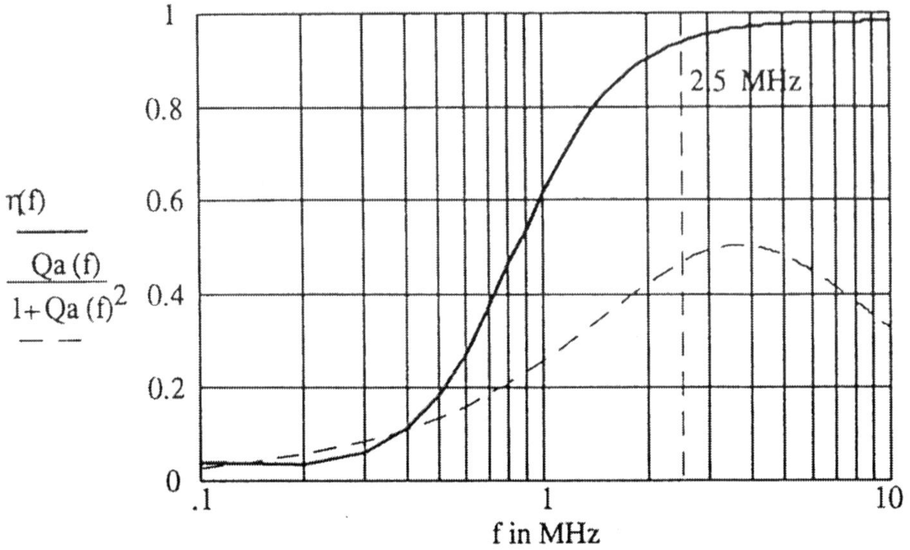

Figure 2 - The coupling efficiency of an electrodeless fluorescent lamp vs frequency (solid line). The arc factor is also plotted vs frequency (dotted line). Acceptable efficiencies are achieved at frequencies above 1 MHz.

Another consideration is the impact of EMI regulations on the system performance. CISPR 15 regulations (2) allow some relaxation of the "magnetic field induced current in a 2m loop antenna" from lamps in a frequency band between 2.2 MHz and 3.0 MHz. Therefore, operating in the middle of this band helps simplify the system design while meeting the coupling efficiency goals.

6. High Frequency Circuit Design

The ballast circuit used to drive the electrodeless lamp is a half bridge circuit as shown in figure 3. The devices in a half-bridge circuit alternately switch on and off with a 50 % duty ratio. This applies a square wave to the load network. The load network consists of the lamp and coil in addition to a matching circuit. The load network is designed so that its resonance frequency is slightly lower than the switching frequency. The load current, i_L, is therefore sinusoidal.

At frequencies in the tens of kilohertz the half-bridge circuit achieves very high efficiencies because the devices are either fully on or fully off.

However at high frequency the circuit suffers from increased switching losses. The primary switching loss is due to the charging and discharging of the output capacitance of the power devices every half cycle. This loss is equal to: CV^2f in the case of a half-bridge circuit [(6)]. This loss is very large at the frequency of operation and significantly affects system efficacy.

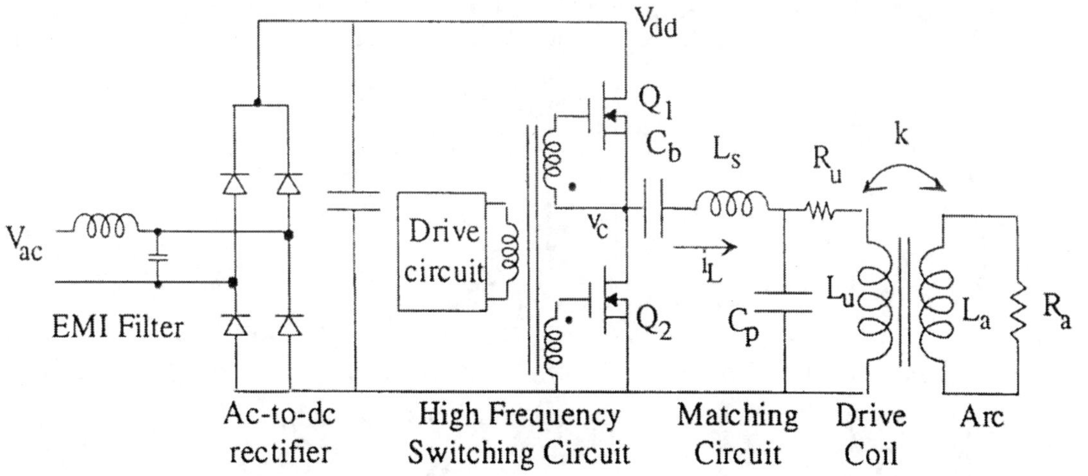

Figure 3 - Schematic of the electrodeless fluorescent lamp ballast showing the EMI filter, the ac-to-dc rectifier, the switching devices, the matching network, the drive coil and the equivalent circuit of the arc.

In order to eliminate the capacitor discharge loss, a zero-voltage switching technique known as transition resonance is used. The load current is allowed to discharge the output capacitor of the devices during a dead time created by keeping both devices off at the transitions. A proper selection of the length of the dead time and the phase and magnitude of the current relative to the midpoint voltage, allows us to meet both the output power requirements of the ballast and to satisfy the zero voltage conditions.

As a result of the switching action of the two devices a trapezoidal voltage appears at the midpoint. The load circuit acts as a band pass filter and allows only a sinusoidal current at the fundamental frequency of the midpoint voltage to flow. Figure 4 shows the midpoint voltage, v_c, the load current, i_L, and the fundamental component of the midpoint voltage, v_{cf}. The timing relationships used in the design are also shown.

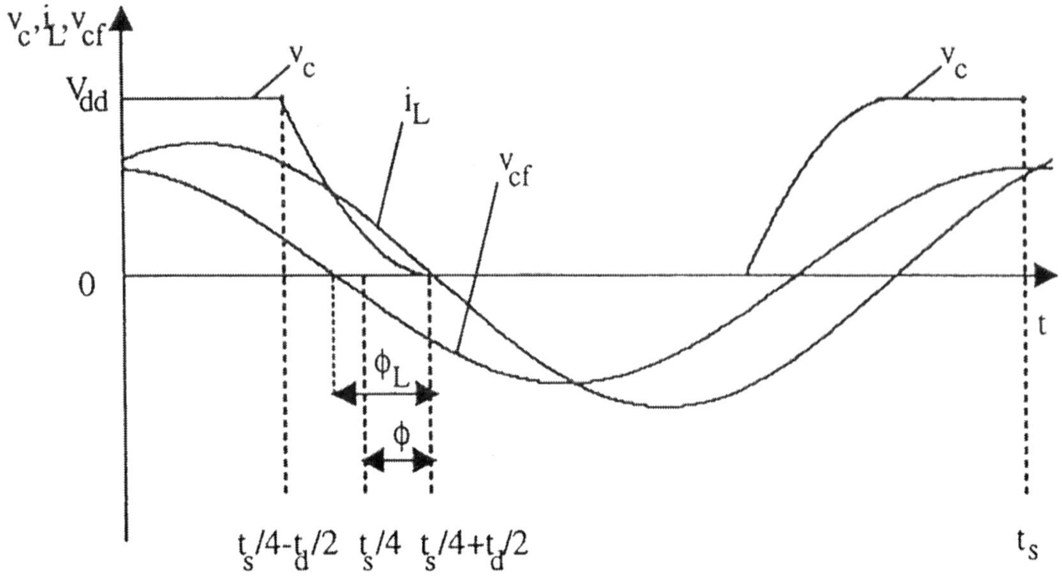

Figure 4- Waveforms of the midpoint voltage v_C, the fundamental component of the midpoint voltage, v_{cf}, and the load current, i_L.

In order to get ZVS, the current amplitude, its phase relative to the fundamental component of the voltage, ϕ_L, and the phase angle of the dead time, ϕ ($\phi = 180(t_d/t_s)$, where t_d is the dead time and t_s the switching period) are related to the frequency of operation, ω, the dc bus voltage, V_{dd} and the output capacitance of the devices, C_{oss}, as follows:

$$V_{dd} = \frac{I_L}{\omega \cdot C_{oss}} \cdot (2\sin(\phi)\sin(\phi_L))$$

The current amplitude and phase must also simultaneously meet the power requirement of the lamp, P_{out}, and the losses in the ballast (including the coil losses), P_{loss}, as follows:

$$I_L = \frac{2 \cdot (P_{out} + P_{loss})}{V_{cf} \cdot \cos(\phi_L)}$$

If we assume that v_C is trapezoidal in shape then the fundamental component of the voltage is given by:

$$V_{cf} = V_{dd} \frac{2 \cdot \sin(\phi)}{\pi \cdot \phi}$$

Simultaneous solution of the three equations above gives the required load complex impedance (for simplicity we use: $Z_{load} = R_{load} + j\,X_{load}$).

7. Matching of the lamp impedance to the ballast

The circuit components, C_p and L_s are used to match the loaded coil impedance to the required impedance Z_{load} for the ballast. The value of the capacitor C_p is selected so that the effective resistance of the parallel combination of the loaded coil and C_p is equal to R_{load}. Then the series inductor L_s is added to make the imaginary part of the combined impedance of C_p and Z_{load} equal to X_{load}.

In the example shown C_b is used as a dc blocking capacitor; its value is large and it has very little effect at the frequency of operation. This combination of component values ensures that the load impedance seen by the switching converter satisfies the output power and the ZVS requirements simultaneously.

8. EMI Mitigation

The electrodeless lamp system meets the CISPR 15 "magnetic field induced current in a 2m loop antenna" requirements [2] because the plasma itself serves as a partial magnetic field shield. Therefore, no additional magnetic field shielding is needed. The CISPR 15 requirements for terminal voltage disturbance TVD (sometimes referred to as conducted interference) are much more difficult to meet. There are several sources of TVD and different approaches are needed with each to meet the regulations.

One source of TVD is the voltage across the coil, which causes capacitively coupled currents to flow in the mains wiring. The internal transparent electrically conducting coating (section 2) is held at a potential close to ground by the capacitative connection to the external frit coating thus reducing TVD. The TVD is further reduced by a second winding on the coil interleaved with the first [7]. Connected only at one end and in the opposite sense, the induced voltage across this second coil partially cancels the electric field of the first winding without affecting the magnetic field of the coil.

A second source of TVD in the system is the ballast itself. The switching devices produce high electric fields with significant harmonics and the currents flowing in the load network can also induce currents in the power line. In order to minimize the TVD from the ballast, the RF section of the ballast is built on a separate board from the input rectifier and the electrolytic capacitor. That RF section is placed in a metal housing that is connected to RF ground to reduce conducted EMI from the ballast. In addition to the screening provided by the metal housing an EMI filter is placed at the input to the ballast (Figure 2). This EMI filter limits the conducted EMI from the ballast to the mains. It also provides a high impedance to any current that may result from residual electric field from the lamp.

Acknowledgements

We want to thank our many colleagues in GE Lighting and GE CR&D who have contributed to this work. We also thank Dr S E Coe for plasma modelling.

References

(1) Wharmby D O, IEE Proc.A, **140**(6),465-473, 1993
(2) IEC, CISPR 15 "Limits and methods of measurement of radio disturbance characteristics of electrical lighting and similar equipment", Edition 4, 1992
(3) Dakin J T, J. Appl. Phys., **60**, 563, 1986
(4) Lister G G and Cox M, Plasma Sources Sci. Technol., **1**, 67-73, 1992
(5) Huddlestone R H and Leonard S L, "Plasma Diagnostic Techniques", (New York: Academic Press), p.8, 1965
(6) Freeland S, "An Introduction to the Principles and Features of Resonant Power Conversion", in "Recent Developments in Resonant Power Conversion ", ed. K K Sum (Intertec Communications Inc.), pp 20-43, 1988
(7) S-A El-Hamamsy, IEEE Trans. Power Elect., **9**(3), 297-308, 1994
(8) U S Patent 4,710,678, 1 December 1987

MEASUREMENTS ON THE PHILIPS QL-LAMP FOR COMPARISON WITH THE DEVELOPED MODEL

J. Jonkers, M. Bakker, J.A.M. van der Mullen, D.A. Benoy, K.T.A. Burm and D.C. Schram
Department of Applied Physics, Eindhoven University of Technology
P.O. Box 513, 5600 MB Eindhoven, the Netherlands

Recently Philips Lighting has developed a new inductively coupled low pressure RF discharge for lighting purposes: the QL-lamp. Its main advantage is the lack of contact between electrodes and plasma, which benefits the life-time. A model to describe this plasma has been developed by Benoy *et.al.* The plasma-parameters and the light emission are calculated locally on base of macroscopic input-parameters (like the number of windings of the coil, input power etc.). Also the impedance of the primary coil and the leakage of electromagnetic (EM) fields can be determined. In this paper all these output-parameters will be compared with measurements to check the validity of the model. The model itself will be discussed in a separate paper.

In Figure 1 the Philips QL-lamp is depicted. The coil is situated in the center of the discharge in order not to obstruct the produced light. Note that the plasma is locked up in the bulb so there is no contact with the coil or other parts of the lamp. The alternating current (2.65 MHz) through the coil produces the EM-field, which is necessary to sustain the discharge. As can be seen in Figure 2 most of the power is absorbed in the torus very close to the coil.

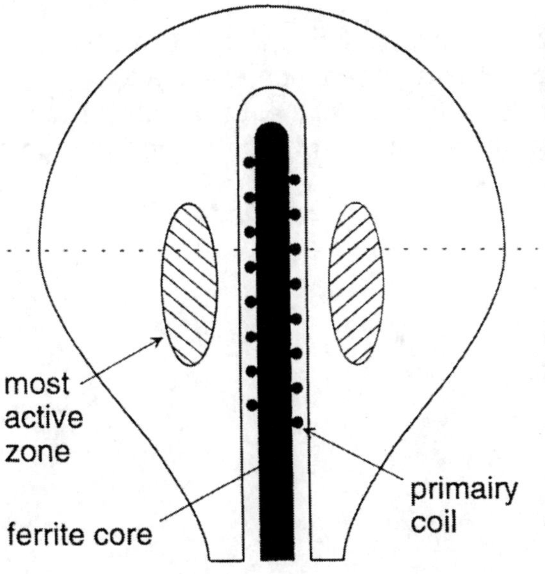

Figure 1: A schematic drawing of the Philips QL-lamp with its most important components. The horizontal dashed line indicates the median plane in which most measurements were performed.

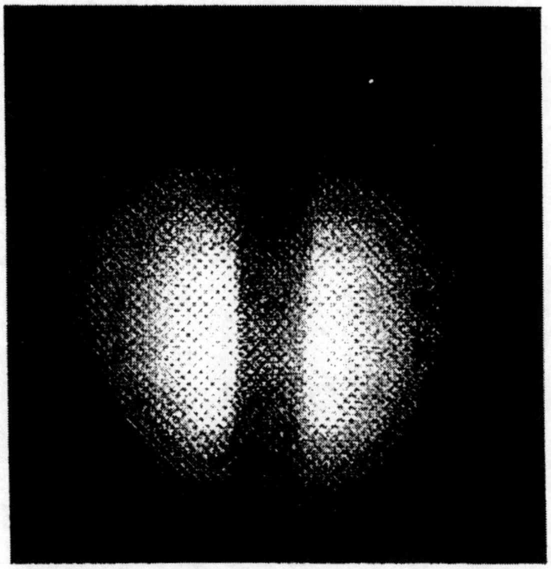

Figure 2: A photograph of the QL-lamp. The most active zones (i.e. the most brightest) are situated near the coil, just below the median plane.

The QL-lamp is in fact a transformer of which the secondary circuit is formed by the discharge. The coupling between both circuits is indicated by the coupling factor. This geometrical factor equals the part of the EM flux produced by the primary coil which is enclosed by the secondary coil. This factor is very important since it determines the leakage of EM energy to the surroundings.

Since the coupling factor depends on the currents in the plasma (i.e. the flow of electrons) it is determined by the spatial distributions of the electron temperature and density. These distributions obtained from the model have been compared with the results of absolute line intensity measurements. The measurements are performed using a combination of monochromator and CCD-camera. The spectrum is depicted in one dimension of the CCD-camera, the other dimension is used to obtain spatial resolution. In this way the absolute density of an excited state can be determined as a function of the radius in one single measurement. A Boltzmann plot of the absolute state densities near the continuum yields the electron temperature (from the slope) and the electron density (from the Saha jump).

Other local information has been obtained by measuring the absorption profile of the argon $4s^3P_2 \rightarrow 4p^3D_3$ transition. The Doppler broadening of this line yields the gas temperature as a function of the radius. Also the density profile of the first excited state of argon has been obtained. Fast fluctuations in this density have been investigated. These fluctuations are caused by the oscillation of the energy coupling (twice the field frequency), which lead to an oscillation in the electron energy distribution function. In this way the quasi-stationary approach of the model can be validated.

An external parameter, which can be measured directly, is the impedance of the primary circuit under working conditions. This impedance will change owing to the presence of the plasma. By using results of the model (i.e. the electrical conductivity and the coupling factor) the resulting impedance can be calculated and compared to the experimental results.

BISTABLE DIMMING BEHAVIOR OF FLUORESCENT LAMPS CAUSED BY ELECTRON-ATTACHING CONTAMINANTS

John F. Waymouth
16 Bennett Road, Marblehead MA 01945 USA

The existence of a bistable mode of dimming fluorescent lamps leading to a hysteresis between light level achieved and dimmer setting is proposed. This mode requires two factors: a normal positive branch of the lamp V-I characteristic, commonly observed in all mercury/rare-gas discharges up to about 5-10% of the typical rated operating currents; and a contaminant gas which attaches electrons to form negative ions. In the attached state, the contaminant molecules are unable to attach further electrons. They are restored to the neutral unattached state by collisional recombination with positive mercury ions.

The conversion of electrons to negative ions effectively removes them from the plasma conductance; therefore the ionization frequency must increase to compensate for this additional electron loss process. I find that for selected values of the ratio of attachment rate constant to recombination rate constant, at full power the contaminant molecules are primarily in the negative-ion state unable to attach electrons. At lower power, they are predominantly in the neutral state, ready to attach electrons. The ionization frequency and corresponding lamp voltage requirement at full power is essentially the same as for an uncontaminated lamp. At low power, however, the ionization frequency required and lamp voltage are sharply increased, to the extent that the lamp V-I characteristic has a second intersection with the ballast load line at low currents in the positive branch. The existence of two intersection points between lamp characteristic and ballast load line results in the bistable dimming operation.

Figure 1 illustrates results calculated from a mathematical model incorporating these concepts. For the sake of simplicity, the model assumes a lamp operating on DC on a resistively-ballasted circuit having a fixed open circuit voltage and a variable resistance. Dimming to low currents is achieved by increasing the value of the ballast resistance at constant open circuit voltage. In this figure, the solid , short-dashed, and long-dashed curves represent the ballast V-I load-lines for full power (normalized ballast resistance = 1), 5 times, and 7 times the full-power ballast resistance, respectively. The curve with solid square points is the V-I characteristic for a good lamp, and the curve with open squares is the V-I characteristic for a contaminated lamp.

At full power, the ballast load-line intersects the good-lamp and contaminated-lamp V-I characteristics at virtually the same current. As the normalized ballast resistance increases to 5, both lamps dim smoothly, the good lamp to a normalized current about 0.15, the contaminated lamp to about 0.07 (Points B and B'). Upon a further increase in ballast resistance to 7, the good lamp dims smoothly to current about 0.1 (Point C). However, for the contaminated lamp, the ballast load line with this resistance no longer intersects the high-current negative-resistance branch of the lamp V-I characteristic. The current drops abruptly to the intersection with the low-current positive-resistance branch (Point C'), current about 0.001, a seventy-fold reduction. Return of the ballast resistance to 5 does not restore the previous current, since the intersection C' is now the stable operating point.

If the characteristic shown for the bad lamp were present at starting, the lamp would never get to full power in the first place, but would remain operating at the low-current intersection C'. However, the contaminants most likely to cause this problem, water vapor and carbon dioxide, will for the most part be adsorbed on the phosphor when the lamp is cold. Thus the lamp would light normally and run up to full power normally. At full power, especially in the hot fixtures customarily found in compact fluorescent applications, the tube walls heat up, desorbing the contaminants. These have little effect upon the V-I characteristic at full power, as noted, but then cause the abrupt jump in current in dimming, as described. Once the lamp is locked in the low current intersection C', the tube walls cool off, the contaminants resorb, the lamp V-I characteristic becomes more "normal", and the intersection with the ballast load-line jumps again to point C, current 0.1. If the lamp power at this higher current is sufficiently high to heat the tube wall to desorb the contaminant gases, the contaminated V-I characteristic is restored, and the current drops to intersection C'. Thus the lamp could thermally cycle repeatedly between these two currents.

One solution to the problem is to increase the ballast open-circuit voltage, so that the ballast load line does not intersect the V-I characteristic of bad lamps at low currents. This is obviously just a crutch to cope with unsatisfactory lamp performance. The best solution is to improve lamp processing to prevent contamination with either water vapor or carbon dioxide.

The poster will provide details of the derivation of the model equations and identify key dimensionless ratios which control system behavior.

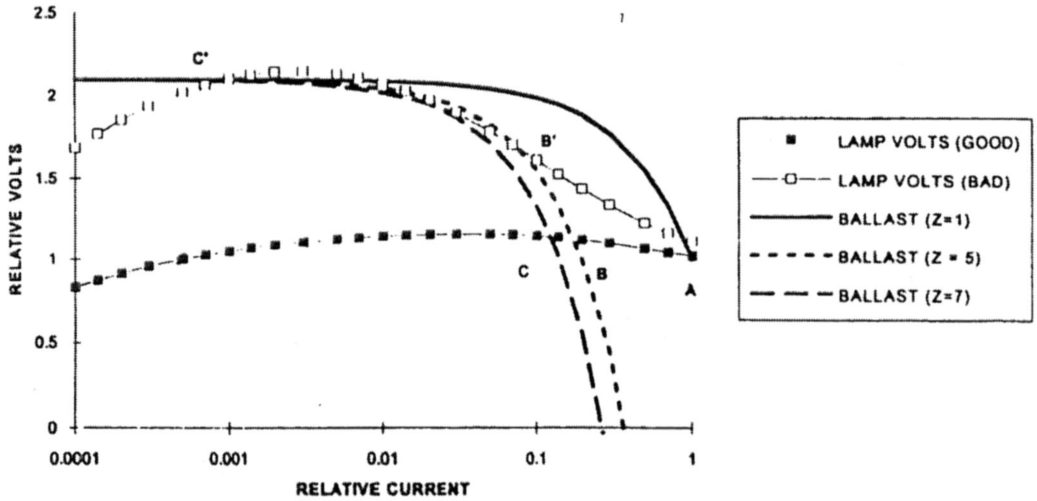

STARTING TIME MEASUREMENT OF NON-PREHEATED ELECTRONIC COMPACT FLUORESCENT LAMPS

TAKEO YASUDA, HIDENORI ITO
FLUORESCENT LAMP ENGINEERING SECTION, LAMP DIVISION
TOSHIBA LIGHTING & TECHNOLOGY
1-201-1 FUNAKOSHI YOKOSUKA 237, JAPAN

Introduction

In recent years, the self-ballasted compact fluorescent lamps (CFL) which have small-diameter discharge tubes and electronic circuits have been used as replacement for incandescent lamps. Simplifying the electronic circuits is important, because CFL is desired to be designed small and highly efficent. Hence, we studied the possibility of non-preheat starting. This enables to omit the preheat-timer components, and get rid of the power loss for heating the electrodes during lamp operation (Fig.1(a),(b).)

Generally, starting is described as a breakdown phenomenon and subsequent glow-to-arc transition. In this work, we measured the starting time of non-preheated CFLs, Ts, as statistical Ts-lag of breakdown (1). All the measurements were carried out after keeping the lamps in the dark.

Measurement

Measurement conditions are as follows.

i) Lamp: having double-"U" shape, arc length of 44cm, inner diameter of 10mm, triply-coiled thermal electrodes (surface area=0.25 cm^2) with oxide emitter, and rare-earth phosphor coating.

ii) Applied Voltage: V(O-P)=1kV, 45kHz, sinusoidal.

iii) Others: Tamb.=25°C, lamp-aging time (Tag)=0 or 300 hours. Before starting; lamps kept in the dark for 4~120 hours (Tk). On starting; lamps kept in the dark or lighted 0.1~10 lx.

The results are shown in Fig.2 and Fig.3.

Figure 2 shows that ‹Ts› gets longer with shorter Tag and longer Tk (‹Ts› is the average of 30 lamps.)

Figure 3 shows that ‹Ts› grows longer in darker enviroment.

Discussion

‹Ts› is calculated by ‹Ts›=1/‹PQ›, where P is the breakdown probability due to one initial electron and Q the number of initial electrons. First, we measured ‹Ts› of the lamps in which a radioisotope (RI):^{147}Pm was filled, and found it to be 390 μs. Since Q in this case was adusted to 3700 1/sec, P is calculated to be 0.7. Using this value, Q is found to change from 1.2 to 72 in the range of ‹Ts› shown in Fig.2. To make ‹Ts› shorter than 0.1 sec, Q has to be greater than 14. However, we must take the distribution of Ts into account. To make Ts longer than 0.1 sec happen less than once per every 10^8 starting, Q›290 is necessary (1). If we want to get this number of electrons from the thermal electrode, it is enough to heat it up to 100°C by the Richardson equation. Fig.1(c) shows a way to realize the minimum preheat starting.

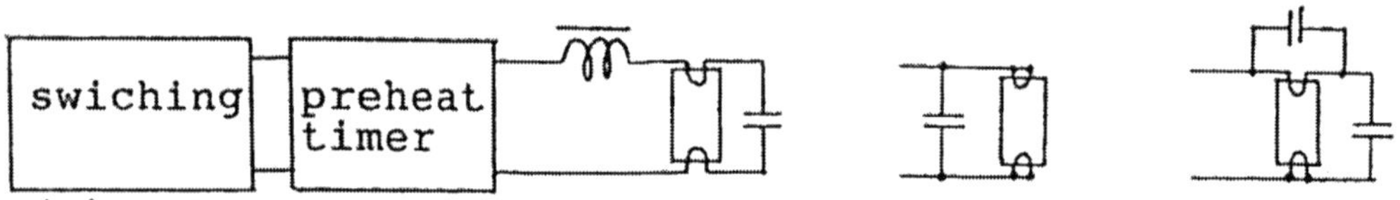

(a)preheat (b)non-preheat (c)minimum preheat

Fig.1 Electronic LC-Resonance Circuits

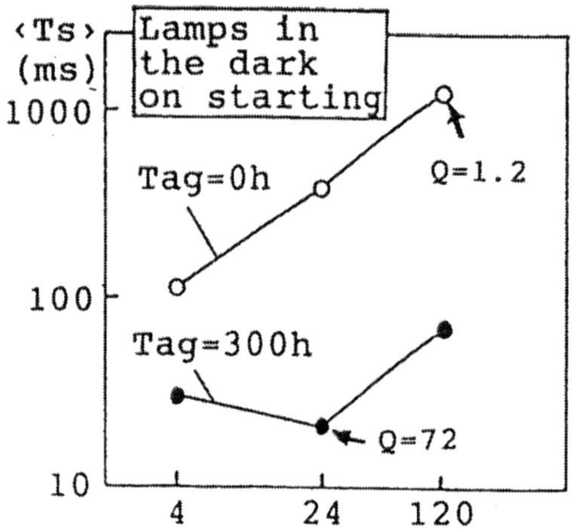

Fig.2

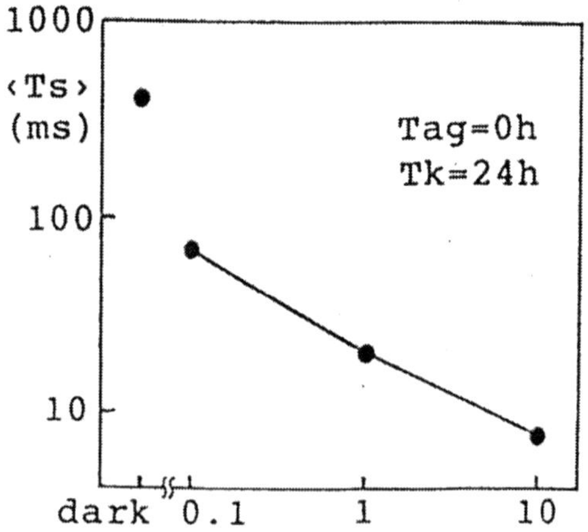

Fig.3

Conclusion

Non-preheat starting is not acceptable for CFL, because it makes Q too small (1.2 in the worst), and consequently Ts-lag too long.

The condition Q›290 is necessary for the brief breakdown. This is achieved by heating the electrode up to only 100°C.

Reference

(1)A Inouye, Y Terashita and K Ogawa,J.Light & Vis.Env. Vol 1,No1 (1977),24.

Thermionic Emission Characteristics of Preheated Cathode Fluorescent Lamps.

Minoru Myojo, Ikuhiro Okuno
Light R&D Center, Light Business Group, Matsushita Electronics Corporation
1-1 Saiwai-cho,Takatsuki,Osaka, 569,Japan

Introduction

The minimum thermionic emission curve for current controlled preheating of fluorescent lamp cathode is specified in IEC standard (International Electrotechnical Commission) [1]. The formula is as given below.

$$i_k = (a/t_e + i_m^2)^{1/2} \quad \cdots\cdots (1)$$

The method for measuring emission time and the constants both "a" and "i_m" for an actual cathode are given by A. Heidemann, et al. [2].

The energy input to a substitution resistor is theoretically compared with that in specified cathode[3].

Experimental

Emission time is measured by detecting the fall of lamp voltage (V_L) with a new method, in which lamp current is kept small and constant during the measurement as shown in **Fig.1** [4]. At this experiment, emission time was mathematically obtained by second derivative of lamp voltage with time.

Results and Discussion

After the examination of thermionic emission characteristics of various types of cathodes, it has become clear that their emission curves are not always similar to the standard as shown in **Fig.2**. Consequently, we present a new emission curve for an actual cathode in the following formula.

$$i_k = (a/t_e + i_m^n)^{1/n} \quad \cdots\cdots (2)$$

Where, exponent n, which is a variable constant and originates in the actual cathode, is decided by using two regression lines from formula(2).

$$i_k^n = a \cdot (1/t_e) + i_m^n \quad \cdots\cdots (3)$$

$$i_k^n t_e = i_m^n \cdot t_e + a \quad \cdots\cdots (4)$$

And, exponent n varies in the range of about 1 to 2 .

Also, after the measurement of the energy input to a cathode and its average resistance by emission time, the relationship between an emission curve and the energy input to a cathode as follows has become clear.

$$\int i_k v_k dt = i_k^2 t_e \cdot (1/t_e) \int r_k dt = (a/t_e + i_m^n)^{2/n} t_e \cdot \langle r_k \rangle \quad \cdots\cdots (5)$$

Where, the integral period is zero to emission time, and $\langle r_k \rangle$ is the average resistance of a cathode. An example is shown in **Fig.3**.

Conclusions

1. An emission curve for an actual cathode was newly presented.
2. The energy input to a cathode has a minimum value with time.
3. Average resistance of a cathode, related to a substitution resistor, changes in a short period of emission time.
4. Results mentioned above show that an Rh/Rc (hot/cold) resistance ratio measurement has a less possibility of describing dynamic emission characteristics.

References

1. 81-IEC-7210, etc.

2. Heidemann, A. ; Denz, W. ; and Roche, W. Specifications for the operation of preheated cathode fluorescent lamps on electronic ballast. *J of the IES* 23(no. 1):115-120 (1994)
3. Myojo, M. ; Yamamoto, K. ; Okuno, I. *National Technical Report* Vol.38 No.1 (Feb.1992) (in Japanese)
4. Miyake, S. ; Myojo, M. ; Okuno, I. Study on non-destractive measuring method for emitter quantity of fluorescent lamps. Proc. 1993 Annual Conf. IEI Jpn. No.3 (1993)

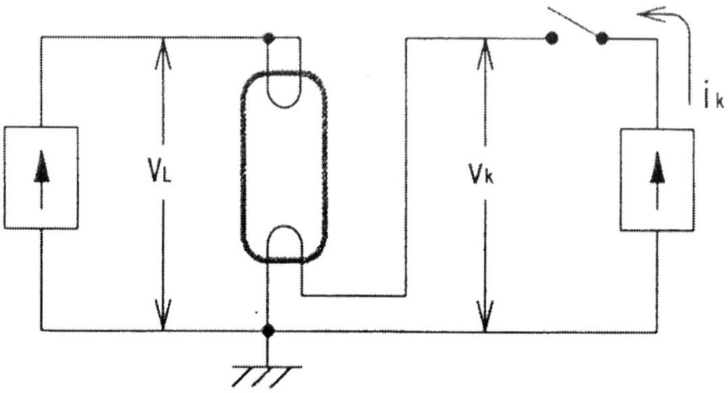

Fig.1 Apparatus for measuring emission time.

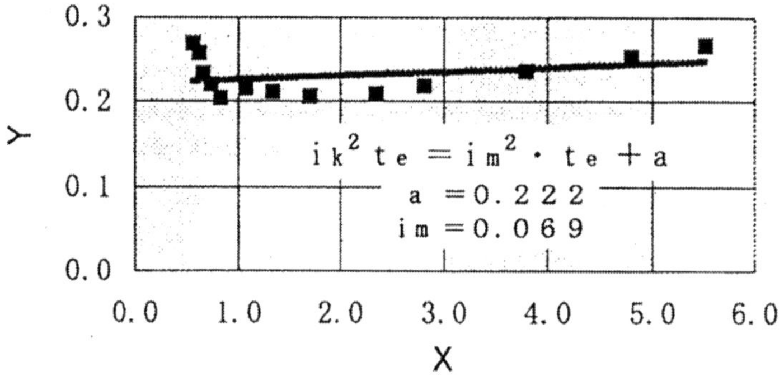

Fig.2 Scatter diagram from formula(1) (compact FL13W).

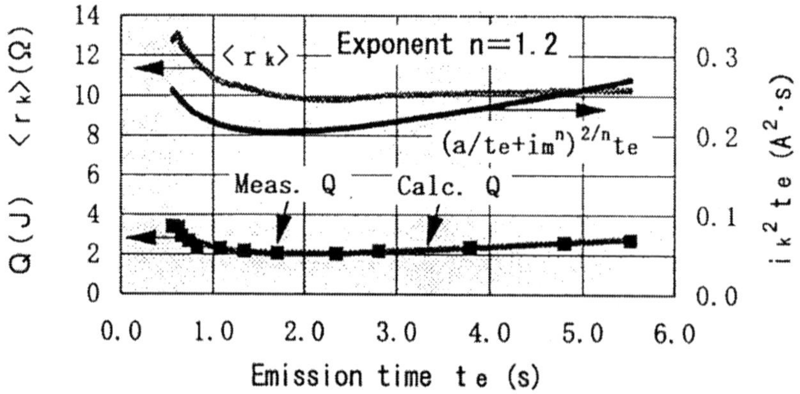

Fig.3 The relationship between an emission curve and the energy(Q) input to a cathode (compact FL13W).

Effect of anode current on the position of cathode hot spot

Li xianli
Nanjing Research Centre of Electric Light Sources, Nanjing, China.

1.Experiments and analyses: Where is the cathode hot spot occurred on the filament? This problem is worth investigation, because it is concerned with the lifetime of fluorescent lamp. Carrefully observing the state of filament when fluorescent lamp is igniting, we can often see that hot spot is occurred at the middle of the filament at first, then it moves to the end B of filament rapidly and is stable there(see Fig1A) . If short -circuit end A with end B of the filament when the fluorescent lamp has ignited, as Fig1c, the temprature of the hot spot B is decreased obviously and at the other end of filament also appears a hot spot A.
What is the main factor determined above positions of cathode hot spots? The following fact attracts us to pay attention. It is well known that in anode cycle the filament receives reached eletrons and collects them to be the anode current passed the filament. The distribution curves of the anode current corresponded with Fig1A and Fig1c are given separately in Fig1B and fig1D. We can see from these curves that places of maximum anode current on the filaments are just the same to the position of cathode hot spots.This seems tell us that effect of anode current on the position occurred cathode hot spot can not be neglected.
It is not enough that we carry out the conclusion only from above experiments and analyses. Because the point B (in Fig1A) and the point A ,B (Fig1c) are not only the points of maximun anode current (for anode-cycle) but also are the points of lowest electric potencial on the filaments (for cathode-cycle). Since the dischage system tends to the lowest energy state, so the pasitions of cathode hot spots also ought be there as above. In order to confirm the effect of anode current on the formation of cathode hot spot we design the experiment as Fig2A. The texted lamp is a fluorescent lamp with dual- filaments electrode. The action of diode D1 is the key of this experiment. It prevents the anode current to pass point B directly from the circuit and forces it to pass the point D at first,and then to pass point C,B. Thus the distribution of anode current become as Fig2B; Diode D2 makes the clectric portential of point D is the same to the point B and C for the cathode-cycle. Diode D3 acts to cancel the rectified action of D2. The experimental result shows: As preheating, the part AB of filament is heated and became red.At first of ignition, the hot spot is occurred at the point B of filament, then the hot spot automatic transfers to the point D from point B and is stable there.This experiment shows again that the cathode hot spot always occurred at the place of the maximun anode current on the filament. It has fully confirmed that important effection of anode current distribution on the fomation of cathode hot spot.
2. Explain:In this paper is carried out a view point that place formed cathode hot spot on the filament is determined by the thermal emission of each point of filament when the cathode-cycle (i.e. the half period that filament acts as cathod) just begin, i.e. the anode-cycle just ended. AT this moment which point of filament have maximun thermal emission, there is the place formed cathode hot spot.

$$J_x = AT_x^2 \exp(-e\Phi_x/KT_x) \qquad (1)$$

where J_x is the thermal emissive current density of each point of filament. T_x is the temperature of each point. Φ_x is the work function of each point. Eq.(1) shows, if the Φ_x of all point of filament are equal. thus the thermal emission of each point of filament depends on the temperature of each point. Ought to point with emphasize, when the cathode-cycle just begin, the temperature of each point of filament is concerned with the obtained energy density of each point of filament in the anode-cycle just ended. this energy density includes two parts:

$$E_x = J_e(V_A+\Phi)+4I_x^2\rho/\pi D^2 \qquad (2)$$

Where E_x is the energy density, J_e is the electron jet density, V_A is the anode fall, Φ is the work function of filament, ρ is the bulk-resistivity of filament, I_x is the anode current passed each point of filament.

The right first part of eq.(2) is the energy density delivered by the electrons arrived at the filament(it acts as anode), In general, it is uniform for each point of filament .The second part is the jole heat produced by the anode current passed each point of filament.Use the Qz for it, so $Qz=4I_x^2\rho/\pi D^2$.We can see (from Fig2B) the Qz of each point of filament are very different.For example,the Qz of point D is four times than point B.This difference of Qz certainly effect the temperature and the thermal emission of each point of filament when the cathode cycle just begin. therefore, the distribution of anode current on the filament is a importent factor determined the position of cathode hot spot.

3.Practical application:In order to redouble the lifetime of fluorescent lamps, the people tested to use the electrode with multi- filaments structure in the past. But the key of the problem is that hot spot emited electrons must can automatic transfer from one filament to another. So that all filments are fully untilized. The practical fluorescent lamps with dual-filaments electrode and ther applied circuit shows in Fig3A. The distribution of anode current shows in Fig3B. if we chose a appropriate R value corresponded appropeiate distribution of anode current, the hot spot can automatic transfer between three points D,B,C. Thus, the lifetime of the fluorescent lamps can be effectively increased.

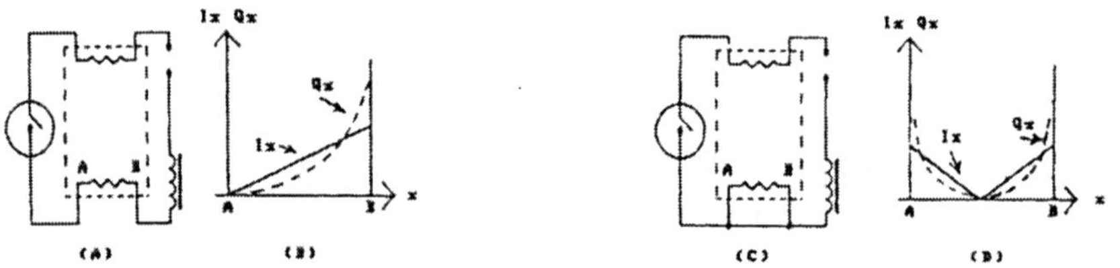

Fig1. The circuit of fluorescent lamp and the corresponding distribution of anode current

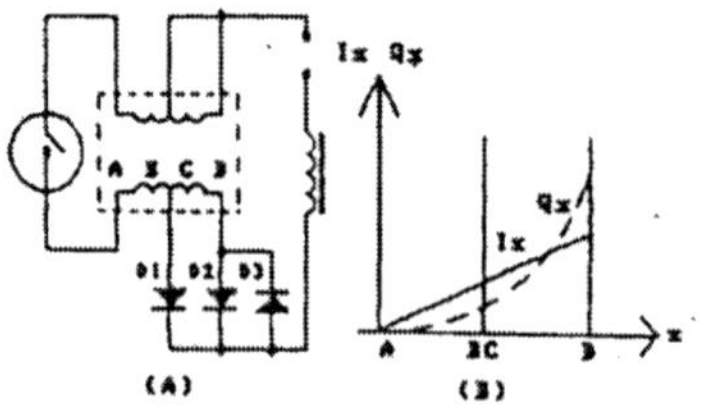

Fig2. The test of hot spot transfer and correspondent distribution of anode current

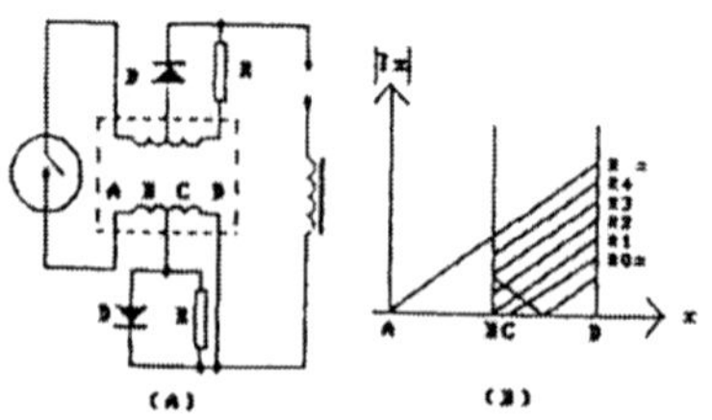

Fig3. Practical circuit of fluorescent lamp with dual-filament structure

INVESTIGATION ON THE ELECTRON ENERGY SPECTRUM OF COMPACT FLUORESCENT LAMPS

N.L.Bashlov, V.M.Milenin, G.Ju.Panasjuk, N.A.Timofeev
St.Petersburg University, Russia.

The modern trend of the development of luminescent light sources filled with the mixture of mercury vapour and noble gases is their minimisation. However, the use of compact tubes (with radius < 0.5 cm) increases wall losses of energy and particles (electrons, ions, excited mercury atoms) making the lamp efficiency worse. To satisfy the requirement of high light efficacy one has to rise the pressure of noble gases. The increasing pressure of a gas (more than 5 Torr) gives rise to some serious experimental difficulties in studying luminescent lamps by probe methods which could give the most complete information about plasma electrokinetic characteristics, and these difficulties are the reason why at present there are not any experimental data of this kind.

The purpose of this work is the experimental study of the electron energy distribution function (EEDF) in the positive column of (Hg+Ar) gas discharge under the increased pressure of the noble gas (up to 20-30 Torr) in compact tubes. The investigation was conducted with the help of a novel probe method which takes into accunt the diffusive motion of electrons in a probe sheath (non-Langmuir case of probe operation [1]). EEDF was derived from the second derivative of the probe current with respect to the plasma potential by solving the integral equation which establishes the connection between EEDF and second derivative of current [2]. The second derivative was measured with the help of the pulse experimental technique [3]. The pressure of mercury vapour was 2-50 mTorr.

The second derivatives and the EEDF obtained with the help of our procedure are presented in the fig.1. As one can see, the distribution functions differ strongly from the second derivatives so that the common treating of probe curves [3] is impossible. The distribution functions normalised on the unit are presented in fig.2. These functions show the probability density for electrons to have the energy. One can see the decrease of the number of fast electrons and the fall of the electron temperature with the argon pressure increasing. The behaviour of EEDF in dependence upon the discharge current and pressure of mercury vapour in the discharge conditions in question, as our experiments shows, is similar to that of usual luminescent lamps ($R \leq 0.5$cm, $p<3$ Torr) [3].

In [3] there was given a theoretical description of the positive column of the discharge in the mixture of mercury vapour and noble gases. As our estimations show, the main assumptions of this description are valid in the discharge in question up to the argon pressure 30 Torr, so one can use the model [3] for calculation of the EEDF. The comparison between calculated EEDF's and experimental one is presented in fig.3. The function (1) is here a function obtained from the second derivative with the help of the common treating, (2) is a function obtained with the help of our procedure, (3) is a result of the Boltzmann equation solution [3]. It is seen that the function (1) differs appreciably from (3). The application of our treating corrects the result and makes the agreement between experimental and calculated functions very good (compare curves (2) and (3)).

References

[1]. Tsendin L.D. -Sov.Phys. JETP 1974 v.39, p.805.
[2].Golubovskii Yu.B., Zakharova V.M., Pasunkin V.N., Tsendin L.D.- Phisika Plasmy 1981, v.7, p.620.
[3]. Milenin V.M., Timofeev N.A. Plasma of low pressure gas discharge light sources - Leningrad, Leningrad Univ. Publ., 1991, 240 p.

Figures

Fig. 1. The second derivatives $j''_a(eV)$ of the probe current with respect to the probe potential (curves 1-3) and electron energy distribution functions $f(eV)$ (curves 4-6) obtained from the curves 1-3 with the help of our treating.
$R = 0.5$ cm, $T_{Hg} = 42$°C, $i = 240$ mA;
1,4 - $p_{Ar} = 5$ Torr, 2,5 - 10 Torr, 3,6 - 15 Torr.

Fig. 2. Electron energy distribution functions $F(\varepsilon) = const \sqrt{\varepsilon} \cdot f(\varepsilon)$ normalized on the unit and measured under different argon pressure.
$R = 0.5$ cm, $T_{Hg} = 42$°C, $i = 240$ mA;
1 - $p_{Ar} = 5$ Torr, 2 - 10 Torr, 3 - 15 Torr.

Fig. 3. Distribution functions $F(\varepsilon)$ in dependence upon mercury temperature T_{Hg}.
$R = 0.4$ cm, $p_{Ar} = 15$ Torr, $i = 250$ mA;
1 - $T_{Hg} = 32$°C, 2 - 40°C, 3 - 50°C, 4 - 60°C, 5 - 66°C.

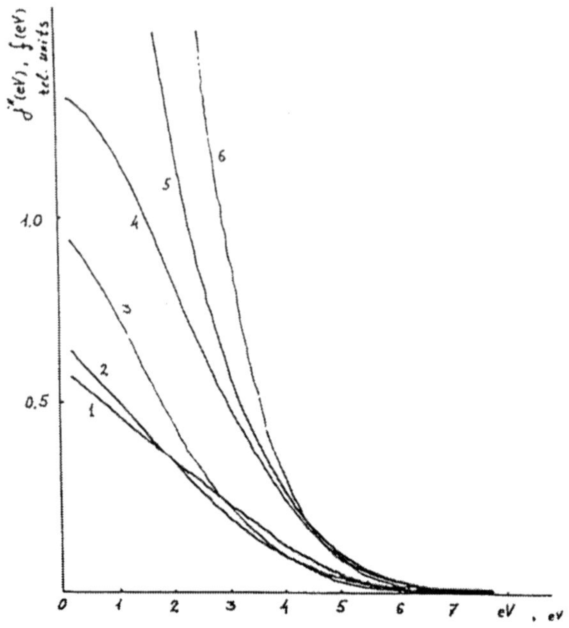

Fig 1.

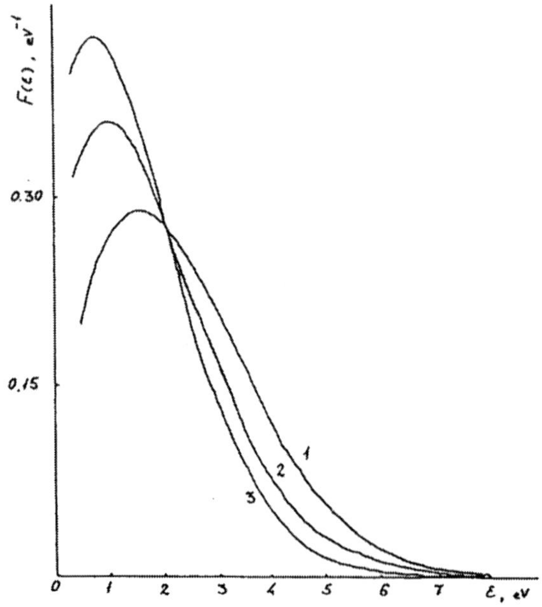

Fig. 2.

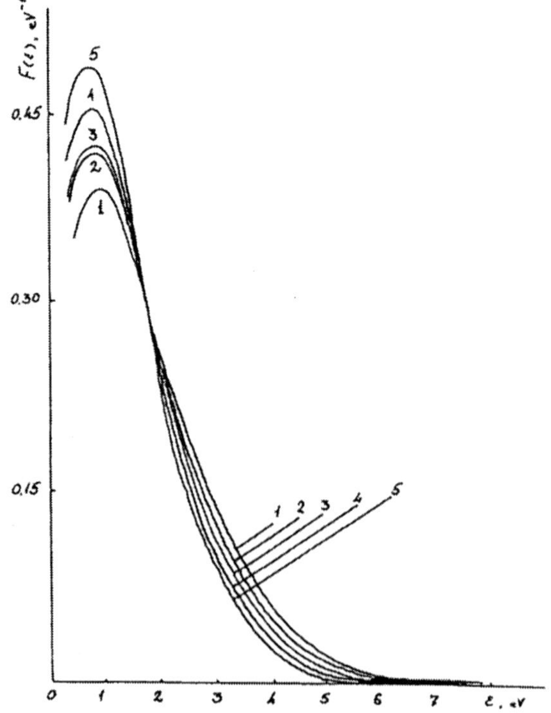

Fig. 3.

INVESTIGATION ON THE ELECTROKINETIC CHARACTERISTICS OF COMPACT FLUORESCENT LAMP

N.L.Bashlov, V.M.Milenin, G.Ju.Panasjuk, N.A.Timofeev
St.Petersburg University, Russia.

The work devoted to the study of the plasma electrokinetic characteristics: the strength of the electric field E, concentration and average energy (or the temperature) of electrons in the positive column of the (Hg+Ar)-discharge in compact tubes (radius ≤ 0.5 cm) under increased pressure of noble gas. The plasma characteristics were obtained :
1) experimentally (described in part 1 of "Compact luminescent lamps...").
2) using a theoretical model as given in reference [1].
3) by discharge similarity principle [2].

The electron temperature measured in the (Hg+Ar) discharge positive column in dependence upon the temperature of mercury vapour and results of calculations are presented in the table 1. Similar data concerning the strength of the longitudinal electric field are also presented here. As one can see, there is satisfactory agreement, but it is to be noted that the measurement of the strength of the electric field in the positive column of the discharge in question is usually troubled by oscillations of the plasma potential [1]. So the experimental errors in determining can reach 10%.

The current dependence is found to be usual: E falls with current increasing. The dependence of E upon the argon pressure and mercury temperature are non-monotone (fig. 1). The measured strength of the electric field as a function of a tube radius [3] and the curve obtained with the help of the similarity laws are presented in fig. 2. There are results of our calculations of temperature and concentration of electrons in this figure as well.

Using the model [2], we calculated temperature and concentration of electrons as functions of external discharge parameters (fig. 3). These data are of great interest because they can be used for the choice of the optimal conditions of fluorescent lamps operation. After moving on to narrow tubes the discharge conditions, which ensure the maximum efficacy of the UV-radiation do not coincide with that of the "classical" luminescent light sources. Our data make it possible to estimate the rates of elementary processes taking place in the plasma in particular, the quenching of the mercury atoms resonance levels in compact tube discharges. The quenching appeared to be rather weak so that it does not influence deeply in the intensity of the mercury resonance radiation. This gives an opportunity to consider the light efficacy of a luminescent lamp as practically an invariant characteristic [2] and to apply to it the approach based on the similarity formalism and find the optimal conditions of compact luminescent lamps operation under the increased pressure of argon. For example, one of the worth-while compact luminescent light sources is the lamp made in a discharge tube with radius 0.5 cm and working at cold spot temperature 50 °C. Using the similarity laws [2], it is easy to obtain the range of the external discharge parameters at which one can expect the maximum of light efficacy: argon pressure 22 Torr, current 28 mA. The similarity formalism can be obviously applied to the investigation of discharges in the mixture of Hg with other noble gases (possibly, expect He). Table 2 illustrates this giving the comparison between experimental values of the strength of E in dependence upon radius in (Hg+Kr) and (Hg+Ne) discharges [3] and results of our calculations based on the similarity formalism.

References

[1]. Milenin V.M. and Timofeev N.A. Plasma of low pressure gas discharge light sources.- Leningrad, Leningrad State Univ.Publ., 1991, 240 p.
[2]. Bashlov N.L., Milenin V.M., Panasjuk G.Ju. and Timofeev N.A.- J.Phys.D: Appl.Phys., 1993, v.26, p.410-417.
[3]. Benetruy P., Zissis G., Damelincourt J.J., A.Por- Proc. 5th Int. Symp. Sci. Tech. Light Sources. York, 1989, p.71.

Figures and Tables

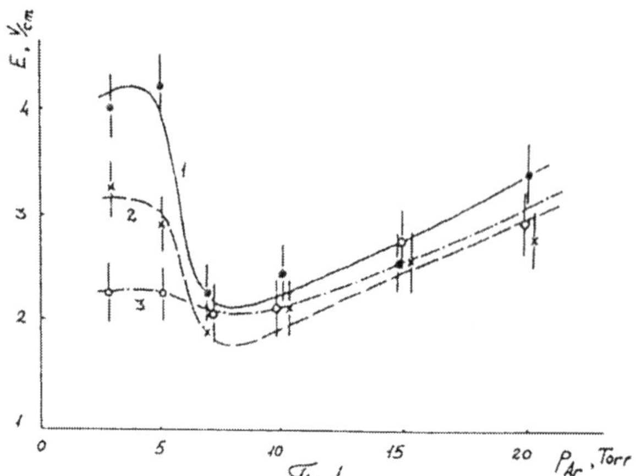

Fig. 1.

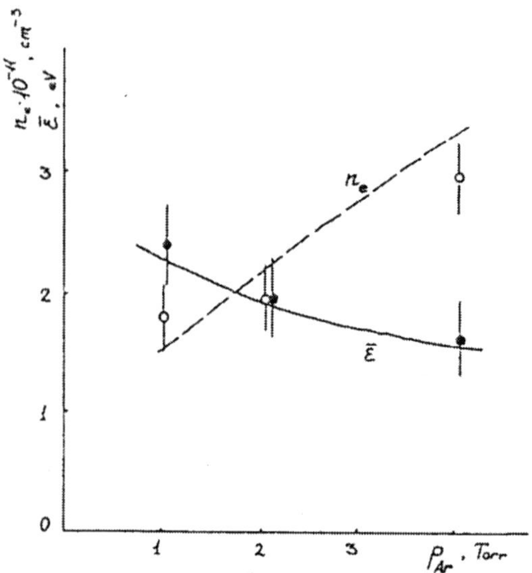

Fig. 2

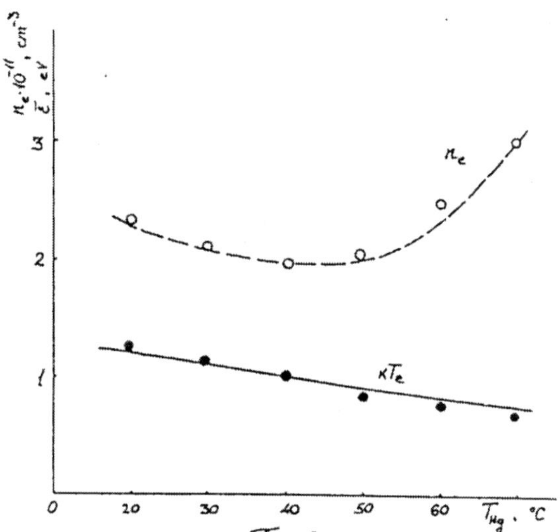

Fig. 3.

Table 1. The electron temperature κT_e in dependence upon the temperature T_{Hg} of mercury vapour. P_{Ar} = 15 Torr, R = 0.4 cm, i = 0.25 A.

T_{Hg}, °C	32	40	50	60	66	
$\kappa T_e^{calc.}$, eV	1.24	1.20	1.12	1.05	1.00	
$\kappa T_e^{exp.}$, eV	1.23	1.14	1.12	1.08	1.02	
$E^{calc.}$, V/cm	-	2.91	2.84	2.80	2.89	3.00
$E^{exp.}$, V/cm	2.3	2.5	2.5	3.0	2.7	-

Table 2 The comparison between experimental values of the strength of the electric field E in dependence upon R in tubes filled with mixtures Hg + Kr and Hg + Ne [3] and results of the calculations made with the help of the similarity formalism.

Mixture	R, cm	i, mA	E, V/cm calculation	E, V/cm experiment
Hg + Kr	0.08	18	12.2	9.1
	0.15	18	6.9	6.4
	0.30	23	5.3	4.6
Hg + Ne	0.08	26	22.7	21.9
	0.15	20	12.1	13.1
	0.30	18	6.8	9.8

LOW POWER SUPER COMPACT FLUORESCENT LAMPS WITH HIGHER LUMINOUS EFFICACY

Dao Yu Fang and Sheng He Song
Institute for Electric Light Sources, Fudan University,
Shanghai 200433, P.R.China

The integral compact fluorescent lamps were introduced in the early eights, as replacements for incandescent lamps. Since that time they have found more and more applications worldwide, because of the energy savings and longer lives. One of the properties of compact fluorescent lamps that limits their widespread use as replacements of incandescent lamps is their dimension. In some cases it is not convenient to use the compact fluorescent lamps as replacements of incandescent lamps, because of the longer dimensions of the compact fluorescent lamps. In this paper we will concentrate on developments for high-voltage (220-240V) low power (3.5-6.5W) electronic compact fluorescent lamps with higher luminous efficacy and super minimal dimensions as replacements for 15-40W incandescent lamps.

The traditional low power compact fluorescent lamps have lower operating voltage and higher operating electric current (because of shorter arc length), the luminous efficacy of these lamps is lower. Therefore, in order to get a higher luminous efficacy, like the higher power compact fluorescent lamps, for the low power lamps a higher lamp operating electric voltage and lower electric current should be into consideration. In order to increase the lamp operating voltage, the most reasonable measurement is to take a glass tube with narrow inner diameter.

For the compact fluorescent lamps with 3.5-6.5W power dissipation, discharge tubes with 7.0 - 7.5 mm inner diameter and 120 - 200 mm arc length would be reasonable. In these cases the operating electric currents are rather low (about 50 - 70 mA), and the lamp voltages are a bit higher (about 80 - 120 V). In view of lower lamp electric current small electronic components can be used and the main part of the electronic ballast can be put into the cup. Bending the glass tube two times and integrating the bended discharge tube with the electronic ballast in the cup, we developed a series of low power super compact fluorescent lamps with higher luminous efficacy. Fig.1 shows the dimensions (in mm) of our super compact integral fluorescent lamps and traditional incandescent lamp. For the super compact fluorescent lamp with 6.2 W power dissipation, the lumen flux is 420 lm and the power factor is 0.72 .

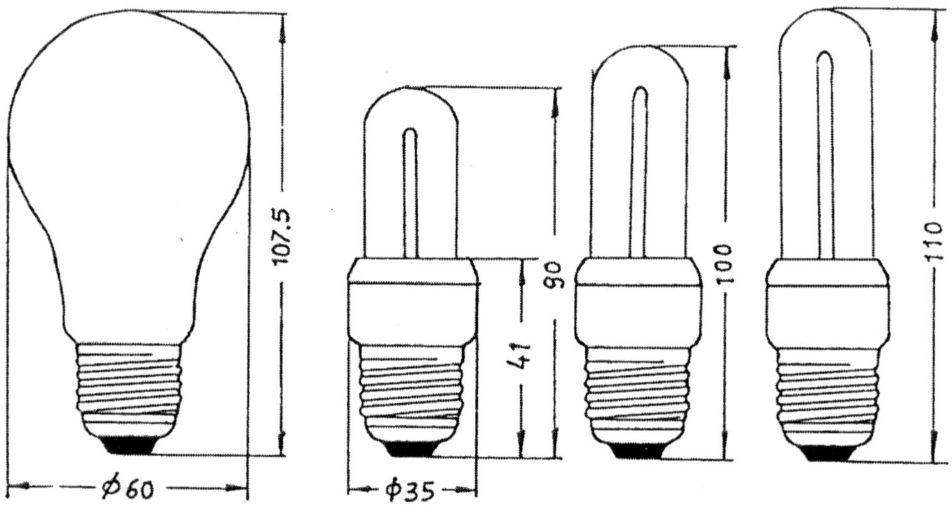

IMPROVEMENT OF THE STARTING PROPERTY OF THE ELECTRODELESS FLUORESCENT LAMPS IN DARK AMBIENCE

Miki Kotani, Masaki Shinomiya, Minoru Myojo*, Toshiyuki Namura*
Light Source R&D Laboratory Matsushita Electric Works, Ltd.
1048, Kadoma, Osaka 571, Japan
***Light R&D Center Light Business Group,Matsushita Electronics Corporation**
Takatsuki, Osaka, 569, Japan

The electrodeless fluorescent lamp was developed as a new compact light source[1]. However this lamp has a disadvantage in the starting performance in dark ambience. After leaving it under the brightness for several hours, it shows much longer delay to start in the dark. This phenomenon has been known in cold cathode fluorescent lamps[2] and high intensity discharge lamps. Previous approaches against this problem were using radio isotope, auxiliary radiation source or cesium oxide impregnated cathode[3]. Our approach is to use cesium oxide coated on the metal mesh located in the lamp bulb.

We tested lamps with the metal mesh coated with cesium oxide as shown in Fig.1,and the standard lamps without cesium oxide as reference.

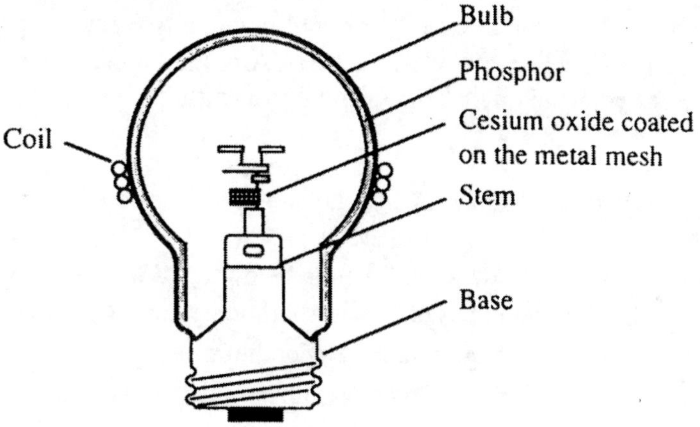

Fig .1 : A Lamp for electrodeless FL

Lamp diameter:45 mm. Ar: 133 Pa.

Lamp starting analysis using statistical method was conducted[4]. The formula (1) presents distribution of starting probability in the dark ambience.
Probability of disable to start is "*Y(t)*".

$$Y(t) = \exp(-\frac{t}{\tau}) \quad \cdots \quad (1)$$

"*t*" is starting delay time, " τ " is average value of starting delay time, and Y(t) is probability of disable to start. Probability of start is *1 - Y(t)*.

The lamps were left under brightness of 1,000 lx for 8 hours, and then they were ignited in the darkness below 0.01 lx. The lamp is spherical and the excitation coil is wound tightly around the lamp. At the starting,input the coil voltage is 500 Vo-p at 13.56MHz. The current into the electric RF power source was measured with an oscilloscope with a current probe. The relation of probability of disable to start and delay time is plotted in Fig.2 .

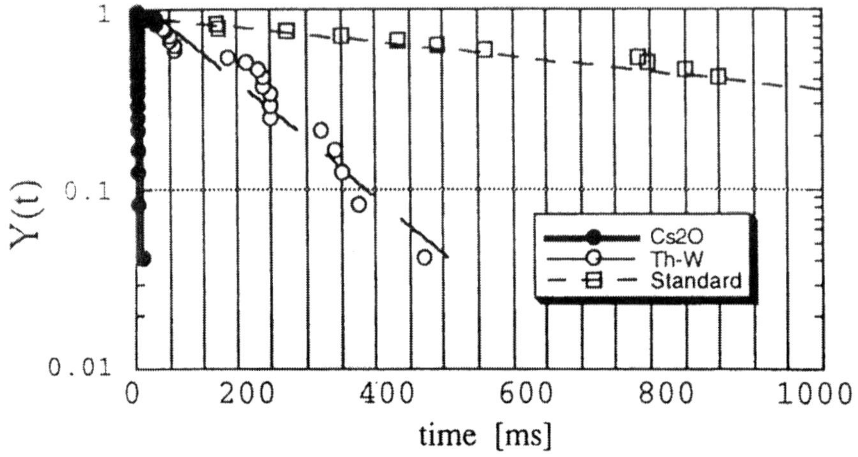

Fig.2: Starting probability in dark ambience as a function of delay time

Excitation coil: 3 turns; wire diameter=1.6 mm; pitch=2mm. Input coil voltage: 500 Vo-p. Operating frequency: CW 13.56 MHz. Ambient temperature: 25 ℃. Cs2O: Cesium oxide coated metal mesh. Th-W: Thorium Tungsten.

Average value of starting delay time of the lamps with cesium oxide was 6 ms. Standard lamp was over 500 ms.

Cesium oxide has very low work function below 1.2 Volts. We suppose that cesium oxide gives some electron by thermal emission under room temperature. As the result, free electron from the cesium oxide assists smooth starting in this condition.

Staring delay time in dark ambience was improved to 1/100 shorter by using cesium oxide coated metal mesh in the lamp.

Reference

1. Shinomiya,M. et al: Development of the electrodeless fluorescent lamp. J of the IES 20(no.1):44 (1991)
2. Inoue,A. et al: Statistical time lag of a cold cathode glow discharge tube. 1989 Annual Conf. IEI Jpn. No.9 (1989)
3. Yamazaki,H. et al: Development of miniature cold-cathode fluorescent lamp for LCD view finder. Proc. 1992 Annual Conf. IEI Jpn. No.42 (1992)
4. S.Cohen, G. Zaslavsky, J.N.Lester: Pulse Starting the 100-W Metal Halide Lamp. J of the IES vol.18:p3 (1989 Summer)

CHARACTERISTICS OF PLASMA IN RF ELECTRODELESS DISCHARGE

I. Mogi, Y. Yabumoto, G. Kanai
R&D Division, Iwasaki Electric Corporation
Gyoda, Saitama, 361 Japan
and
T. Kawabe
Institute of Physics, University of Tsukuba
Tsukuba, Ibaraki, 305 Japan

Recently, efforts of development of radio frequency(rf) discharge electrodeless fluorescent lamp have been increased. This lamp encloses Mercury and rare gases in the bulb, which is set in a rf coil. RF electric current flowing through the coil causes discharge in the bulb. It is important to know the relationship between the plasma parameters in the bulb and the light emission, particularly, in the range of the wave length of Ultra-Violet for fluorescent lamp. In this paper we report the experimental results on the measurement of the plasma parameters, such as the electron density(N_e) and electron temperature(T_e) by use of double probes, and on the comparison of the parameters with the light emission with the wave length of 254nm from the plasma in the bulb.

Experimental arrangement is shown in Fig. 1. Several types of the discharge tubes have been used in the present experiment. The diameter and the length of the bulb are same, however, the location of the tips of the probe are different to measure the spatial distribution of N_e and T_e both in radial and axially. The discharge tube is made of quartz glass in order to measure Ultra-Violet spectra from the plasma. The spectrum and spatial distribution of the 254nm radiation which excites phosphor were measured by the spectrum analyser and slit.

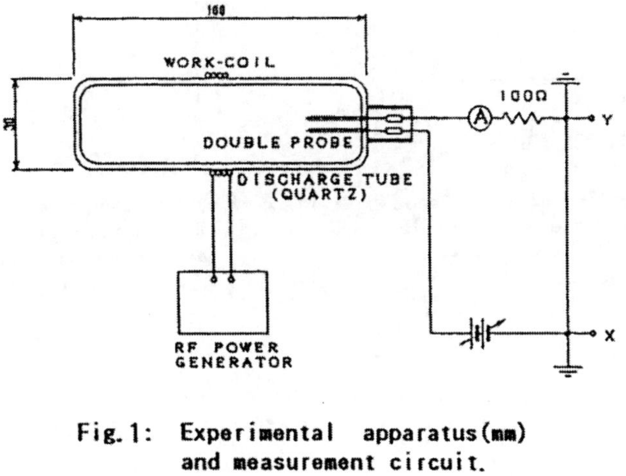

Fig. 1: Experimental apparatus(mm) and measurement circuit.

bulb diameter:30×100mm
pressure of Ar:133Pa
number of turn of coil:4
probe diameter:0.35mm
probe length:1.0mm
frequency of rf power:13.56MHz
input power:10–40watts

T_e and N_e have been measured by changing the discharge parameters, such as gas pressure, rf power. Simultaneously, 254nm radiation intensity was measured at each location of the bulb. By Abel transformation of 254nm radiation, distribution of the local intensity of the radiation was obtained.

Fig. 2. shows the axial distributions of N_e, T_e and integrated intensity of UV radiation in radial direction in a range of Z=0–40mm from the rf coil when the rf power was 10watts. From this figure, one can find that N_e decreased strongly along the axis, while T_e did not vary much. The UV emission decreased, which is considered to be due to decrease of N_e, mainly. It was found that the most of the characteristic of the plasma parameters in the discharge tube is same as the ones in standered fluorescent lamps[1]. Fig. 3. shows radial distribution of the local intensity of UV emission from the plasma. It was found that the local intensity has hollow structure near the coil.

Further experimental investigations on the dependence of the UV emission on the plasma parameters in the rf discharge are going on to obtain the scaling law.

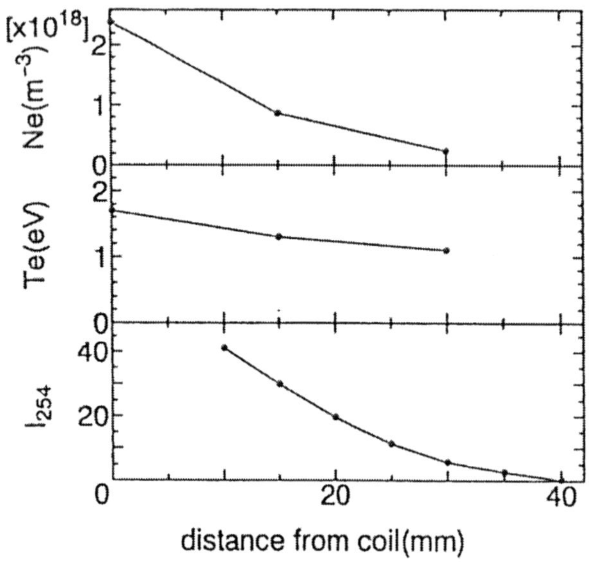

Fig. 2: The axial distribution of N_e, T_e and intergrated intensity of UV radiation from rf coil.

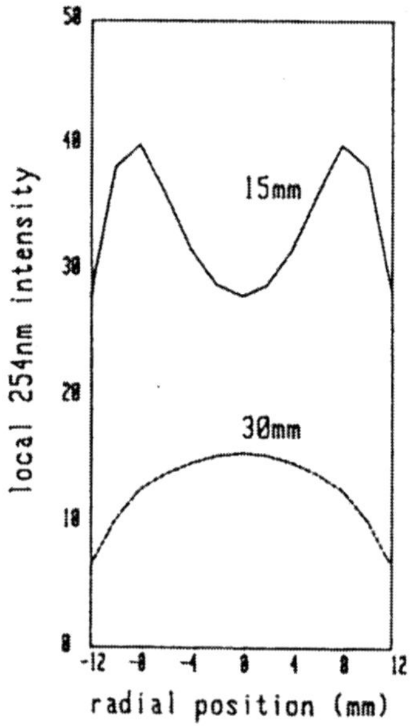

Fig. 3: Radial distribution of the local intensity by Abel transformation. Parameter is distance from coil.

Reference

1) J. F. Waymouth and F. Bitters, J. appl. Phys. Vol. 27 No. 2, 122–131(1956).

CONTROL OF UV RADIATION EMITTED FROM FLUORESCENT LAMPS

Johnson Kachidza and Pamela K. Whitman
GE Lighting, Nela Park, Cleveland, OH 44112

Introduction

There has been growing interest in measurement and control of ultraviolet energy emissions from lighting sources as more is learned about the biological impact of UV. In particular, new interest has been generated in the US as a new UV standard, BSR/IESNA RP-27[1], proposed by the IESNA Committee on Photobiology on the "Photobiological Safety of Lamps and Lamp Systems" is now being circulated for discussion.

Bergman, et. al recently presented a comprehensive review of measurement techniques and calculation of UV emissions from a variety of lighting sources.[2] This paper focuses on current state of the art for fluorescent lamps and methods for improved (reduced) UV emissions from these sources.

Comparison of Commercially Available Fluorescent Lamp Sources

In this paper, we present data obtained from measured UV spectral irradiance using the guidelines recommended by the proposed IESNA/RP-27 standard. We compute the Permissible Exposure Time (PET) value for each source at a distance from the source where the illuminance is 1000 lux and using the methodology recommended by and described in detail in (1).

The PET for common fluorescent lamps can vary widely based upon either lamp type or materials of construction. For instance, two 32 Watt T8 fluorescent lamps having the same nominal performance (lumen output and color) produce PET times from 15 to 54 hours depending upon the glass thickness, impurities, and the phosphor composition. The difference in UV emissions is primarily due to differences in the iron impurity levels in the soda lime glass envelopes; the T8 lamp made in the UK having around 500 ppm iron vs. the North American glass tubing containing around 200 ppm iron. Some difference in UV emissions can be attributed to phosphor type, for instance the common Ce-containing green triphosphors (CAT, LAP, CBT) generate differing quantities of UV-B in addition to visible radiation. Finally, increasing either phosphor coating or envelope glass thickness can reduce UV emissions from fluorescent lamps.

Controlling UV Emissions Through Glass Composition

UV absorbing glasses for various uses have been manufactured utilizing small amounts of UV absorbing material such as oxides of titanium, cerium, iron, vanadium, and manganese. Cerium oxide has been used in combination with vanadium oxide and titanium oxide[3] for reducing UV emissions from fluorescent lamps. Still, the vanadium oxide can produce a greenish yellow or amber coloration, which limits its acceptance. Iron oxide is a common impurity found in fluorescent lamp glass and will itself absorb UV-B radiation in the 280 - 320 nm range. However, the oxidation state of the iron oxide must be carefully controlled during glass melting to avoid discoloration of the glass and subsequently adversely affecting lamp performance.

UV emissions can also be effectively suppressed by doping the envelope glass with cerium oxide in combination with iron oxide.[4] Utilizing 0.03 - 0.07 wt. % cerium oxide in combination with 0.02 - 0.06 wt.% iron oxide will provide a 5 to 10-fold increase in while remaining essentially transparent to visible light.

Controlling UV Emission through Utilization of Internal and External Coatings

UV emissions can be effectively reduced by introducing a non-luminescent layer chosen to absorb or back-scatter UV which is either generated or not absorbed by the phosphor layers. This UV absorbing or reflecting layer can be located either internally between the glass envelope and the phosphor layers or externally. Materials which are contained inside the lamp must be able to survive in the mercury discharge without developing visible absorption and without contaminating the lamp discharge atmosphere. External coatings need sufficient toughness to survive handling and also must not develop visible absorption or otherwise affect the lamp color.

Internal coatings have long been explored for improved lumen maintenance, phosphor adhesion, and even to protect internal conductive tin oxide films from darkening[5]. One such film is a protective coating of titanium dioxide which is formed on the inner glass surface of a fluorescent lamp by applying a solution of tetraisopropyl titanate or tetrabutyl titanate[6]. Subsequent conversion of the organometalllic compound to the oxide yields a titania film. However, due to the high refractive index of titania, interference colors and high reflectivity limit the use of such a film in a fluorescent lamp. To minimize these problems, other organic metal compounds, such as magnesium acetate can be added to the mixture to produce a mixed titania magnesia film with reduced refractive index and hence lower lumen loss.

Exterior coatings have been developed which utilize combinations of colloidal titania, ceria, or zinc oxide[7]. In this case, the UV absorbing particles are dispersed in an organo-silicon sol gel which forms in an abrasion-resistant protective silicone matrix upon subsequent drying and heat treatment. Use of either ceria or zinc oxide in combination with titania allows reduction in the quantity of titania necessary to achieve the desired reduction in UV. This is desirable since the titania thus applied can become a reflector of visible radiation, and thus a layer which is thick enough to produce the desired reduction in UV emission will also reduce lamp performance.

We have developed internal coatings which utilize 10 - 30 nm colloidal powders applied to the interior glass surface of the lamp. These powders are dispersed in water using appropriate dispersants which will burn cleanly upon lehring the lamp. Ultra-fine titania powders have been developed to minimize the back-scatter of visible radiation while maximizing the reduction in transmitted UV. When well dispersed, these particles are too small to be effective scatterers of radiation above 400 nm, but are large enough to effectively scatter UV radiation below 360 nm. Hence, the reduction in emitted UV radiation is enhanced beyond that expected for an equivalent thickness of coherent titania film but visible radiation is not trapped in the lamp by excessive back-scattering. Coatings thus applied in thicknesses ranging from 0.1 - 0.5 microns were able to increase PET times by an order of magnitude without significant impact on lamp performance -- lumens or color. Similarly, precoat films formed from commercially available colloidal cerium oxide sols were able to increase PET by without substantively impacting lamp performance. One advantage of these colloidal films over the doped glasses discussed previously is that these films also reduce UV-A emissions from 320 - 360 nm, which can lead reduced color fading [8] of organic dyes found in fabrics and paintings. In addition, these internal colloidal films act as a protective layer to minimize the interactions of the highly energetic Hg discharge with the soda lime glass envelope and hence can replace the colloidal alumina or silica precoat layers which are commonly utilized to improve lumen maintenance.

Conclusion

A variety of methods have been demonstrated to reduce the UV radiation emitted from fluorescent lamps below currently proposed legislative standards. Doping of the envelope glass with cerium, iron, or a combination of these elements in the proper valence can provide up to a 10-fold increase in the PET without loss of lamp performance. Internal and external coatings containing colloidal titania, ceria, zinc oxide, Internal and external coatings containing colloidal titania, ceria, zinc oxide, or a combination of these materials can produce similar increases in PET while also reducing UV A emissions below 360 nm.

References:

1. "Photobiological Safety of Lamps and Lamp Systems", BSR/IESNA RP-27 (formerly Z311.1), 193, Illuminating Engineering Society of North America, New York.
2. R.S. Bergman, T.G. Parham, T.K. McGowan, "UV emissions from general lighting lamps, " proceedings of the IESNA Annual Conference, 1993.
3. Escher-Desrivieres, U. S. Patent 2,860,059.
4. G.L. Bucher, C.H. Welker, E.E. Hammer, C.E. Scott, T.F. Soules, " UV absorbing lamp glass", US Patent 5,350,972.
5. E.E. Hammer, US Patent 3,845,343; J.B. Jansma, T.G. Parham, P.K. Whitman, US Patent 5,258,689; K. Shiraiwa, EP 540 788 A1 (application); K. Arai, H. Sumoto, K. Ichinomiya, U. S Patent 5,008,789; K. Narisei, JP 60-250550 (application); S. Hiramatsu, JP 61-91848 (application).
6. W.C. Martyny, " Fluorescent lamp envelope with transparent protective coating", US Patent 3, 599,029.
7. A. Tay, "Ultraviolet ray suppressive light source, application agent for it and its manufacture, JP Patent 64-37686 (application); Y. Sakakibara, " Fluorescent Lamp", JP Patent 2-84346.

Performance Advantages of a Helical Compact Fluorescent Lamp

Thomas F. Soules, Jennifer I. Barry, Erwin G. Steinbrenner and Laura Kicher
GE Lighting
NELA Park, Cleveland, Ohio 44112
and
Timothy Sommerer
GE Corporate Research and Development
Schenectady, New York

The problem for the lamp designer making compact fluorescent lamps (CFL's) is how to configure a sufficiently long arc length so that for a current giving reasonably high efficiency, 100-320 mA, enough light output is achieved within a small package size. In particular, since lamp harps and fixture reflectors are usually designed around the shape of incandescent lamps, the ideal CFL would fit within the envelope of the incandescent lamp of corresponding light output. CFL's today made from connected vertical bent tubes are significantly taller than their incandescent counterparts and since their base must accommodate six or eight tubes, the base size is larger than that of an incandescent lamp.

The purpose of this report is to discuss performance advantages of a double helix CFL design formed from a single piece of glass tubing. In this design the arc travels from one electrode up in a helical fashion, crosses over the top and then travels back down between the coils of the first helix 180 degrees out of phase with the arc point directly across. Our helical lamp designs are about 1/3 shorter than comparable CFL's with vertical legs and fit within the envelope of the corresponding wattage incandescent lamps. Examples are shown in Figure 1.

Compact shape

Several factors contribute to the more compact shape of the helical design. A single turn of a helix contains more arc length than a comparable height of 6 or 8 tubes in the same outer radius. More importantly the double helix continuously trades off diameter for height so that whereas CFL's made of 3 bent tubes spaced about 2 mm apart are 37-39 mm in width and those with 4 bent tubes 47 mm, the double helix can be made to have the same diameter as the corresponding GLS incandescent: 60, 67 or 73 mm and as shown in Fig. 1 can have a changing diameter along the axis of the lamp thus filling in the top portion of the lamp with arc length. A further decrease in overall lamp height is achieved by mounting the electrodes generally in a radial direction bent off a tangent of the helix. In conventional CFL designs the electrodes are mounted vertically so that their height adds directly to the height of the lamp.

Reduced visible light loss

In addition to its compact shape the helical lamp design results in less light trapped and lost in the structure. Some light is lost when a CFL lamp is connected to its base due to light generated inside of the lamp being multiply reflected and "light piped" down into the base and light radiated by the structure which bounces off the base housing. See Figure 2. This light loss was determined by measuring the peak lumens of lamps connected to their bases and then measuring the lumens when the lamps were separated different distances from their bases with white tape placed over the holes in the base. Lamps were also operated us'ng a high frequency power supply with narrow white wires running through small holes in the ballast housing with the lamps connected and then separated. Although the light lost in this way depends on the length of the tubes, the visible reflectivity of the coating and the proximity of the lamp to the base, in most cases it amounts to about 1.5 % of the total light output for each tube-base connection. Hence a lamp made using three bent vertical tubes with 6 ends in the base gains about 9 % in light output when the base is removed. A helical lamp with only 2 ends connected to the base looses only 3 % when connected to its base. If the lamp is inserted in the base up to the first turn in the coil another 1 % of the light is lost.

The second source of visible light loss in CFL's discussed by Wesselink (Philips J. Res. 35, 337-353,1980) and Ingold (GE Report, 1994) is due to light radiated into the interior of a CFL structure which undergoes multiple reflections with other tubes with some light transmitted through openings between the tubes and through the tubes themselves and some lost by absorption by the phosphor coating and glass of other tubes (Figure 2). This loss was measured by placing an optical fiber at various points in the different CFL structures. The average escape probability for light from the fiber when corrected for light lost due to reflection back to the fiber is given by

$$P_{esc} = \frac{(1-P_s)+P_s(1-R-A)}{1-P_sR}, \qquad (1)$$

where P_s is the ratio of the cross sectional area of tubing to the area of the lamp envelope and is a measure of the probability that light radiating into the structure will strike another tube. R is the reflectivity of the tubes and A is the tube absorptivity. The light radiated by the CFL relative to the straight lamp is given by

$$L = S_{out} + S_{ins}P_{esc}\ , \qquad (2)$$

where S_{out} is the fraction of the tubes surface area which radiates outward and S_{ins} is the fraction which radiates into the lamp structure. S_{ins} can be calculated as was done by Wesselink or it can be approximately measured using black felt inside the lamp. Typically for lamps with 3 or 4 bent tubes 40-45 % of the light radiates inward and 12-17 % of that light is lost by absorption depending on phosphor thickness giving 5-7 % of the light radiated being lost. The light loss in this way for the more open helical structures can be calculated as a function of coil spacing and is typically around 3 %.

Better light distribution

From Figure 1 it is clear that the light distribution from a helical lamp is uniform around the axis of the lamp. It is also quite uniform from the top of the lamp down to 90 degrees. In this way the light distribution from the helical lamp closely approximates a GLS incandescent lamp. Typical CFL lamps with vertical tubes radiate much more light sideways and very little light in the forward direction of the lamp. They may or may not radiate uniformly around the axis of the lamp depending on the arrangement of the tubes.

Design flexibility

Helical lamps can be made in much smaller package sizes than shown in Figure 1 by decreasing the diameter of the tubing. This increases the impedance or watts dissipated per unit length of the arc and in the helical design further gives a proportionately smaller lamp since the lead or pitch of helix can be decreased. To take full advantage of narrower tube diameters, CFL's with vertical tubes would have to add additional tubes connected into the base with concomitant increase in light loss discussed above.

With tubes narrower than about 12 mm fluorescent lamps are less efficient due to increased wall losses. This can be compensated for by running the lamp at a lower current which increases efficacy. Hence if lamp current and voltage can be controlled, a helical lamp can alternatively be designed to fit within a specified design package and the tube diameter and lamp current and voltage chosen for maximum luminous efficacy.

Rugged construction

Finally since helical lamps are made from a single tube these lamps are actually the most physically rugged on the market.

Figure 1. Current HELIAX helical plug-in compact fluorescent lamps

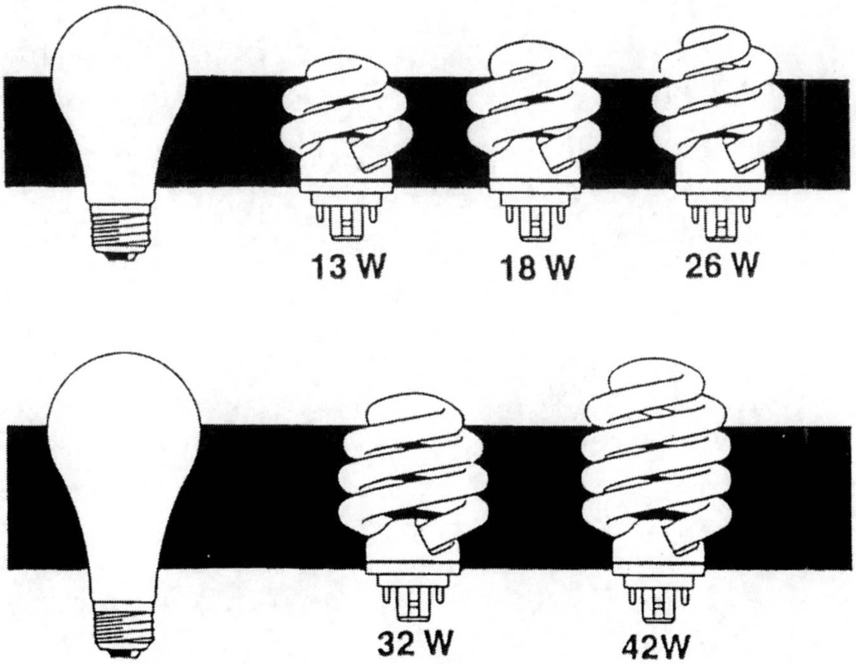

Figure 2. Schematic showing light radiated by a CFL including (A) light radiated from outer phosphor coated surfaces, (B) light generated on the inside of the tube which is multiply reflected and light-piped toward the end of the tubing, (C) light bouncing off the base of the lamp, (D) light radiated into the interior of the lamp which escapes through open areas in the lamp and (E) the same but transmitted through other phosphor coated tubes.

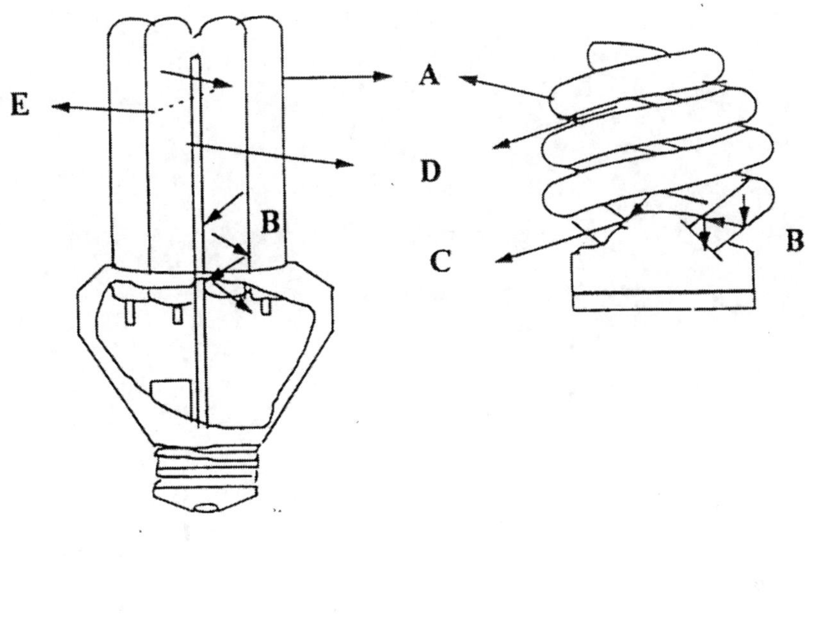

PROGRESS IN ELECTRONIC DEVICES FOR LAMP OPERATION

H. Nishimura
Matsushita Electric Works, Ltd.
1048, Kadoma, Osaka 571 Japan

1. INTRODUCTION

Among electronic devices for lamp operation, electronic ballast for fluorescent lamps is currently in the spotlight.
In 1994 in Japan the percentage of application of the electronic ballast for fluorescent lamps, among all kinds of ballasts, reached 26% in the area of residential use. In the United States, where the energy-saving "Green Light Program" is being promoted, the percentage has reached 24% in office and commercial use. Functions and performance demanded for the electronic ballast differ in applications. This paper discusses the progress of the electronic ballast in terms of applications to office/commercial and residential use, use for replacing incandescent lamps, and applied systems.

The electronic ballast for HID lamps was first marketed in 1986, for use as a general illumination. This appearance came 10 years after that of the electronic ballast for fluorescent lamps. The basic difference between the electronic ballast for HID lamps and that for fluorescent lamps is that the former provides no distinctive improvements in luminous efficiency.
Due to this difference, advantages other than luminous efficiency were pursued in the electronic ballast for HID lamps. This paper also discusses developments focusing on advantages obtained from the combination of HID lamp and electronics.

A number of electrodeless lamps and lighting systems have been developed recently. An electrodeless lamp system can be constructed only in combination with electronics. However, much of the technique is still developing with regard to the total efficiency, dimming capability, and development into electrodeless HID lamps. The progress of the technique is discussed also from such viewpoints.

2. ELECTRONIC DEVICES FOR FLUORESCENT LAMPS

2-1. ELECTRONIC BALLAST

Fig. 1 shows change by years in the popularization of the electronic ballast;

(a) illustrates the progress in Japan, (b) in the United States. Noting the popularization in that manner of the electronic ballast, this paper tries to account for its manner of popularization and to review the progress.

a) The electronic ballast for office and commercial use (typically, 40-W class and greater)

The electronic ballast for office and commercial use has been employed for energy-saving purposes.[1] Energy saving is accomplished by improvements in luminous efficiency, operating fluorescent lamps with a high frequency of several tens of kHz. Accordingly, the core of the electronic ballast is an inverter for generating a high frequency. One of the lessconspicuous advantages of the inverter is that it permits variations to a degree in the electrical characteristics of a fluorescent lamp, so that the design can accommodate varied characteristics. In other words, in a operating circuit using conventional ballast, the value of the lamp voltage for the fluorescent lamp is determined by the value of the supply voltage, while in the case of a operating circuit using an inverter, the value of the lamp voltage of the fluorescent lamp can be selected from relatively optional values. Reviewing lamp design from this viewpoint and combining an electronic ballast and a lamp of higher luminous efficiency (for example, a "T-8" size lamp operated solely by a high frequency), we attained greater energy-saving effects.[2]
In addition, as the inverter makes it possible to employ a fluorescent lamp of a smaller diameter, it became possible to design a thinner luminaire, as shown in Fig. 2, giving higher added value.[2][3][4]

b) The electronic ballast for residential use (typically, 40-W class and smaller)

More than 6 million units of residential luminaire with electronic ballast are shipped annually in Japan. The electronic ballast for residential applications should satisfy the following demands (1) to (3), as well as the energy saving demand.

(1) In a multi-lamp operating unit, removing one lamp does not cause any of the remainder to go off.

(2) When a failed lamp is replaced with a normal one, the newly installed lamp should light even if replacement is carried out with power supplied. (The lamp should light up independently of turning the power switch ON/OFF.)

(3) The luminous output should be dimmable by approximately 50%.

That these demands were met by several years ago has contributed to the popularization today of the electronic ballast.[5 - 22]

c) The electronic ballast for compact fluorescent lamps

The main application of the compact fluorescent lamp is as a substitute for the incandescent lamp. Consequently, one of the most important technical problems of the electronic ballast for compact fluorescent lamps is compatibility between consideration for temperature increase of electronic circuit components and their mounting into a small space. Contrivance for that purpose has been made, including the automatic cutting of output under high temperature condition of components, employment of a structure insulating against heat produced by the lamp, and development of heat-resistant components.[23 - 25] As an alternative to incandescent lamps, compact fluorescent lamps should basically be inexpensive and also should permit dimming with a dimmer designed for incandescent lamps.[25 - 28]

2-2. DEVELOPMENT AND IMPROVEMENT OF THE ELECTRONIC BALLAST

a) New method for the improved remedying of input current distortion

In general, the electronic ballast is constructed by combining an AC/DC converter and high-frequency inverter. A common way to make improvements in remedying input current distortion has been the use of an AC/DC converter. Two types have been used. One is known as the passive filter type and the other, the active filter type. These two types of filters are the result of trade-offs between price and size/weight. As a way to utilize the advantages of these two types, consideration was given to sharing of circuit components between the active filter and inverter as shown in Fig. 3. [27][29 - 34] The method has been partly already put into practical use.

b) The dimming method applicable in the range of very low luminous output using the electronic ballast for fluorescent lamps

The electronic ballast for fluorescent lamps is capable of dimming by varying the operating frequency. In the range of small ratios of luminous output (ex. 10%), however, it suffered problems such as unstable discharge or going off. In order to attain reduction in the luminous output ratio safely (ex. to approximately 0.1%), the circuit configuration as shown in Fig. 4 was examined.[35] Fig. 4 illustrates the addition of two circuit elements;

(1) supply of a very small current to the lamp under low luminous output condition as a measure against unstable discharge and (2) voltage compensation under low luminous output condition as a measure against accidental going off. An electronic dimming ballast employing this method is already in practical use.

c) Electronic ballast applied lighting system

A typical example of systemic use of the electronic ballast is the color temperature controlling lighting system. The color temperature controlling lighting system uses RGB fluorescent lamps, in which an electronic ballast capable of dimming in the range of low luminous output is used to operate the lamps. A light diffusing panel is placed under each fluorescent lamp. The color of each fluorescent lamp is blended on the panel. It is possible to obtain a radical change in the color temperature on the light diffusing panel by controlling the dimming ratio for each fluorescent lamp.[36] Tab. 1 shows an example of specifications used in such configuration.

3. ELECTRONIC DEVICES FOR HID LAMPS

3-1. ELECTRONIC BALLAST

Differing from the case of fluorescent lamps, luminous efficiency attained by operating an HID lamp with a high frequency cannot be expected to increase. Conversely, as is commonly known, operating an HID lamp with a high frequency may cause "acoustic resonance." For this reason, many other kinds of methods for operating HID lamps have been discussed in addition to the use of a high frequency.[37 - 45] A typical method except the high-frequency operating method is a low-frequency square-wave operating method, and both methods are already in practical use.

Since operating HID lamps with the electronic ballast cannot be expected to increase luminous efficiency, this has led to demand for other advantages from the electronic ballast.
As a result, development for the following applications has been attempted.

- Spotlights and downlights using compact HID lamps utilizing the advantage of light weight
- Specific applications such as the backlight for liquid crystal projector, utilizing the characteristic of flicker-free luminous output.
- Headlights on automobiles utilizing the characteristic that the electronic ballast can operate on a battery

Recent papers report on the possibility of longer service life of HID lamps, noting that the electronic ballast offers easy control of output characteristics.[46][47]

3-2. SUPPLEMENT ON THE INTRINSIC CHARACTERISTICS OF LAMPS

a) Addition of the hot-restart function

Hot-restart of an HID lamp requires a voltage of several tens of kV. It is desirable to decrease such a high level of required voltage in terms of safety and product reliability. Fig. 5 shows an example of methods applicable to this voltage reduction, illustrating that it becomes possible to reduce the pulse energy necessary for hot-restart, by superimposing two kinds of pulse voltages differing in width and height and applying them to the lamp.[48][49]

b) Stabilization of pulse voltage for lamps with a built-in bimetal starter

Operating systems that combine a ballast for mercury vapor lamps and a bimetal switch installed in the outer bulb of a high pressure sodium lamp are popular in Japan. The bimetal switch is used to generate starting pulses. However, as the timing in the opening of the bimetal varies, the system has some delay after turning the power switch on until the starting of the lamp, and the pulse voltage generated is not uniform.

Various discussions were given to the HID lamp igniter.[50][51] In particular, one possibility that is drawing attention is that which has, instead of the bimetal switch, a non-linear capacitor in the outer bulb of a high pressure sodium lamp. The non-linear capacitor becomes operable immediately when the power switch is turned on. It also offers stability in the value of generated pulse voltage.

4. ELECTRONIC DEVICES FOR ELECTRODELESS DISCHARGE LAMPS

Electronics are necessary to operate an electrodeless discharge lamp.

In commercialized products, 2.65 MHz or 13.56 MHz is used as the operating frequency. Both frequencies have advantages and disadvantages when viewed from the aspects of circuit efficiency, price, RF power transfer to lamp, and controllability.[53 - 55]

The first commercial electrodeless fluorescent lamp and its applied fitting were marketed for decoration and display purposes. One of the characteristics of the electrodeless fluorescent lamp is crisp turning on/off of luminous output.

Fig. 6 compares its crispness with that of common fluorescent lamps and incandescent lamps. This characteristic is desirable for decoration/display applications. After the decoration/display applications, electrodeless fluorescent lamps producing under 6000 lm of luminous output were put to practical use for general illumination.

The electrodeless discharge lamp has had only a brief history since it was first marketed. Accordingly, it still has a number of problems in the field of electronics. One of these problems is the addition of a dimming function. In the case of a operating circuit using 13.56 MHz, the luminous output can be dimmed, varying the duty of RF power supplied to the lamp, with the operating frequency being fixed. That which has a lower limit of 5% in luminous output has been put to practical use.[56]

5. SUMMARY

Progress in electronic devices for lamp operation is strongly correlated with demands arising from lamps themselves and circumstances in which they are used. These factors are quite significant in the discussion. Consequently, I have reported the status of the progress of each kind of lamp in each area of application.

In the area of the fluorescent lamp, where its popularity is great, the development of electronic devices should be aimed at through energy saving, enhancement of convenience in residential applications, and realization of compact units substituting for incandescent lamps. In the case of the HID lamp, objectives for employing electronic devices are diversified, ranging from the development of light-weight and compact luminaire combining an electronic device with a compact HID lamp to the use of HID lamps not designed for illumination purposes. In the area of the electrodeless discharge lamp, the history of practical electronic devices has only just begun. Accordingly, the future direction of progress seems to be toward improvements in the basic performance and addition of lighting unit functions.

REFERENCES

[1] S.Takahashi et al : J. Illum. Engng. Inst. Jpn. [77], 7, 394 - 398, (1993)
[2] T.Kakitani et al : Proc. Illum. Engng. Soc. NA. Annual Conf. 759 - 782, (1993)
[3] K.Takahashi et al : Proc. Illum. Engng. Inst. Jpn. Annual Conf. #16, (1992)
[4] T.Kobayashi et al : Proc. Illum. Engng. Inst. Jpn. Annual Conf. #17, (1992)
[5] K.Kousa et al : Proc. Illum. Engng. Inst. Jpn. Annual Conf. #12, (1992)
[6] Y.Nakazawa et al : Proc. Illum. Engng. Inst. Jpn. Annual Conf. #15, (1992)
[7] K.Tanaka et al : Proc. Illum. Engng. Inst. Jpn. Annual Conf. #22, (1991)
[8] F.Nakaya et al : Proc. Illum. Engng. Inst. Jpn. Annual Conf. #24, (1991)
[9] Y.Matsui et al : Proc. Illum. Engng. Inst. Jpn. Annual Conf. #26, (1991)
[10] F.Nakaya et al : Proc. Illum. Engng. Soc. NA. Annual Conf. 493 - 516, (1990)
[11] Y.Hiratomo et al : Proc. Illum. Engng. Inst. Jpn. Annual Conf. #24, (1990)
[12] Y.Yamanaka et al : Proc. Illum. Engng. Inst. Jpn. Annual Conf. #25, (1990)
[13] H.Ohsaki et al : Proc. Illum. Engng. Inst. Jpn. Annual Conf. #27, (1990)
[14] H.Ohsaki et al : Proc. Illum. Engng. Inst. Jpn. Annual Conf. #18, (1989)
[15] K.Atagi et al : Proc. Illum. Engng. Inst. Jpn. Annual Conf. #19, (1989)
[16] Y.Yamanaka et al : Proc. Illum. Engng. Inst. Jpn. Annual Conf. #20, (1989)
[17] N.Kishimoto et al : Proc. Illum. Engng. Inst. Jpn. Annual Conf. #21, (1989)
[18] Y.Igarashi et al : Proc. Illum. Engng. Inst. Jpn. Annual Conf. #24, (1989)
[19] K.Yamada et al : Proc. Illum. Engng. Inst. Jpn. Annual Conf. #34, (1988)
[20] H.Komatsu et al : Proc. Illum. Engng. Inst. Jpn. Annual Conf. #36, (1988)
[21] F.Nakaya et al : Proc. Illum. Engng. Inst. Jpn. Annual Conf. #37, (1988)
[22] C.Ohki et al : Proc. Illum. Engng. Inst. Jpn. Annual Conf. #38, (1988)
[23] M.Kika et al : Proc. Illum. Engng. Inst. Jpn. Annual Conf. #7, (1991)
[24] Y.Hirao et al : Proc. Illum. Engng. Inst. Jpn. Annual Conf. #19, (1990)
[25] K.Kawabata et al : Proc. Illum. Engng. Inst. Jpn. Annual Conf. #18, (1990)
[26] J.J.Groot et al : Proc. Illum. Engng. Soc. NA. Annual Conf. 55 - 73, (1993)
[27] E.E.Hammer et al : Proc. Illum. Engng. Soc. NA. Annual Conf. 49 - 65, (1992)
[28] H.Fujii et al : Proc. Illum. Engng. Inst. Jpn. Annual Conf. #23, (1987)
[29] NIKKEI ELECTRONICS (journal of Japan) ,no.622, 161 - 168
[30] A.Hisako et al : Proc. Illum. Engng. Inst. Jpn. Annual Conf. #13, (1994)
[31] M.Yamada et al : Proc. Illum. Engng. Inst. Jpn. Annual Conf. #15, (1993)
[32] A.Hiramatsu et al : Proc. Illum. Engng. Soc. NA. Annual Conf. 341 - 360, (1992)
[33] H.Mannami et al : Proc. Illum. Engng. Inst. Jpn. Annual Conf. #18, (1992)
[34] A.Hiramatsu et al : Proc. Illum. Engng. Inst. Jpn. Annual Conf. #19, (1992)
[35] A.Okude et al : Proc. Illum. Engng. Soc. NA. Annual Conf. 140 - 164, (1991)

[36] H.Takeuchi et al : Proc. Illum. Engng. Soc. NA. Annual Conf. 700 - 728, (1993)
[37] S.Nakanishi et al : Proc. Illum. Engng. Inst. Jpn. Annual Conf. #35, (1989)
[38] K.Arai et al : Proc. Illum. Engng. Inst. Jpn. Annual Conf. #37, (1989)
[39] T.Oumi et al : Proc. Illum. Engng. Inst. Jpn. Annual Conf. #39, (1987)
[40] K.Shimizu et al : Proc. Illum. Engng. Inst. Jpn. Annual Conf. #40, (1987)
[41] H.Nagase et al : Proc. Illum. Engng. Inst. Jpn. Annual Conf. #41, (1987)
[42] K.Shimizu et al : Proc. Illum. Engng. Inst. Jpn. Annual Conf. #29, (1986)
[43] H.Nagase et al : Proc. Illum. Engng. Inst. Jpn. Annual Conf. #37, (1986)
[44] H.Nishimura et al : Proc. Illum. Engng. Soc. NA. Annual Conf. 392 - 410, (1987)
[45] H.J.Faenrich et al : Proc. Illum. Engng. Soc. NA. Annual Conf. 411 - 441, (1987)
[46] N.Fukumori et al : Proc. Illum. Engng. Soc. NA. Annual Conf. 343 - 361, (1994)
[47] T.Higashi et al : Proc. Illum. Engng. Inst. Jpn. Annual Conf. #24, (1993)
[48] M.Kika et al : Proc. Illum. Engng. Inst. Jpn. Annual Conf. #30, (1992)
[49] N.Arakawauchi et al : Proc. Illum. Engng. Inst. Jpn. Annual Conf. #24, (1993)
[50] G.Zaslavsky et al : Proc. Illum. Engng. Soc. NA. Annual Conf. 402 - 428, (1989)
[51] J.M.Proud et al : Proc. Illum. Engng. Soc. NA. Annual Conf. 636 - 650, (1986)
[52] R.Mackenzie et al : Lighting in Auatralia [15], 1, 23 - 25, (1995)
[53] K.Vegter et al : Proc. Illum. Engng. Soc. NA. Annual Conf. 641 - 653, (1993)
[54] V.A.Godyak et al : Proc. Illum. Engng. Soc. NA. Annual Conf. 654 - 674, (1993)
[55] M.Shinomiya et al : Proc. Illum. Engng. Soc. NA. Annual Conf. 138 - 158, (1990)
[56] T.Nakano et al : Proc. Illum. Engng. Inst. Jpn. Annual Conf. #20, (1993)

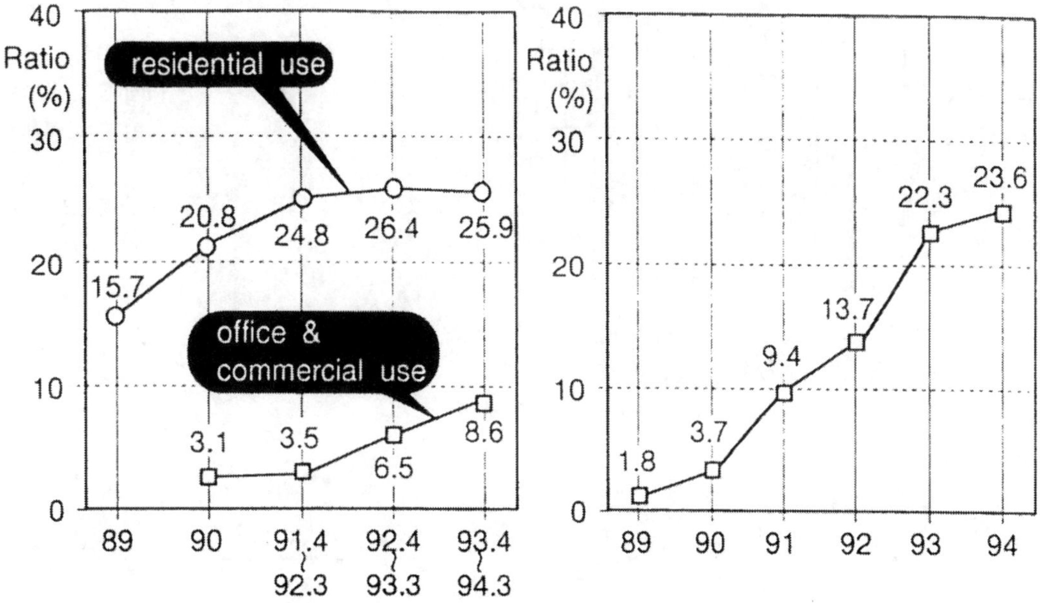

Fig.1 Popularization ratio of the electronic ballast v.s. Year
(Electronic／All Kind = shipment number respectively)

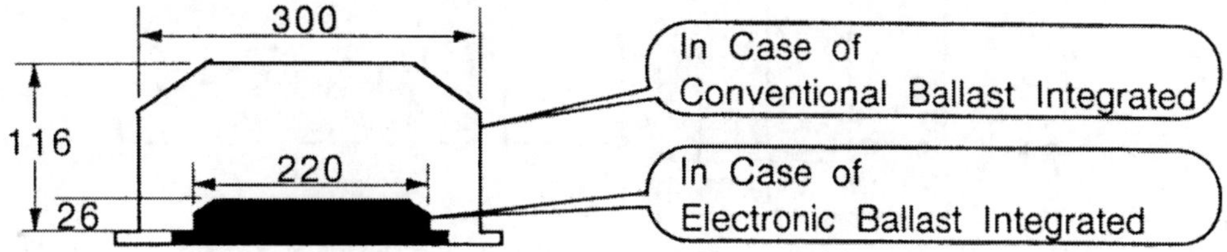

Fig.2 Sectional plan of the luminaire (Recessed type)

Luminaire	Recessed (with Light Diffuser)
Lamps	36W Fluorescent Lamp Red(2), Green(3), Blue(2)
Ballasts (for each lamp)	Electronic Dimming Ballast =Dimmable continuous 0 - 100%
Control Signal for Dimming	1kHz, Duty-Control Control each ballast separately
Variable Color Temp. Range	2800 - 30000K (Arbitrary)
Variable Luminous Output Range	20 -100% (Arbitrary)

Tab.1 Specification example of color temperature control lighting system

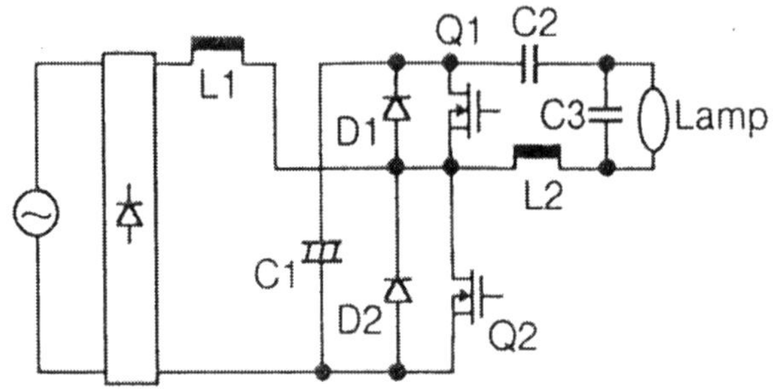

Fig.3 Circuit example (improved THD)
Q2 & D1 is served both as
boost converter and inverter

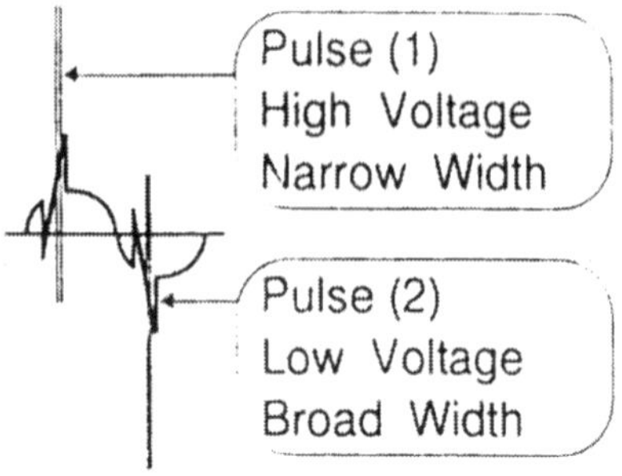

Fig.5 Open circuit voltage
for hot-restrike

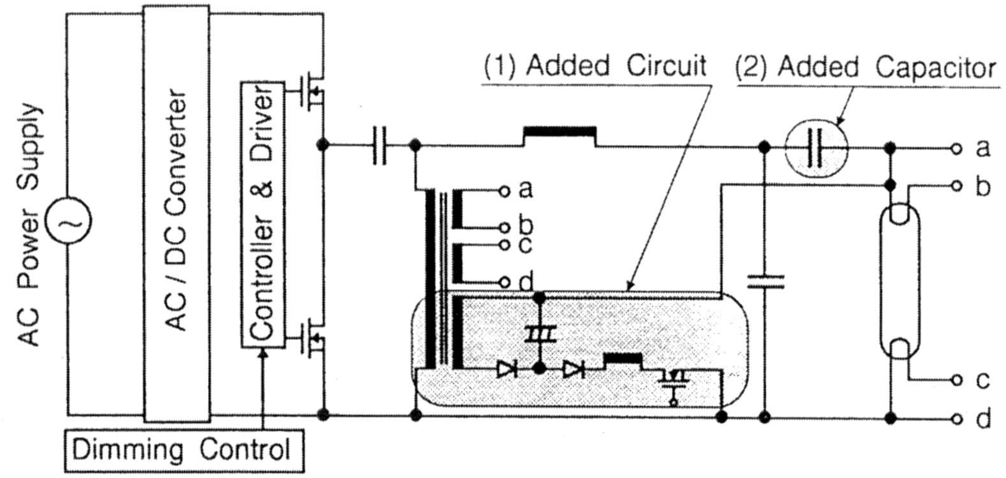

Fig.4 Electronic dimming ballast for fluorescent lamp

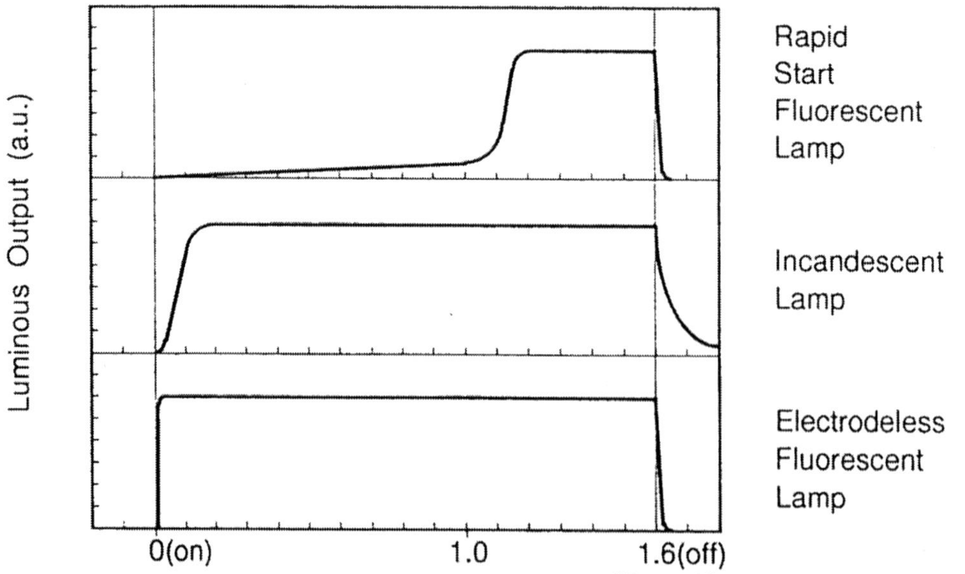

Fig.6 Responsibility of luminous output

EXPERIMENTAL STUDY OF THE FLUORESCENT LAMP WITH AN ELECTRONIC DIMMER CONTROLLER

Guan-Chyun Hsieh
National Taiwan Institute of Technology
Taipei, Taiwan, R.O.C.

Eel-Peeng Tsai
China Electric Mfg. Corp.
Taipei, Taiwan, R.O.C.

Ping-Shen Sung
Taiwan Electric Research & Testing Center, Inc.
Taipei, Taiwan, R.O.C.

Recently, the dimming control incorporated in the high-frequency lighting systems has been widely developed to compensate for the daylight coming in through the outdoor. In this paper, a series-resonant half-bridge inverter (SR-HBI) with dimmer controller is proposed for dimming the fluorescent lamp in a constant-frequency and constant-duty pulse-width control mode. The presented SR-HBI inverter operates in the discontinuous-conduction mode (DCM) with three launching modes for energizing the fluorescent lamp including heating filament, ignition for gas discharge, and lighting. Experimental investigation for the dynamic characteristic of the fluorescent lamp has explored the important parameter, lamp capacitance, which is more significant in high frequency operation [3]. It is obvious from the test result that the lamp resistance and lamp capacitance are both functions of the lamp current and operating frequency. In the design of SR-HBI, the fluorescent lamp is installed in parallel with the starting capacitor. Remarkably, the starting capacitor presents a low reactance before firing the lamp and becomes a high impedance after igniting the lamp. An accurately modeling of the dimmer system in series resonance is then achieved according to the derived dynamic characteristic of the fluorescent lamp [3]. A 40-Watt fluorescent lamp equipment including a dimmer controller with three-section controls is realized for verifying the theoretical prediction. In this design, the resonant frequency is set at 62 kHz and the three-dimming sections are typically specified as luminance's of 100%, 66%, and 30%. The controlled parameters for the three-dimming sections are respectively measured and shown in Table 1. The use of constant switching frequency and constant duty cycle for controlling the fluorescent lamp can overcome the EMI and lower the temperature rise of the system. Fig.1 is the measured lamp voltage in series-resonant for energizing the fluorescent lamp in heating, ignition, and lighting. It is seen that the cold-starting voltage is 1.3 kV_{p-p}, the cold-starting current is 1.24 A_{p-p}, and the heating time for the filament is 1.5 second. Fig. 2 shows the series-resonant lamp voltage and lamp current for fully lighting of 100% luminance at switching frequency of 50.37 kHz. The proposed SR-HBI for energizing the fluorescent lamp leads to significant advantages in lighting performance. This design results in high power factor of 0.997, lower temperature rise of 25°C, energy savings of near 8 watts, and 10% increase of lighting efficiency.

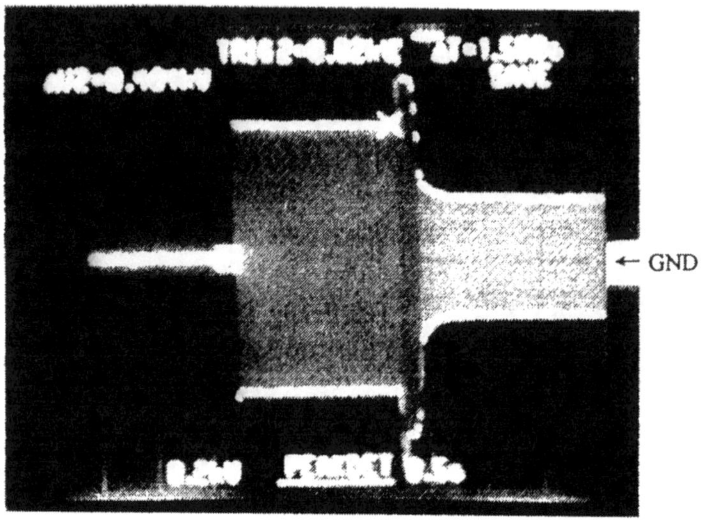

Fig. 1 Measured lamp voltage profile for energizing the fluorescent lamp (Ver.: 200 V/div., Hor.: 0.5 sec/div)

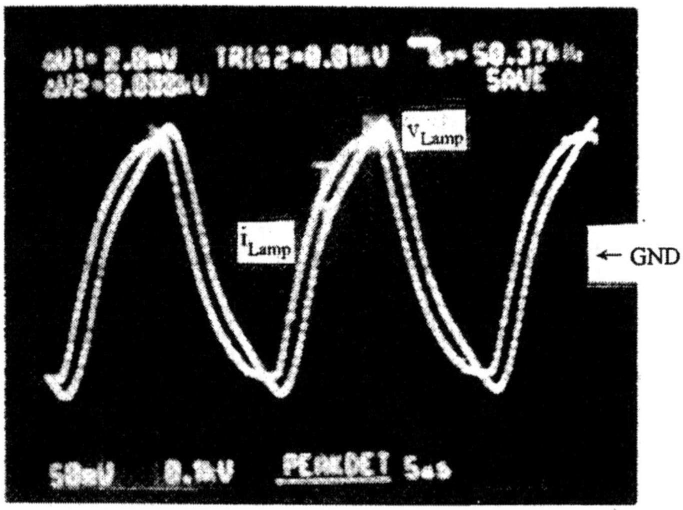

Fig. 2 Series-resonant lamp voltage (Ver.: 100 V/div.) and lamp current (Ver.: 0.2 A/div.) for fully lighting at the switching frequency of 50.37 kHz. (Hor.: 5 μs/div.)

Dimming sections	1	2	3
Dimming percentage	100%	66%	30%
Resonant frequency	62 kHz	62 kHz	62 kHz
Switching frequency	51.4 kHz	51.4 kHz	69.2 kHz
Duty cycle (%)	37	37	37
PFC output voltage	365 V	250 V	250 V
Input power	43.8W	·31.1W	24.5 W
Lamp output power	32 W	22 W	12.3 W

Table 1 Measured data for a 40-W fluorescent lamp energized by SR-HBI with dimmer control

THE STARTING SITUATION OF THE FLUORESCENT LAMP OPERATED BY THE ELECTRONIC BALLAST

Tadao Uetsuki, and Nobuhiro Hayashida
Matsushita Electric Works LTD.

Introduction

Electronic ballasts operating fluorescent lamps at relatively high frequency have been used practically. The high frequency operation seems to cause some different performance of the lamp from the low frequency operation in the starting process of the lamp. If there is the significant difference in the starting, it would influence on the lamp life. The purpose of this paper is to report the starting situation focusing on the vicinity of the electrode (filament coil) operated with high frequency, which would be different from the case of low frequency operation.

Experiment

One of commercially available ballasts was used to operate the lamp. Firstly, we observed the vicinity of the electrode (filament coil) during the starting period under the normal operation, in which the ballast supply some preheat current to the lamp. Secondary it was observed under the condition that the electrode (filament coil) was heated without voltage applied to the lamp (between electrodes). Lastly it was observed under the instant start condition.

The specification of the lamp used in this experiment is same as regular FL30 on the market, and the electronic ballast was designed for it which is also on the market.

1st experiment

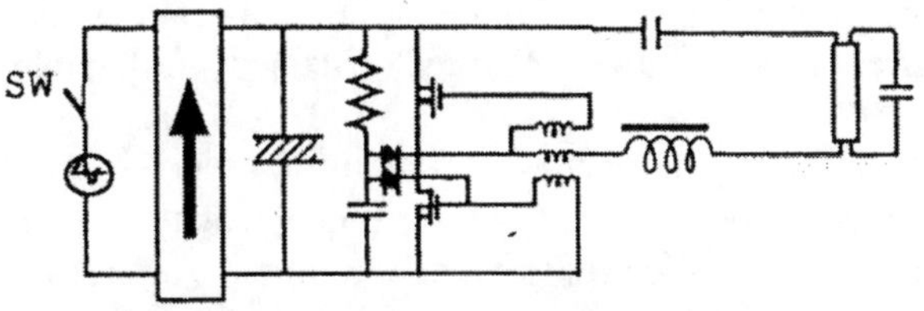

Result

As soon as the SW is turned on, the high intensity plasma occurred inside of the electrode (filament coil)._ PHOTO-1

After that the lamp current increased to the rated current._ PHOTO-2

Both the time until the lamp current reaches the rated current and the intensity of this plasma depend on the voltage applied to the lamp.

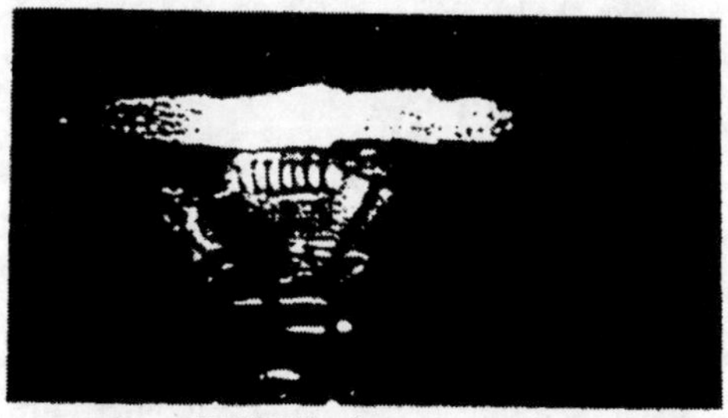

PHOTO-1

PHOTO-2

2nd experiment

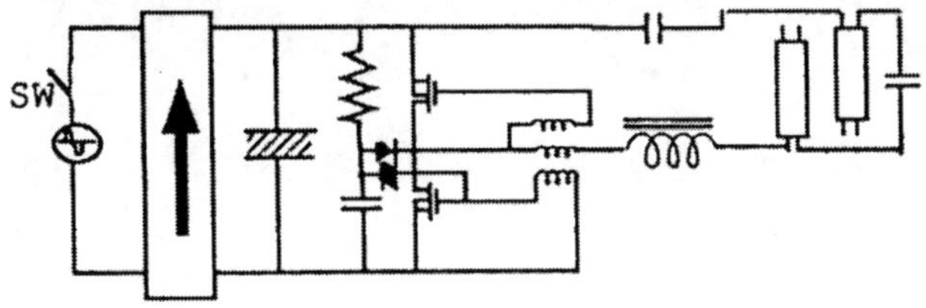

Result

The electrode (filament coil) was heated to red in around 1 sec. after turning on the SW ._ PHOTO-3

And then (so-called) afterglow was observed._ PHOTO-4

The plasma inside of the electrode (filament coil) shown in PHOTO-1 was not observed.

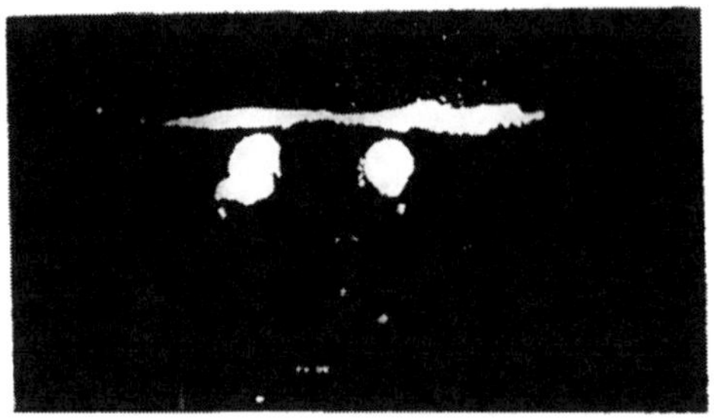

PHOTO-3

PHOTO-4

3rd experiment

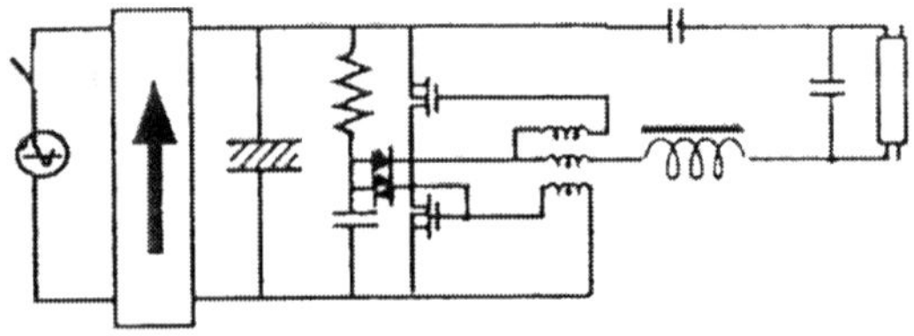

Result

Just after the SW was turned on, the high intensity plasma occurred inside of the electrode (filament coil). After that the lamp current increases to the rated current. These situations were same as those shown in PHOTO-1 and PHOTO-2.
Both the time until the lamp current reaches the rated current and the intensity of this plasma depend on the voltage applied to the lamp strongly compared with the case of 1st experiment.

Discussion

The plasma (having really high intensity) occurs inside of the electrode (filament coil) when the lamp is operated by an electronic ballast. From the result of three experiments, this plasma is considered to be caused not by the current flowing into the electrode (filament coil), but by the voltage applied to the lamp.
Desirable conditions of the preheating of the high frequency operation for a specific lamp are recommended by Dr.Albert Heidemann and Werner Denz.(1) But the high intensity plasma has been observed in the vicinity of the electrode under the electronic ballast operation. The electric conductivity of the high intensity plasma is really large compared with that of the electrode (filament coil). So there should be a possibility that some of the preheating current (to heat filament coil) pass through the plasma. As the result, there should be the possibility that the electrode temperature is not sufficiently high for the lamp starting.

Conclusion

When the lamp is operated by an electronic ballast, the high intensity plasma which was not been observed under low frequency operation occurs inside of the electrode (filament coil).
There should be a possibility, under the high frequency operation, that some of the preheating current (to heat filament coil) pass through the plasma, so that the electrode temperature is not sufficiently high when the lamp starting.
The relation between this plasma characteristics and the electrode temperature during the starting process should be researched quantitatively.

Reference

(1) Albert Heidemann and Werner Denz, IESNA ANNUAL CONFERENCE, P.751, 1993

THE CIRCUIT MODEL OF FLUORESCENT LAMPS UNDER HIGH FREQUENCY OPERATING

M. Taihei, T. Urayama
Department of Electrical and Electronic Engineering
Faculty of Science and Engineering, Ritsumeikan University

1916 Noji-cho, Kusatsu, Siga, 525 Japan

The mathematical models of discharge lamps are theoretical models introduced from the plasma modeling and experimental models expressed by equivalent conductance equation. These numerical models are complex and required for solution many hours of time on computer.

In recent years, the Computer Aided Design(CAD) technology is widely used in the design and analysis of various kinds of electronic circuit.

Applying the CAD to the discharge lamps operating circuits, the models must be presented by the circuit Component describing the current and voltage characteristics of discharge lamps.

Fig. 1 shows the typical voltage and current of fluorescent lamps under high frequency operation. The current and voltage characteristics of fluorescent lamp depends primarily on the state of ionization of electric discharge and also the state of ionization are affected largely by the operating frequency, Namely, at low operating frequency of less than several hundred hertz, The lamp voltage waveforms indicates a similar rectangular shape and can be divided clearly into reignition mode, negative resistance mode and distinguish mode in every half cycle.

A reignition voltage reduces at high frequency operation of greater than 1 kHz the boundaries of those three modes become indistinct and the voltage and current waveforms have a similar shape in each other.

Although it is mentioned in generally that a relationship between current and voltage of discharge lamps becomes linear under high frequency operation. Its the current and voltage characteristics are slightly non-linear as shown in Fig. 2.

The numerical models of non-linear property are explained by a number of equation such as nth power equation, polynomial equation.

In order to examine the degree of non-linearity under high frequency operation, we present the current and voltage characteristics as shown in Fig. 2 by nth power equation as follows

$$v_L(t) = k\ i_L(t)^n \qquad (1)$$

Where, $V_L(t)$ and $i_L(t)$ are simultaneous values of lamp current and voltage, k is the coefficient that the dimension is resistance.

As the fluorescent lamps are operated on the sinusoidal source and inductive ballast, a typical value of the experimental results of coefficient k and power n are shown in Fig. 3. It is clear from experimental results that the value of n changes in range of 2.0 to 1.0 within frequency of 10 to 100 Hz .

REFERENCE

V.J.Francis; Fundamental of Discharge Tube Circuits, Methuen & Co. Ltd.., England, ('48)
K.Wani; J. of IES of J., 73-9 ('89)
J.H.Campbell, et al. ; Illuminating Engineering, Vol. XLVIII, No.2 p.95 ('53)
U.Mader, et al. ; Proc. IEEE, APEC '93, P.418 ('93)

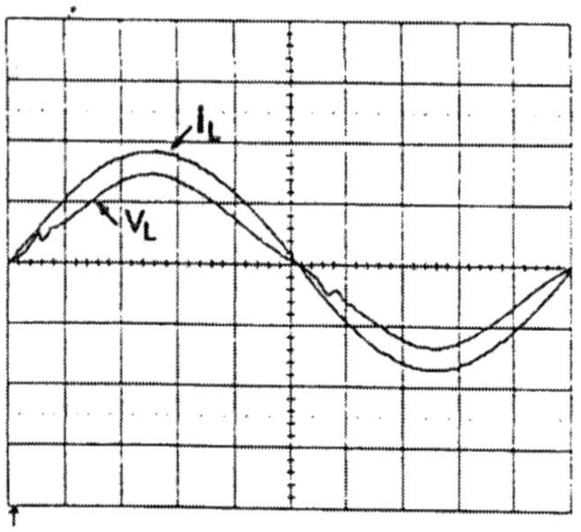

fig. 1 Lamp current and voltage waveform under operating frequency of 10kHz

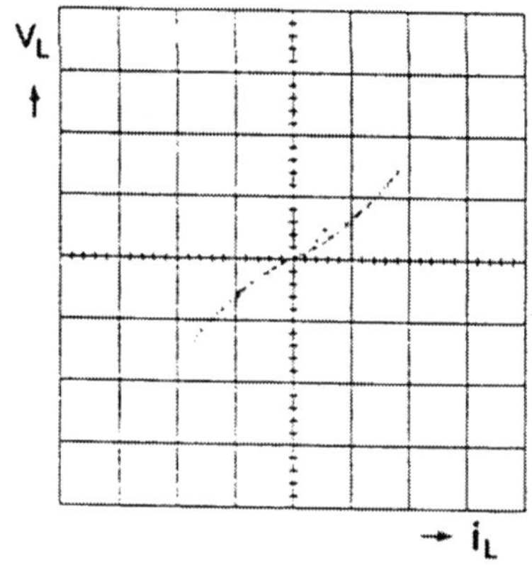

fig. 2 Voltage and current characteristic under operating frequency of 10kHz

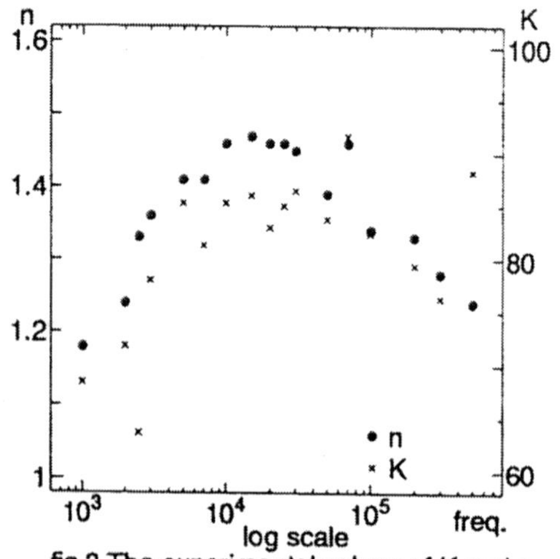

fig.3 The experimental values of K and n

THE EFFECTS OF THE OPERATING FREQUENCY ON THE DYNAMICS OF THE BARIUM EMISSIONS AND THE ELECTRODE TEMPERATURE OF A FLUORESCENT LAMP

Katsuhide Misono(1) Joseph T. Verdeyen(2)
(1) Toshiba Lighting & Technology Corporation
1-201-1, Funakoshi-cho, Yokosuka-shi, Kanagawa 237 JAPAN
(2) University of Illinois at Urbana-Champaign
607 East Healey Street, Champaign, IL 61820 USA

1. Introduction

The electrode phenomena of fluorescent lamps have been widely studied recently because of the practical importance of this subject.[1]-[4] In this work, we will discuss the dynamic behavior of the barium emissions (BaI:553.5nm, BaII:455.4nm) and the electrode temperature under the square wave AC operation ranging from 6Hz to 6,000Hz, and propose the guidelines to design the electrode suitable for high frequency operation.

2. Experiments and Discussion

The lamps investigated were usual F20T10 with double coils coated with alkaline earth oxides (emission mix). These lamps were aged for more than 100 hours to stabilize the characteristics. The line emissions from the barium atoms and the thermal emissions from the hot spot (882.5nm) were isolated by a monochromator (HEATH; EU-700), detected by a photomultiplier (RCA; 31034A), and recorded on a digital oscilloscope (TEKTRONIX; 11402A).

Fig.1 and Fig.2 show the dynamics of the hot spot temperature and ionized barium emission for different operating frequencies. These figures demonstrate that the barium emissions increase during the cathode cycle even though the spot temperature decreased (at ~6Hz) implying that sputtering of the emission mix is a major source of barium in vapor phase. At higher operating frequency (~600Hz), the barium emissions and the time derivative of the spot temperature increased during the cathode cycle even though the time averaged temperature was lower than at 6Hz. After the anode oscillation disappeared (>600Hz), the barium emissions as well as the spot temperature saturated (Fig.3). This suggests the existence of dense plasma near the electrode at high frequency operation.

We explain these observations in the following manner: First, the higher electron density enhances the ionization of the mercury atoms, which are then accelerated back to the cathode and this, in turn, increases the spot temperature as well as increases the sputtering rate. Finally, the latter increases the density of barium in the vapor and thus the barium emissions increase.

We used the diamagnetic characteristics of a coil inductively coupled to the tube to measure the plasma density. From the impedance characteristics of the coil with and without plasma, the plasma parameters (electric conductivity, effective collision frequency and plasma electron density) were determined. Fig.4 shows the electron density as a function of the operating frequency. This revealed the existence of more dense plasma at high frequency operation compared with the density at low frequency operation.

A mathematical model based on the continuity equations for neutral and ionized barium atoms (Ba,Ba*,Ba^{+},Ba^{+*}) qualitatively explained the experimental results.

3. Conclusion

Sputtering cannot be negligible to discuss the loss mechanism of the emission mix during operation. The difference of the electrode phenomena between low and high frequency operations comes from the change of the plasma density near the electrode as well as the disappearance of the anode fall. The design of the electrode for high frequency operation should be modified from that for low frequency operation to improve the life performance. The fundamental ideas are the reduction of the plasma density near the electrode and the optimization of the electrode temperature distribution. One of the solutions is to control the auxiliary heating power.

References

[1] Bhattacharya,A.K.,*J.Appl.Phys.*,65-12(1989)4595,65-12(1989)4603

[2] Watanabe,Y.,*LS-6*(1992, Budapest, Hungary)14:L

[3] Moskowitz,P.,*LS-6*(1992, Budapest, Hungary)25:P

[4] Misono,K.,Verdeyen,J.T.,*46th Gaseous Electronics Conference* (1993, Montreal, Canada)LA-2,QB-6

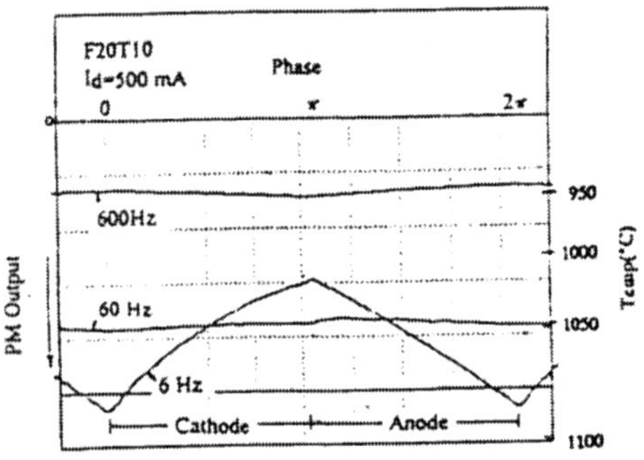

Fig.1 The Hot Spot Temperature

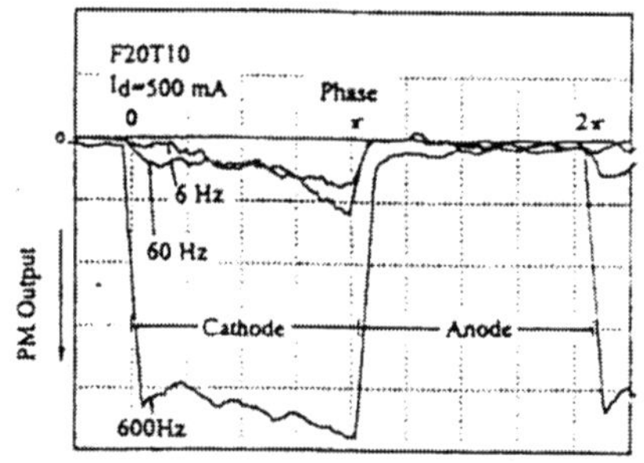

Fig.2 The Ionized Barium Emission

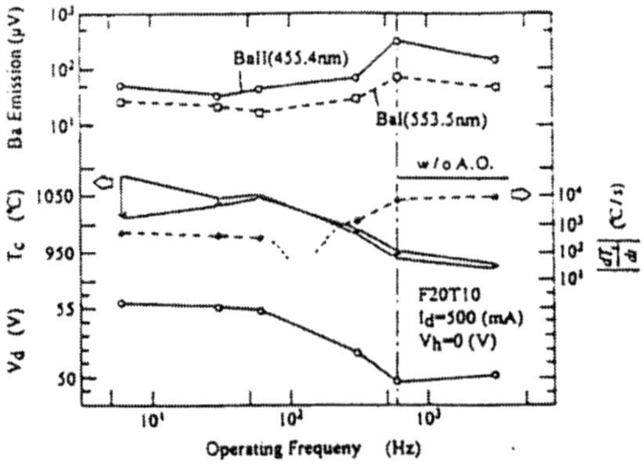

Fig.3 Effects of the Operating Frequency

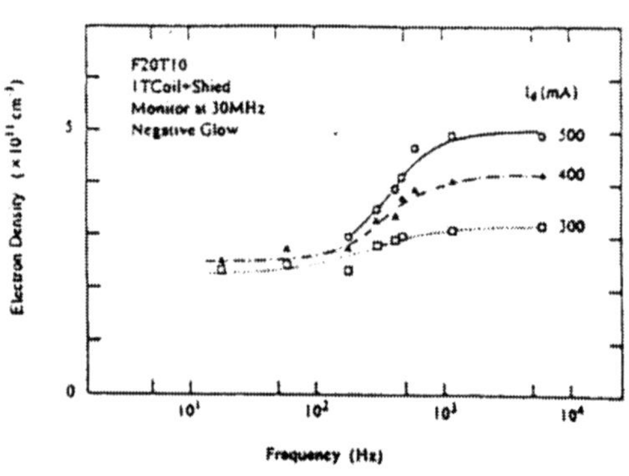

Fig.4 Electron Density as a Function of the Operating Frequency

Color Control of Fluorescent Lamps

Masaharu AONO[A], Masafumi JINNO[A], Makoto KUBO[B]
and Ryohei ITATANI[C]

A:Department of Eloctrical and Electronic Engineering, Ehime University
3 Bunkyo-cho, Matsuyama, Ehime 790, JAPAN
B:Department of Electronics Engineering, Kyoto University
Yoshidahonmachi, Sakyo-ku, Kyoto 606-01, JAPAN
C:Niihama College of Technology
7-1 Yagumo-cho, Niihama, Ehime 792, JAPAN

Lighting creates the comfortable room atmosphere. Five colors of *Lamps*, *Warm White*, *White*, *Neutral* and *Daylight* are specified for the color of the "White" fluorescent lamps by the Japanese standard system (JIS Z 9112). One fluorescent lamp produces only one color of the light. Therefore, in order to change the color of the room light, one must exchange the fluorescent lamp for another one of the desired color. That is impracticable.

Authors have developed the color-changeable discharge lamps using pulsed discharge[1)2)]. In this system, decreasing pulse width and frequency, the electron temperature in the positive column increases and in turns, radiation from atoms of higher upper state energy levels increases.

This method was applied to the operation of fluorescent lamps. A mixture of phosphors excited by the mercury resonance lines and by the xenon resonance lines is coated on the inner wall of a fluorescent lamp filled with mercury and 200Pa xenon. The excitation levels of xenon atom are higher than those of the mercury atom (as shown in Fig.1). In the pulsed discharge with the short duration and the low frequency the spectra of the xenon are emitted more intensely than those of the mercury. Then radiation from the phosphors for xenon is emitted more intense than that from the phosphors for mercury.

The phosphors Y_2SiO_5:Ce (A) absorbs xenon resonance lines and emits the blue light (shown as ○ on the CIE chromaticity diagram in Fig.2). On the other hand, the phosphors 3.5MgO-0.5MgF_2:GeO_2:Mn (B) absorbs mercury resonance lines and emits the red light (◆ in Fig.2). The blend (C) of the above two kinds of phosphors (A) and (B) mixed at the weight ratio of 1 to 1 emits the light of the chromaticity coordinate (0.277, 0.176) under the pulsed operation (pulse width = 10μs, repetition frequency = 3kHz, peak current = 75mA) (□ in Fig.2) and the light of the coordinates (0.470, 0.241) on the 28.6kHz AC operation (■ in Fig.2).

The blend (D) of two kinds of phosphors $Ca_{10}(PO_4)_6$FCl:Sb, Mn and $Ca_{10}(PO_4)_6F_2$:Sb,Mn mixed at the weight ratio of 1 to 2 absorbs mercury resonance lines and emits the orange colored light (● in Fig.2).

Increasing the repetition frequency of the pulsed operation, the color of the blend (E) comprised of (A) and (D) at the weight ratio 1 to 2 shifts from the blue side to the orange colored side (△→▽→◇ in Fig.2).

By shifting the color of the raw phosphor for mercury toward the right-hand side in the chromaticity diagram, the color-variable range of a fluorescent lamp will be expanded. To attain the fine control of the color, the contributions of the color produced by the gaseous atoms within the lamp must be estimated.

When the ionization potential of one kind of gaseous atom is lower than the resonance potential of the other, the latter does not emit its radiation under AC or DC operation. Since the ionization potential of mercury exists between the resonance and ionization potentials of xenon, xenon always emits its radiation under AC or DC operation.

The most of phosphors for mercury radiation available now are excited by xenon radiation and the converse is also true. These two facts reduce the control range of the color. Another important problem to be solved is the low intensity on the pulsed operation with short duration and low frequency.

The authors wish to express their gratitude to Mr. H.Kurowaka for his many contributions to this experimental work.

References

1) M.Aono, H.Kurokawa, M.Kubo and R.Itatani: J. Light & Visual Environment, Vol.3,No.1,pp1-9 (1989)

2) R.Itatani, M.Kubo, M.Aono and M.Ryoko: Proc. 6th Int. Symp. Science & Tech. Light Sources (LS-6), Budapest, pp.85-86 (1992)

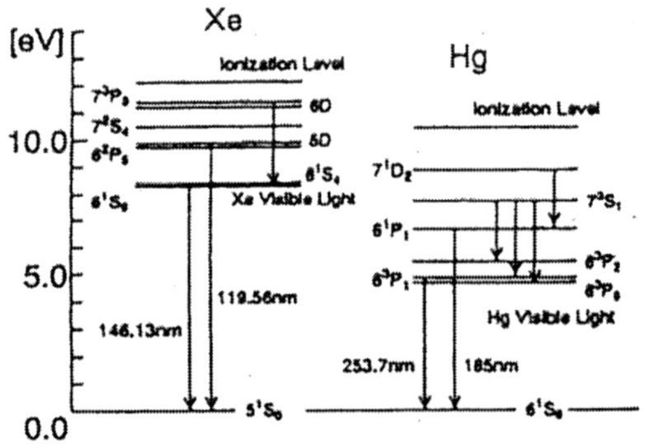

Fig.1 Exciting energy levels of Hg and Xe

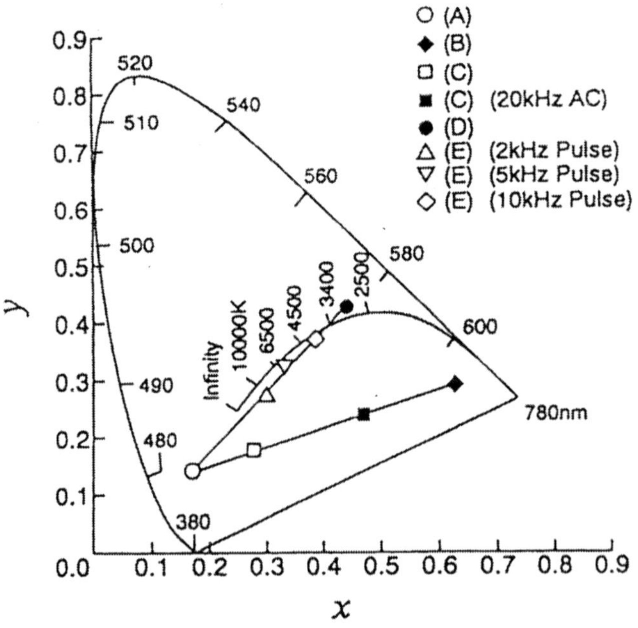

Fig.2 The colors of fluorescent lamps on CIE chromaticity diagram

DESIGN CONSIDERATIONS FOR COLOR TEMPERATURE VARIABLE FLUORESCENT LAMPS

Shinkichi Tanimizu * and Akihiko Kougami **

Hitachi Ltd.,
* Heating and Lighting Division, 888 Fujihashi, Ome, Tokyo 198, Japan
** Central Research Laboratory, Kokubunji, Tokyo 185, Japan

This paper describes design considerations and experimental results for the color temperature variable discharge lamp, designed to operate on a pulsed circuit which renders the lamp capable of changing from about 3000K to 6500K.

The color of light from a fluorescent lamp has two components: that from the mercury vapor discharge and that from the phosphor coatings. The 436nm blue line stands for the former color, and the color shift for Hg-Ne lamp has been reported 1). The latter color is produced by different phosphors or by varying the proportions of RGB phosphors in the blend.

In order to realize large color shift along the Planckian locus under pulsed operation, it is very effective to select at least a complementary pair of phosphors such as (Type I) having extraordinary different decay times (Table 1 and Fig. 1) and/or (Type II) having different saturation characteristics of luminescence for the same increase of pulse voltage (Fig. 2). Here, the strong 436nm line ($7^3S_1 \rightarrow 6^3P_1$) may be selected as a component of the former pair, initial decay time of which is from 6 to 9 μs (Fig. 3).

As for the tricolor phosphor lamps with color points falling on the Planckian locus, each of RGB light output can be calculated, assuming total light outputs of RGB to be constant. Both of R and B outputs vary drastically depending on the color temperature, while G output is nearly constant for the range of 3000K to 7000K. Therefore, the RB pair selection becomes important.

For the low pressure mercury discharge lamp (F8T5 tube) coated with long decay Mn^{2+} red phosphor, which belongs to Type I, the largest color shift of ΔT = 3170K was obtained by changing duty factor of 1 μs pulse (Fig. 4). The observed color differences (Δuv) were within 0.005. This color shift means that single fluorescent lamp can provide a variety of colors, the most common being Warm White, Cool White and Daylight. Another approach using the saturation effects of phosphors has been carried out by a pulse height modulation method for the low pressure Xe lamp, i.e., no visible radiation lamp. The RB pairs belonging to Type II were selected from the curves as shown in Fig. 2. At a fixed duty factor of 12.5%, the observed color shifts were limited to ΔT = 600K.

Reference

1) R. Itatani et al : LS-6, Budapest, Hungary, 1992, Paper 15:L, p. 85.

Table 1 Typical fluorescent decay times of various activator ions embeded into oxygen-dominated host lattices.

Emission color	Activator ion	1/e decay time
Blue	Ce^{3+}	~ 80 ns
	Eu^{2+}	0. 8 ~ 1 μs
	Sb^{3+}	~ 6 μs
	VO_4^{3-}	~15 μs
Green	Mn^{2+}	5 ~15 ms
	Tb^{3+}	1 ~ 4 ms
Red	Eu^{3+}	0. 7 ~ 1 ms
	Mn^{2+}	30 ~50 ms
	Mn^{4+}	~ 3 ms

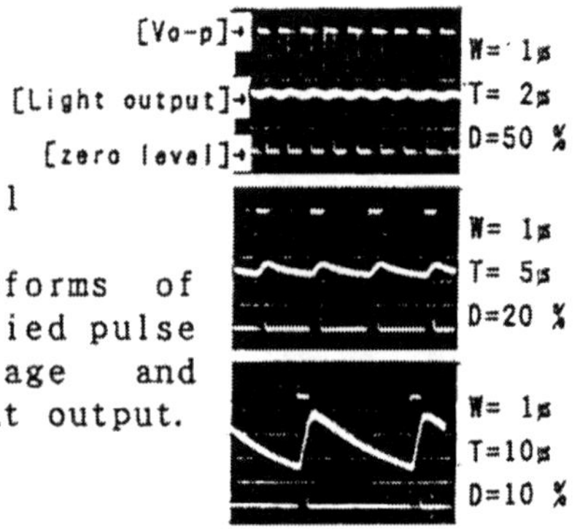

Fig. 1

Waveforms of applied pulse voltage and light output.

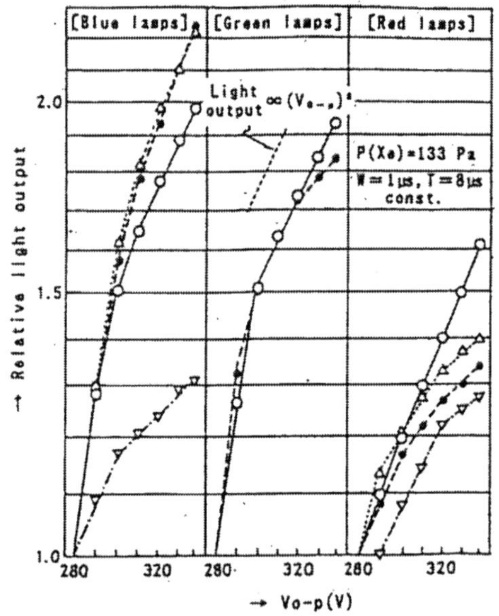

Fig. 2 Log-log plot of relative light output vs dc-pulse voltage for low pressure Xe lamps.

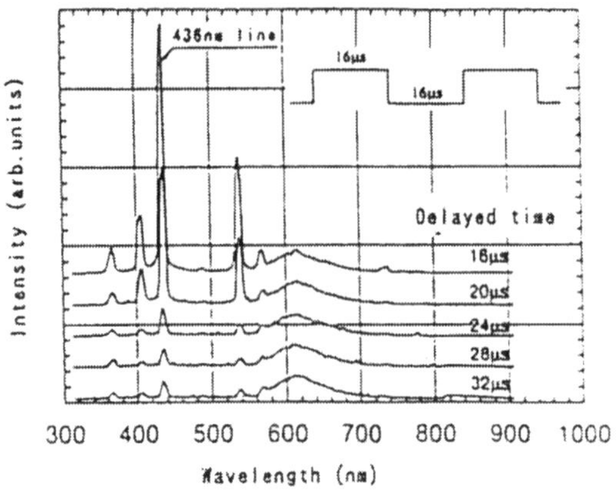

Fig. 3 Time-resolved spectra for a low pressure Ar-Hg discharge lamp coated with Mn^{2+} red phosphor.

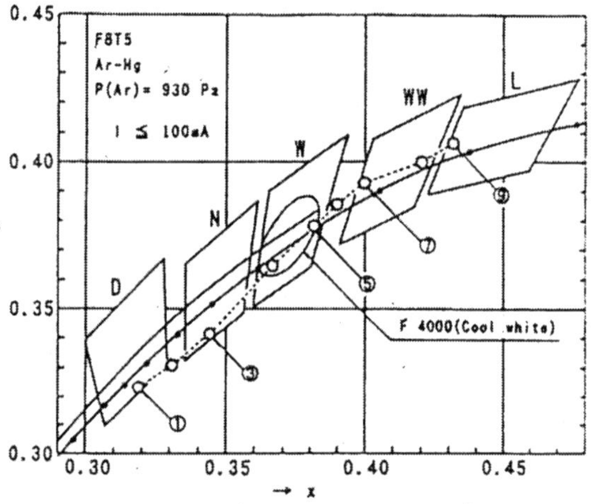

No.	Duty factor D (%)	Color temp. T (K)	Color difference Δuv	Classification JIS Z-9112
①	50.0	6280	-0.0032	D
②	33.3	5620	-0.0050	—
③	25.0	5010	-0.0046	N
④	16.7	4330	-0.0013	W
⑤	12.5	3980	+0.0005	W
⑥	10.0	3830	+0.0018	—
⑦	8.3	3650	+0.0023	WW
⑧	6.3	3280	+0.0020	WW
⑨	3.1	3110	+0.0024	L

ΔT=3170 K

Fig. 4 Chromaticity diagram showing the effect of change in duty factor for a low pressure Ar-Hg discharge lamp.

MODE TRANSITION IN AN INDUCTIVELY COUPLED DISCHARGE

Masaaki Monte, Mamoru Matsuoka and Motoichi Kawaguchi
Department of Industrial and Technical Education, Mie University, Tsu, Mie 514, Japan

1. Introduction

In this report we describe the experimental result on the mode transition in an argon plasma sphere excited inductively with a one-turn rf current coil, which has been often used for the purpose of lighting recently. The transition looks not one between E discharge and H discharge,[1-4] but one between two modes both dominated by an electromagnetic field.

2. Experimental arrangement

An argon plasma sphere of 13 cm in diameter and of pressure of 1 Torr was excited by a one-turn rf coil wound on its equator. The frequency of the coil current was 13.56 MHz. The input power to the coil was up to 500W, which was provided by an oscillator through a matching circuit. The spatial profiles of plasma density and electron temperature were measured by a movable Langmuir double probe connected to a carefully designed rf blocking circuit. The spatial profile of radiated light was detected from a video camera, whose output was processed by a computer by using the method being developed by one of the authors.

3. Experimental result

With increase of rf input power, an axisymmetrical-shaped plasma sphere started (mode 1). The sphere in this mode was bright, basically throughout the plasma. This mode sustained with more increase of input power up to 320W, followed by increase of brightness. When the input reached that value, the discharge suddenly changed to a ring-like one, together with much more brightness (mode 2). Also in this mode, the shape of the plasma column was axisymmetrical.

3.1 Plasma density profile

The spatial profile of plasma density on the coil surface is shown in Fig.1, where the lower line is the density profile just below the mode transition of input power of 320W, and the upper line shows the profile just after the transition (input power of 380W). According to the transition from mode 1 toward mode 2, the plasma density abruptly rose up by two order of magnitude near the coil, while its change at the center of the sphere was much less. Therefore, the plasma sphere in mode 2 looked to be constricted toward the coil. This was also quite clear from the data of the video camera. On the other hand, the coil current at the transition (about

10A) did not change so much as the plasma density. Such situation suggests that a strong variation of the impedance of the plasma sphere occurred at the transition. In each mode, the plasma density increased slowly with the input power being increased.

3.2 Hysteresis characteristics

The dependence of the peak value of plasma density on the input power showed a hysteresisloop according to the direction of variation of input. This implies that the mode transition was driven by some nonlinear mechanism.

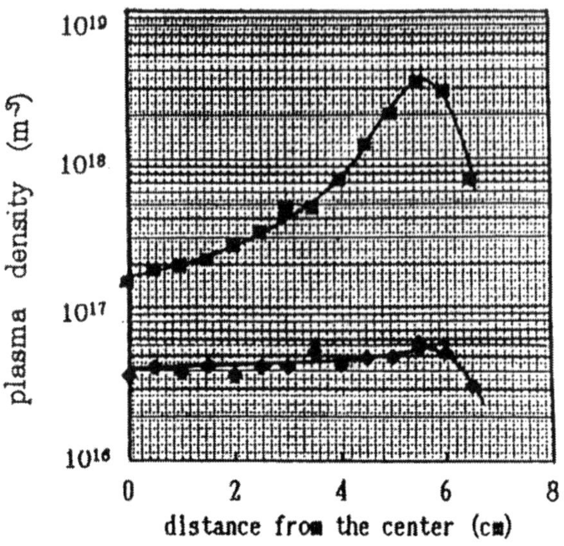

Fig.1. Radial variation of plasma density on the coil surface.
Lower curve: just before transition.
Upper curve: just after transition.

4. Discussion

From the shape of the discharge column, even the mode 1 seemed not an electorostatically generated plasma, but electromagnetically generated. One possibility for the key to the mode transition in the electromagnetically excited plasma is the skin depth δ of the electromagnetic wave emitted from the coil. When the input power reaches the threshold of 320W, δ becomes shortened to about the sphere radius. Then, the wave is prevented to intrude at the center, which makes the equivalent radius of the plasma current ring larger and strengthens the coupling between plasma current and the coil current. This results in increase of the plasma current and plasma density. Thus, skin depth becomes shorter, which causes more increase of plasma density and the constriction of plasma ring toward the coil. Such a nonlinear positive feedback mechanism is supposed to cause the transition.

references

1) T.Takamoto et.al.: J.Inst.Elec.Eng.Jpn., vol.91, No.4, 173 (1971).
2) M.Shinomiya et.al.: J.Illum.Eng.Soc., vol.29, No.1, 46 (1991).
3) H.U.Eckert: High Temperature Soc., vol.6, 36 (1974).
4) D.O.Wharmby: 47th Gaseous Electronics Conf., NB-3 (1994).

A Study On Triphase Source For Mass-Using Discharge Lamps

Wang Liancai Yang Lixi Zeng Qingshan
Pang Yongzhi Wang Zhezhang
(Zhengzhou Institute of Technology, China)

Abstract:

Mass-using discharge lamps in the triphase illumination will be cause some problem for the electric source, therefore some triphase symmetrical circuit and varied unsymmetrical circuit have been studied. According to the analyses of expermental datd and the neutral current waves, some results are discussed in this paper.

When discharge lamp is driven by industry frequency sine- volt and works on steady state, its voltage waveform is approximate to triagle -wave. Its harmonic is composed of odd-harmonoc and most of it is third harmonic. In three-phase four-wire power supply system, the third harmonic which angular phase difference is 120 will superpose precisely. The waveform showed on the top is the neutral current waveform, another is the lamp current waveform. The waves which are showed in Fig.1 ~ Fig.6 are shooted by 7201-A storage oscilloscope.

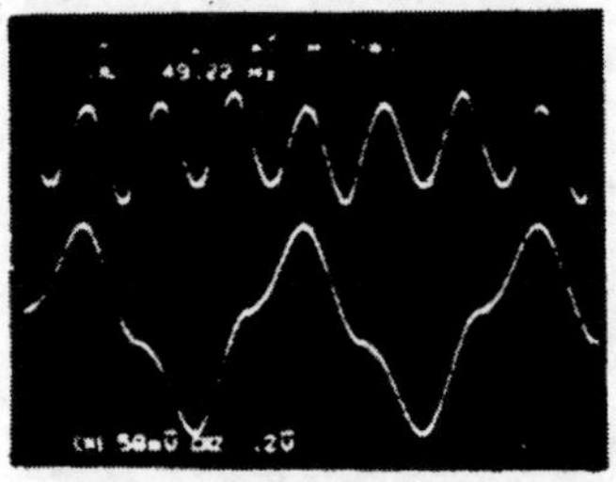

Fig.1

Neutral and Phase Current Waves

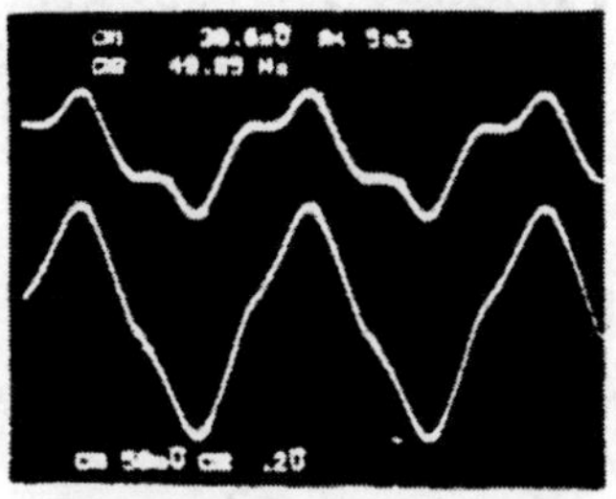

Fig.2

Nonsymmetry 421

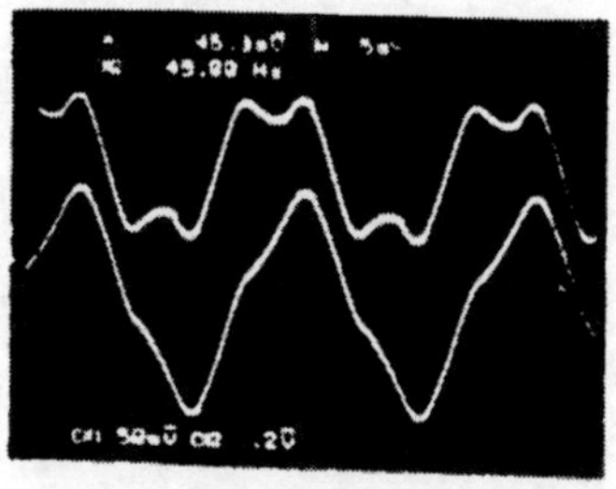

Fig.3

Nonsymmetry 431

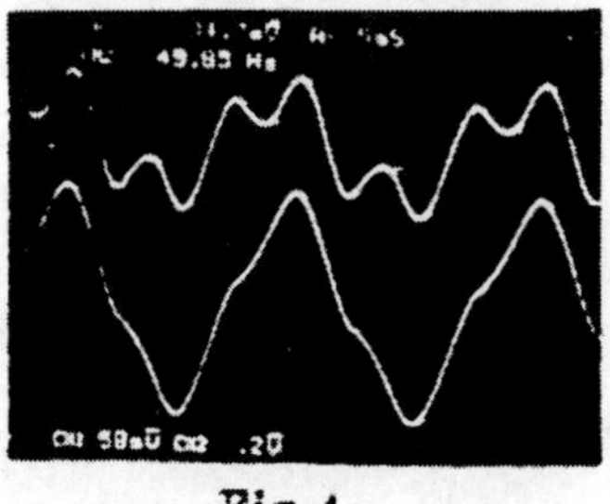

Fig.4

Nonsymmetry 432

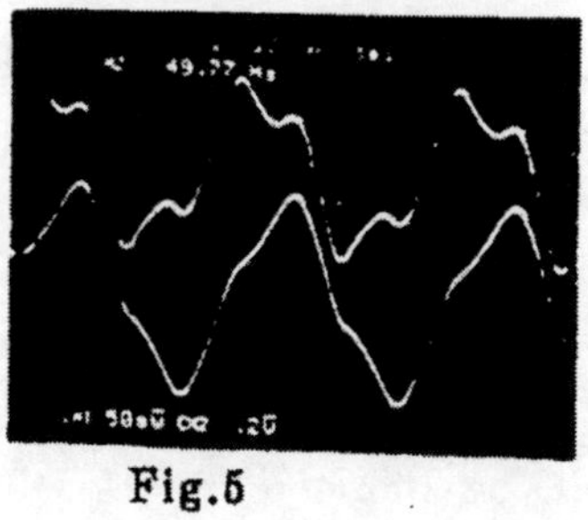

Fig.5

Nonsymmetry 441

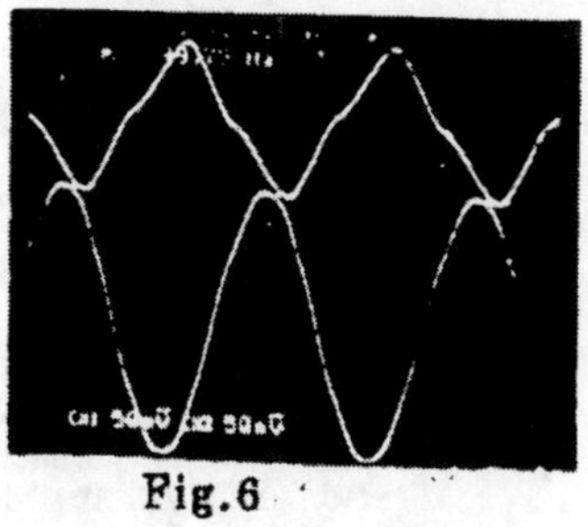

Fig.6

Detal- shaped Connected with Load

The waves of the neutral current and the full-load phase current are showed from Fig.2 to Fig.5 when the circuit is at the several typical cases of nonsymmetry.

As expressed in discharge lamp circuit, the circuit of discharge lamp is a inductive circuit, its power factor is usually 0.9, but it will be a strongly inductive or capacitive character circuit in the special cases, the factor may be littler than 0.6. We can consult the phasor analysis, if the phase B is capacitive load and the phase C is inductive load, the neutral current $\dot{I}_0=\dot{I}_a+\dot{I}_b+\dot{I}_c$ will be larger than the full-load phase current, even twices as the later.

The relation between the voltage and the current of the discharge lamp circuit on the steady-state working condition is a downward volt-ampere character because of the nonlinear property of the lamp, the principal harmonic is triple harmonic wave. If we change the mode of the load connection from Y - shaped to delta-shaped, the triple harmonic waves will be superimopsed and compensated each other, it will form a circulation because their phase difference are 120 , the triple harmonic waves neither effect the electric source system nor the working condition of the discharge lamp.

When the lamps are delta-shaped connected, the waves are showed on the Fig.6, the phase current wave is above the line current, even if the former has some distoetion, the later is still a sine- wave. If the load is connected with V- shaped, one of the line current wave is still a sine wave. Therefore, if the electric discharge lamp are used in large area illumination and are connected with delta-shaped, the higher harmonic which is brought by the nonlinear property of the lamp could be made up. Therefore the conclusions as follows:

(1) The neutral current of the triphase circuit compose mainly with the triple harmonic if the load is the discharge lamp, the ratios of the virtual value of the neutral current to the full-load phase current is 32.7%, even if the triphase load is totally symmetry.

(2) The neutral current of unsymmetrical triphase load is larger than that of the corresponding linear load, it will be laeger than the full-load phase current in the special condition. There are plentiful harmonices in the neuteal current, the discharge lamp circuit becomes an new harmonic source.

(3) If the lamp load are connected with the delta-shape, the higher harmonic caused by the lamps could be compensated, the waveform of the power system could be improved.

These analyses and the expermental data are of great value to the illuminative power supply design and the cable manufacture.

HPS-Lamps without Mercury in Pulse-Current Operation

F. Serick, H. Kaase
Institut of Lighting Engineering TU Berlin
Einsteinufer 19, D-10587 Berlin

The safe plasma temperatures in wall-stabilized HP-discharge-lamps are limited by the thermal rating of the tube. Therefore, standard HPS-lamps with stationary power supply only work with center plasma temperatures of about 4000 K ($CRI \approx 20$, $CCT \approx 2000$ K) /1/. At enhanced vapour pressure and extensive broadening of the self reversed Na-D-profile the *CRI* can be increased considerably, but the limit for *CCT* is about 2500 K.

It is possible to overcome these technical restrictions by pulse-current operation of the discharge /2/. The thermal inertia of the tube wall is about four orders of magnitude higher than that of the plasma. Therefore, at the same averaged power input $\bar{P}$, with short gate times one can work with considerably higher plasma temperatures without thermal overloading of the wall. With help of different pulse currents and suitable duty factors it is possible to vary the spectral distribution of the plasma radiation in wide limits.

To explore the photometric parameters of the pulsed HPS-system we used 70 W PCA arc tubes (3.5-mm-i.d., arc length 35 mm). After a long time 50 Hz AC operation to reach a stable state, the discharge was provided with two rectangular pulses by thyristor-controlled pulse forming networks. A simmer current keeps the plasma conducting during the period between the pulses.

We could vary the following parameters well aimed:

1. Preheating power $\tilde{P}$ (responsible for the vaporized Na-quantity)
2. Height and width of the first pulse P_1, t_1
3. Time duration between first and second pulse Δt
4. Power conversion of the simmer current P_s, ($A_s = P_s * t_s$)
5. Height and width of the second pulse P_2, t_2, ($A_2 = P_2 * t_2$)

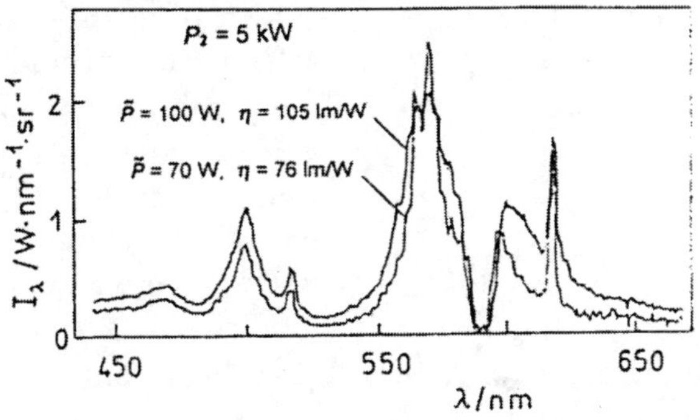

For electrical data recording we used storage oszilloscopes. The spectral emission in special time windows was analyzed by a OMA-system with MCP. Fig. 1 shows pulse spectra for P_2 = 5 kW and two preheating powers.

From the data at very different plasma parameters it is possible to form analytical approximations giving connections to photometric parameters; as for example: η_s [lm/W] = .77P_s[W]+ 17.7. The total luminous efficacy for a multiple pulsing operation results in: $\bar{\eta} = (\eta_2 A_2 + \eta_s A_s) / (A_2 + A_s)$

By using the different approximations one can perform a mathematical simulation to discover the best photometric parameters for predeterminated boundary conditions. Fig. 2 shows $\bar{\eta}$-curves versus P_s for several P_2 (---); conditions: Δt = 4 ms, t_2 = 180 μs. According to the relation $\bar{P} = [P_s(\Delta t - t_2) + P_2 t_2] / \Delta t$

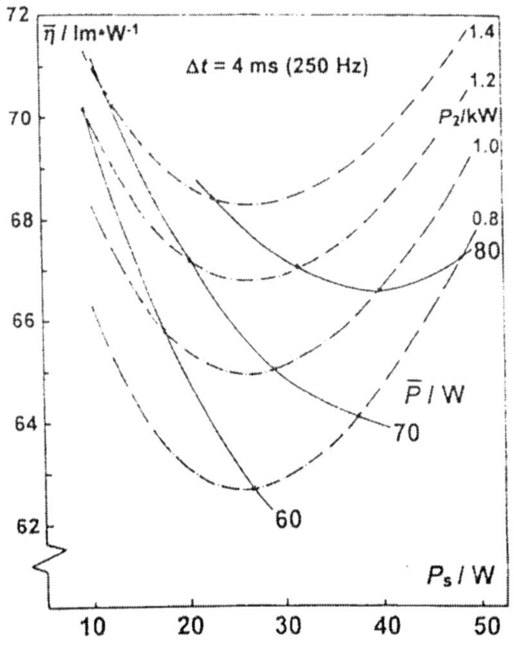

we calculated contours (——). The crossings represent operating points of this system; luminous efficacies $\bar{\eta}$ > 70 lm/W can be reached. Similar investigations at different pulse parameters were carried out for CRI and CCT too.

In commercial pulse ballasts the rectangular pulses are realized by oscillations of about 50 kHz. This effects in a small plasma modulation and slight deviations in comparison with our results. Nowadays the new DSX-system (OSRAM) for 77 W lamp power achieves:

$\eta \approx 58$ lm/W; CRI = 80...85;
CCT = 2650...3050 K.

Tube wall temperature profiles of HPS-lamps with pulse ballasts and with conventional 50 Hz sinus supply respectively were measured by D-line pyrometrie /4/. At the same power $\bar{P} = 70$ W, the temperature in the middle between the electrodes E is lower by 120 K for the pulse regime (Fig. 3). That means, the wall power input by radiation and heat conduction is smaller. Therefore, the power conversion of the pulsed discharge may be raised without overloading the wall. The wall temperature increases by ≈ 5 K per W ($\bar{P}$). For a similar temperature distribution as for $\bar{P} = 70$ W in conventional mode, the pulse power conversion may be > 90 W. This finding allows to improve the photometric parameters of the pulsed HPS-system.

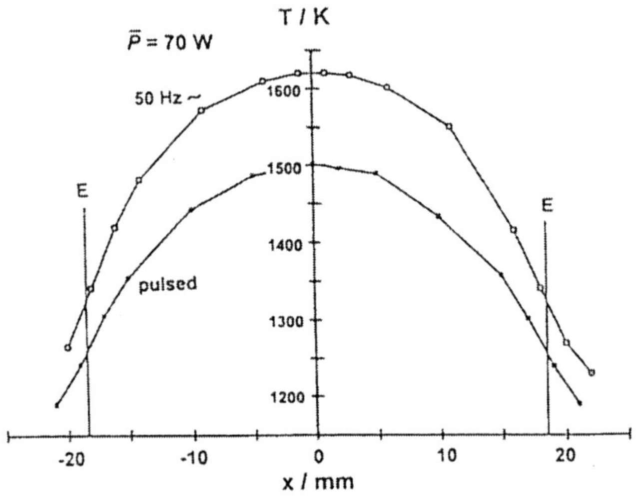

References

/1/ De Groot,J.J.; J.A.J.M. van Vliet; The High-Pressure Sodium Lamp; Philips (1986)
/2/ Osteen, M.M.; US Patent no. 4.137.484 (1979)
/3/ Günther, K. et all; Contrib. Plasma Phys.; 30(1990)6; 715
/4/ Kloss S.-H.: Course work; Inst. Lichttechnik TU Berlin (1993)

BUILT-IN STARTER WITH PULSE CUT-OUT FOR HID LAMPS

Shunichi Sasaki*, Takenobu Iida* and Russell Loane**
*: Iwasaki Electric Co.,Ltd. Saitama plant
1-1 Ichiriyama machi Gyoda city Saitama pref. 361 Japan
**: Eye Lighting Industries Pty Limited
157 Wellington Road,East Brisbane,Qld4169,Australia

Mercury vapour lamps (M.V.L.), although relatively inexpensive, lack the high efficacy of High Pressure Sodium lamps (H.P.S.L.) and the improved colour quality of Metal Halide lamps (M.H.L.). As a result, many users have decided to use H.P.S.L. and M.H.L. instead of M.V.L.

But it is necessary to attach an external starter to a ballast or an internal (built-in) starter to a lamp for starting of H.P.S.L. or M.H.L., and there are some problems caused by the starters at lamp failures.

For example;

1) in the case of lamp starting failure:
 * radio frequency noise generation
 * insulation failure in the circuit

2) in the case of the arc tube leakage in the vacuum outer jacket as H.P.S.L.:
 * abnormal arcing in the outer jacket
 * abnormal arc current

These phenomena may be the causes of the ballast failure. One of the solutions is to cut out pulse when the lamp fails.

We developed a new built-in starter that can cut out the pulse when the lamp fails.

--Structure and outline of the starter--

This starter is constructed by Ferro-Electric Capacitor device (FEC) , semiconductor switching device (SIDAC) and resistors. We call this starter "FEC starter".

FEC's electric aspect is disclosed as a "Hysterisis Locus" as shown in Fig.1. It is the non linearity characteristics of the FEC in conjunction with the ballast that generated the high voltage starting pulse at the same position on each half cycle as shown in Fig.2. This characteristics are appeared below the curie point (90℃) of the FEC and disappeared above it.

--Pulse cut-out #1 (at lamp starting failure)--

The circuit diagram of FEC starter is shown in Fig.3.
The resistor (5) in the circuit is mounted to keep the electrical characteristics of the FEC during the lamp life. And also this resistor works as a heater when the lamp can't start, because very small current flows through the resistor. This current heats up not only the resister but also the FEC. When the FEC's temperature rises above it's curie point, the pulse is cut out.

--Pulse cut-out #2 (at arc tube leakage in a vacuum outer jacket such as H.P.S.L.)--

If xenon gas leaks from the arc tube, arcing will be happened at somewhere in the outer jacket. In this case, if the lamp's starter is the FEC starter, the arcing will be formed between the electrode's surfaces or their edges of the FEC, because the FEC is very thin disc. Then the FEC will be fired, and the pulse voltage will be cut out permanently.

We can get the built-in starter with pulse cut-out for not only H.P.S.L. but also M.H.L. to select a resistance of the resistor and a positioning between the FEC and the resistor correctly.

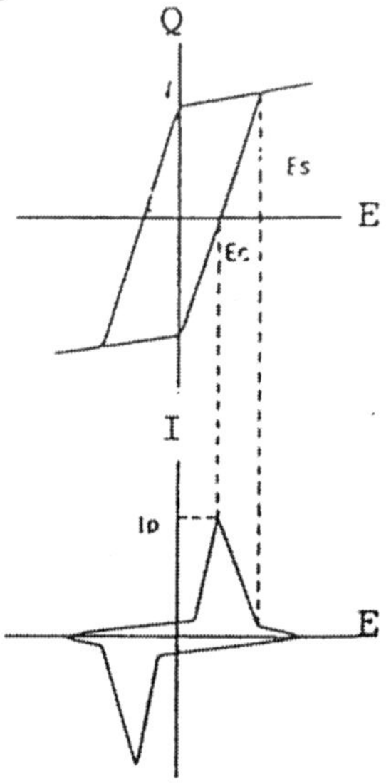

Fig.1 Hysterisis Locus & Volt-Current characteristic

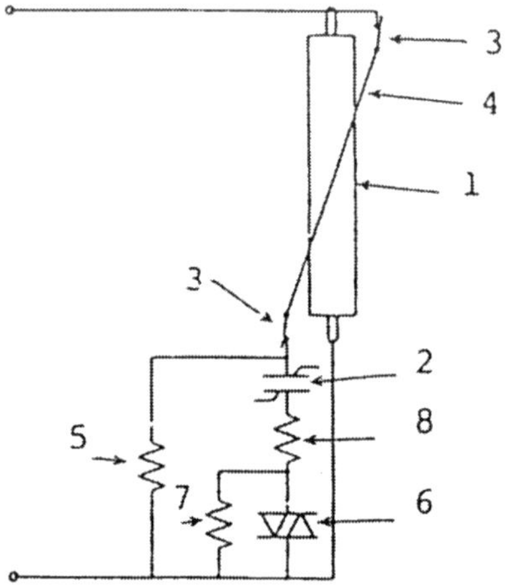

Fig.3 Schematic Diagram of FEC starter

1:ARC TUBE 2:FEC
3:BIMETAL SWITCH
4:STARTING ACCELERATOR
5:RESISTOR(Rc)
6:SEMICONDUCTOR SW.
7:RESISTOR(Rs)
8:CURRENT DAMPER

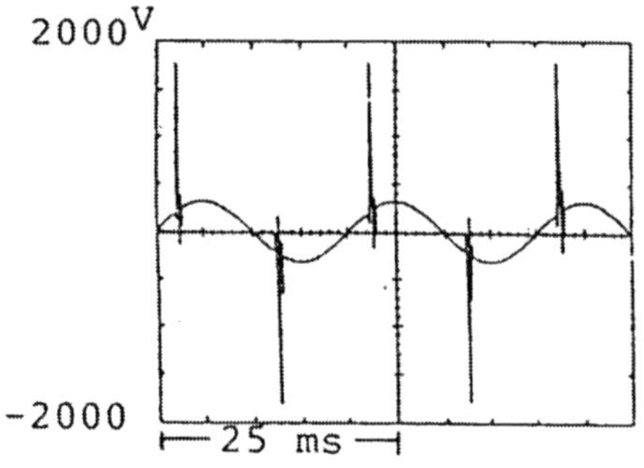

Fig.2 Pulse Profile

BIBLIOGRAPHY:
1. T.Iida et al.: Ceramic Pulse Generating Capacitor and Built-In Starter for High Pressure Sodium Lamps.
34th IES National Convention, Sydney, 1988
2. S.Sasaki et al.: Metal Halide Lamps Designed for use with Standard High Pressure Sodium Ballast.
IES of Australia and New Zealand, Gold Coast, 1992

ELECTRONIC OPTIMIZATION OF HID LAMPS

K. Günther

OSRAM GmbH, Nonnendammallee 49-61, D 13625 Berlin

Electronic operation of gas discharge lamps offers the possibility to provide the electrical power as sinusoidal waveform in the low, medium and high frequency range, as square wave, or as a sequence of pulses at a wide range of duty cycle and peak powers. The response of the lamp depends on the gas composition and on the geometry of the arc and the discharge vessel. It is possible to keep constant one of the electrical quantities or a composition of them to stabilize the lamp power, the arc position, or an other feature of importance. The basic problem to do so is to find an information which is representative for the actual condition of the lamp, and to translate it into an electrical signal which can be understood by the electronic power supply. In the case of pulsed mode operated HPS lamps it is possible to modify the colour temperature and the colour rendering index independently by means of controlling the peak power and the repetition frequency of the pulse series. The possibilities are considered to compensate manufacturing deviations, lamp degradation during life, and effects of environment by means of an electronic feed back: the lamp condition is checked from the electrical data and will be corrected, if necessary, by a change of the electrical input. In this way it is possible to control critical situations of the lamps including starting and end of life effects.

1. Introduction

The customer expects from a discharge lamp a proper luminous flux and, especially, a stable colour temperature and colour rendition corresponding to the figures in the data shields. The actual data, however, are the result of a complicated arrangement between lamp and ballast. Each deviation of the lamp properties, e.g. manufacturing deviations, changes of the lamp environment, aging of the lamp, as well as changes of the power supply, will influence the sensitive equilibrium of heating and dissipation of the gas discharge and the heat balance of the arc tube, and will result in a more or less serious change of the photometric properties. On the other hand, a well thought-out control of the power supply parameters can modify the lamp parameters or stabilize them against external influences and aging.

2. Destabilizing interactions in the system

Fig. 1 shows a schematic interaction pattern considering external and internal influences on a HID system containing a power supply and a high pressure discharge lamp. Here we discuss the typical case of a fill system containing incompletely evapourating additives, the partial pressures of which depending upon the composition of the fill and upon the tempe-

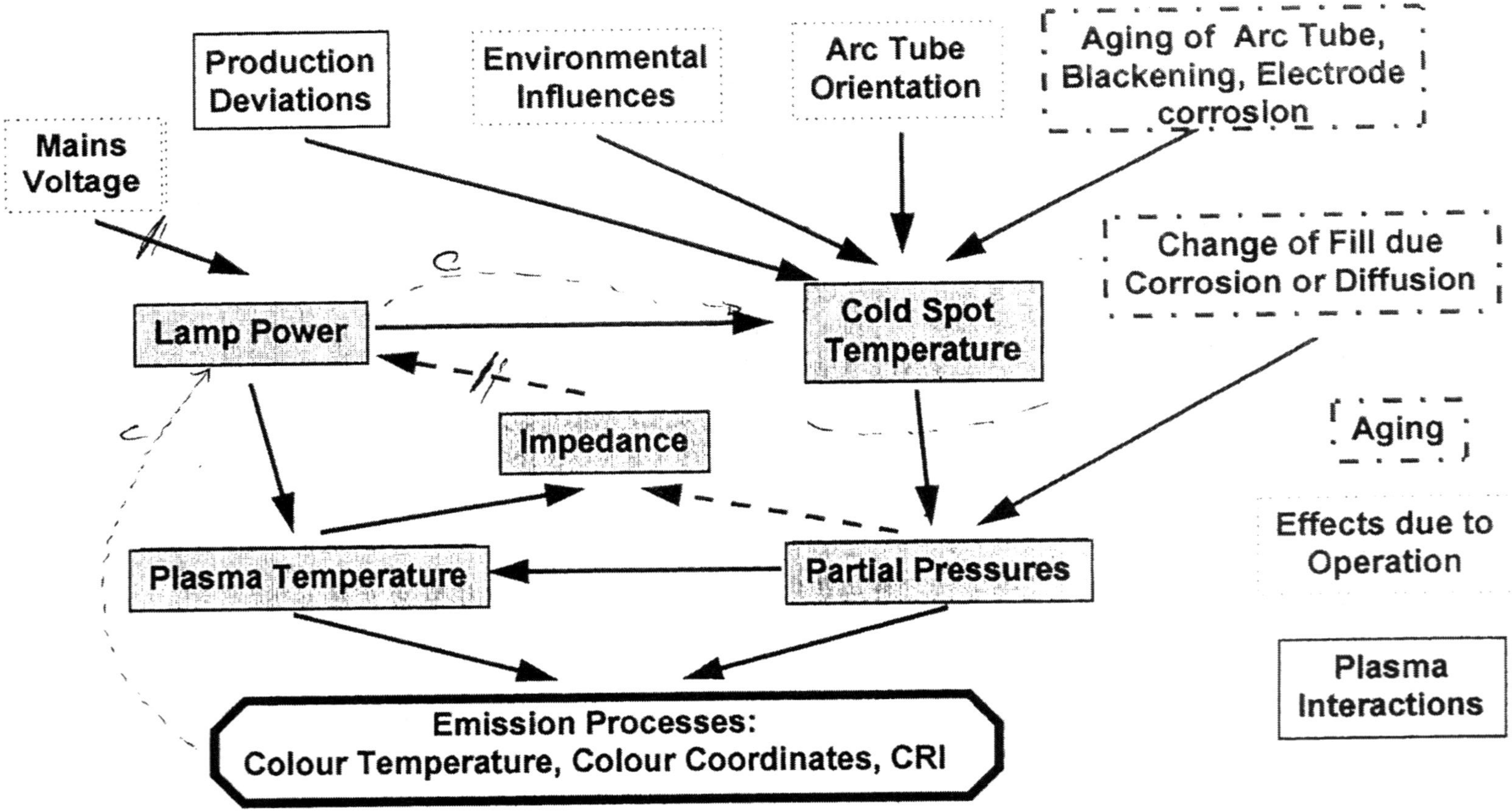

Fig. 1. System response on external influences and internal interactions

rature of the coldest area in the arc tube. This extremely important temperature is affected by a series of influences:

- deviations in manufacturing,
- external influences: environment (e.g. tightly closed luminaire), arc tube orientation (can shift the temperature distribution of the arc tube), mains voltage...
- lamp aging: absorbing layers near the electrodes, electrode corrosion, change of fill composition due to diffusion or corrosion...
- lamp power.

By means of a suited lamp design these effects can be diminished but not eradicated; so the partial pressures are influenced, and the power balance of the discharge changes the plasma temperature and the emission processes, i.e. the photometric properties. The numerous feedback loops in the diagram can enforce or suppress these shifts in dependence on the actual plasma conditions. The corresponding change of the electrical conductivity σ may even influence the power fed from the power supply into the lamp building an additional feedback loop. The result of these processes may be described in the form of the voltage-current characteristics of the discharge. This characteristics must be distinguished from the lamp characteristics which includes the heat balance of the arc tube and especially the cold spot behaviour (the time constants for the establishment of these balances differ by more than four orders of magnitude). The final result of these processes in interaction with the ballast characteristics is the operation point characterized by current, voltage , and power of the lamp and partial pressures and temperature of the discharge determining the photometric properties.

What can be done to protect the system against the results of these unavoidable mechanisms? Three basic classes of power supply should be defined according to their control function for the system:

1. **The magnetic ballast**: It shows the well known ANSI trapezoids as result of the deviation range of lamp lines and mains voltage range.
2. **The conventional electronic ballast**: It compensates mains voltage deviations and different plasma impedances and forces the system to a constant lamp power regardless of the actual power needs of the lamp to realize the designed operation point.
3. **The intelligent electronic ballast**: It derives a signal from the lamp containing information on the photometric condition and controls the lamp power in such a way that the designed colour temperature and colour rendition are established.

3. Control signals from HID systems

To stabilize a HID system we have to find a signal from the lamp which is correlated to its photometric properties. Because it is not a possibility to use any optical or spectroscopical signal, a connection must be found between the electrical behaviour and the spectral distribution of radiation. This problem cannot be solved generally, and we try to find rules for different classes of HID lamps. Let us choose a generalized line strength as proposed by SCHIRMER (1960):

$$P_{rad} = n_M \cdot \overline{K}_M \cdot e^{-\frac{\overline{E}_M}{kT}} \qquad (1)$$

where n_M is the number density of the radiating particles, and $\overline{K}_M, \overline{E}_M$ are an averaged line constant and excitation energy, respectively.

As a representative for the electrical properties we consider the electrical conductivity. To be more transparent we use the simple formula of Lorentz (1905) for partially ionized plasmas:

$$\sigma = C \cdot \frac{n_e}{\sqrt{T}} \frac{1}{n_{Hg}Q_{Hg} + n_M Q_M + n_e Q_I} \qquad (2)$$

with C = const, T the plasma temperature, and n_{Hg}, n_M, n_e are the particle number densities of mercury, of the component M, and of the electrons. The Q are the transport cross sections for electron collisions of the indicated species. Let us apply these expressions on different lamp types. Table 1 shows the application of equations (1) and (2) to the special conditions and density relations of the listed lamps. It can be seen that there is a strong correlation between the electric and the photometric properties in the case of **high pressure xenon** lamps and **Hg free HPS** lamps, i.e. the radiation properties of the lamps can be detected by the power supply by means of a test of voltage and/or current. For **standard HPS** lamps this correlation holds also when a constant relation between the vapour pressures of mercury and sodium in the case of a given amalgam can be assumed. Obviously it must fail when the amalgam ratio changes due to sodium corrosion or diffusion. The correlation is completely lost in the case of **metal halide** lamps where the electrical properties are determined nearly exclusively by the mercury and the radiation distribution by the halide additives. Because they coexist without any interaction it is extremely difficult to get an information about the radiation from electrical tests.
A chance to carry out the task even for metal halide lamps may be a dynamic test by applying a fast power step to the lamp: The discharge can be considered as a energy reservoir whose energy is mainly determined by the temperature and density of mercury vapour. The relaxation of the system is given mainly by the radiation which dependends on the particle number densities of the metal vapour. In this way the decay or rise time contains information about the radiation distribution. Dynamic tests of this kind may provide also other useful information about the lamps as e.g. excess of halogen in case of increased reignition peaks.

4. Signal processing

It is obvious that depending on the lamp type under consideration electrical signals of different kind must be monitored by the power supply and processed in different way. An intelligent power supply should carry out at least the following tasks:

- control of the ignition
- control of the warm up phase from glow to arc
- stop the power supply after a given time if there is no ignition
- test the lamp condition using the signals discussed in the last section and compare them with reference parameters
- readjust the waveform of the electric power supplied to the lamp until the reference data are realized
- stop the power supply if the allowed control range of the reference parameters is exceeded and indicate end of life for the lamp.

It turned out that the most economic and flexible way to meet these different tasks is to use a microcontroller. The following example is reported to demonstrate this idea.

Lamp	Conditions	Conductivity	Radiation
High pressure xenon	$n_{Hg} = 0,\ n_{Xc}Q_{Xe} << n_eQ_i$	$\sigma = C\frac{1}{\sqrt{T}Q_i} \propto T^{3/2}$	$P_{rad} = n_{Xe}\overline{K}_{Xe}e^{-\frac{E_{Xe}}{kT}}$
Standard HPS	$n_{Hg} >> n_{Na},\ n_{Hg}Q_{Hg} >> n_eQ_i$	$\sigma = C\frac{n_e}{\sqrt{T}}\frac{1}{n_{Hg}Q_{Hg}}$	$P_{rad} = n_{Na}\overline{K}_{Na}e^{-\frac{\overline{E}_{Na}}{kT}}$
Hg free HPS	$n_{Hg} = 0,\ n_{Na}Q_{Na} >> n_eQ_i$	$\sigma = C\frac{n_e}{\sqrt{T}}\frac{1}{n_{Na}Q_{Na}}$	$P_{rad} = n_{Na}\overline{K}_{Na}e^{-\frac{\overline{E}_{Na}}{kT}}$
Metal halide	$n_{Hg} >> n_M,\ n_{Hg}Q_{Hg} >> n_eQ_i$	$\sigma = C\frac{n_e}{\sqrt{T}}\frac{1}{n_{Hg}Q_{Hg}}$	$P_{rad} = n_M\overline{K}_Me^{-\frac{\overline{E}_M}{kT}}$

Table 1. Dependence of the electrical and optical properties of the different lamp types on the plasma temperature and the particle number densities.

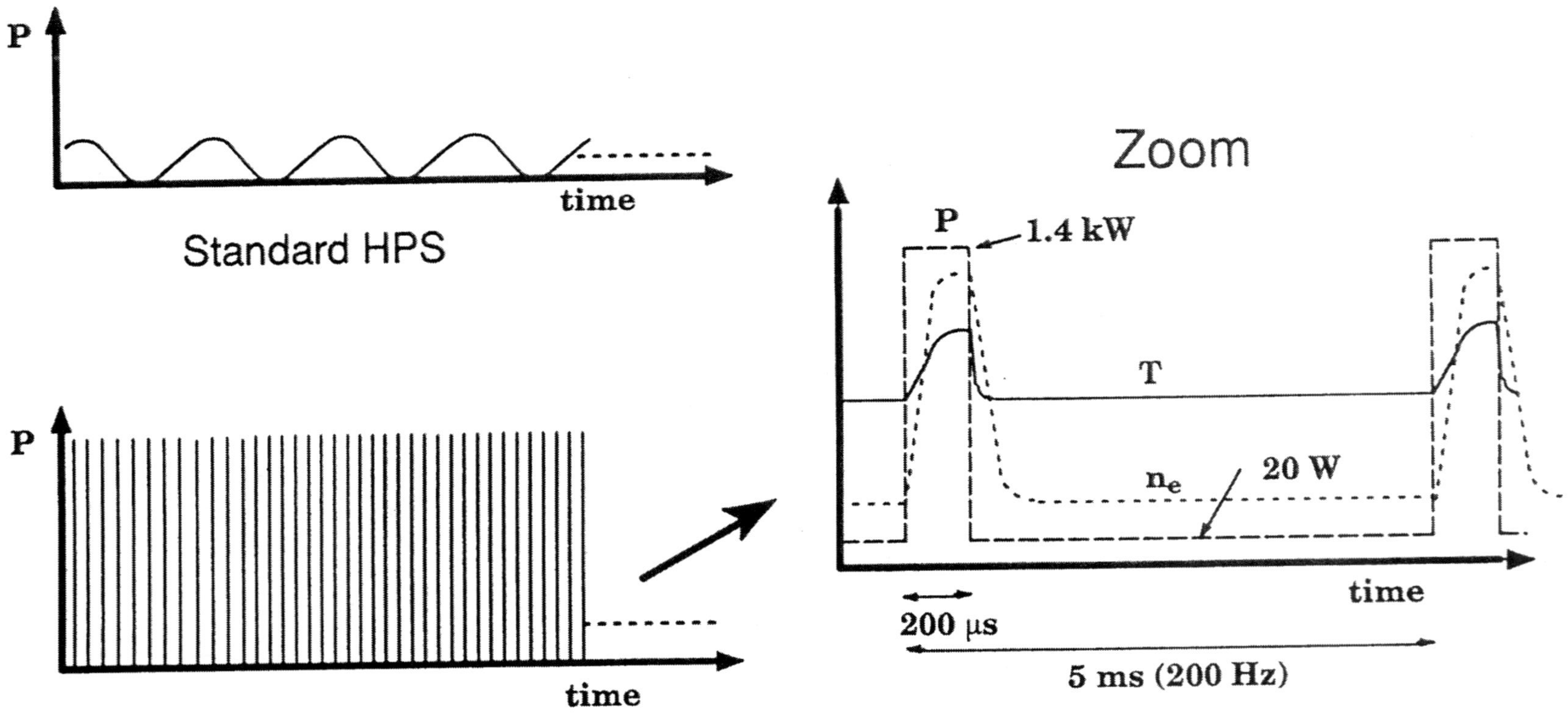

Fig.2. Comparison of the electric power waveform P(t) for magnetic ballast operation and pulsed mode operation

n_e: Electron density T: Plasma temperature

5. Pulsed HPS lamp

Pulsed operation is an effective means to raise the correlated colour temperature of a HPS lamp by means of a transient increase of the plasma temperature of the discharge [1-4]. During the pulse phase the population of higher lying energy levels of the sodium system is increased, and spectral transitions including the recombination continuum of the 2 ^{2}P level in the blue region are excited resulting in an elevated colour temperature. Thus, we use the fact that the time constant of the plasma is much shorter than that of the arc tube which feels the average power only. The lamp is fed from an electronic control gear containing a microprocessor which monitors the electrical behaviour including ignition, warm up and switch off. All data of the power waveform (see Fig. 2) as pulse length, pulse height, repetition frequency and keep alive power are computed and controlled to ensure the designed correlated colour temperature and colour rendering index kept within very narrow limits regardless of individual deviations of the lamp, environmental influences and aging effects. It was found that this system shows much better efficacy and colour rendering when operated completely without mercury: at the high power levels there would be the danger that the mercury is excited and contributes to the power balance. The results are well known since fifty years: the efficacy does not exceed 50 LPW, and the colour rendering index is about 55. A second benefit is the simplicity of the system and the related strong correlation between the electrical and spectroscopical data. As an example, Fig. 3 shows the correlation between lamp voltage and colour rendering index for a set of lamps.

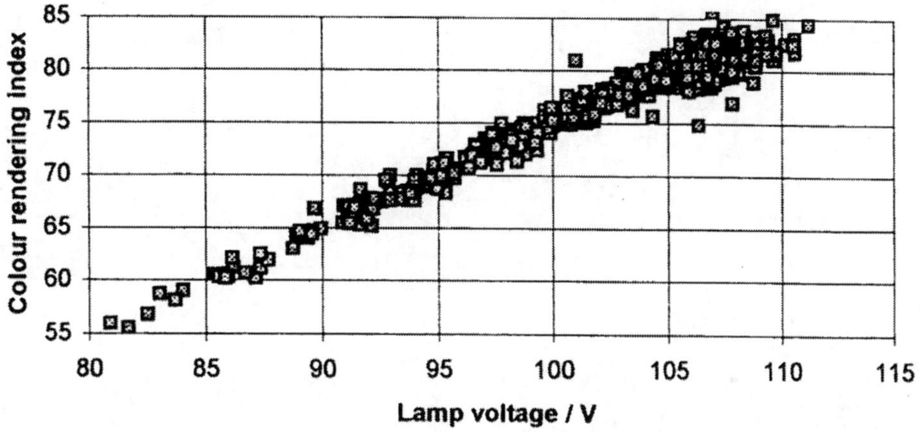

Fig. 3. Correlation between lamp voltage and colour rendering index for a set of lamps

The design data of the system can be modified by program, of course; so the colour temperature can be switched between two fixed levels at 2700 and 3100 K without changing the lumen output or the colour rendering index to adapt the colour rendering to the character of the objects to be illuminated. For outdoor illumination the lumen efficacy of the lamp can be increased at the expense of the colour rendering index without changing the hardware.

References

[1] Johnson, P.D., T.H. Rautenberg; J. Appl. Phys. **50** (1979) 3207

[2] De Groot, J.J., J.A.J.M. van Vliet; The High Pressure Sodium Lamp, Philips Techn. Library, Kluwer Techn. Boeken B.V. Deventer, 1986

[3] Brates,N., E.F. Wyner, J. Illum. Engn. Soc. **16** (1987) 50

[4] Günther,K., H.G. Kloss, T. Lehmann, R.Radtke, Contrib. Plasmaphys. **30** (1990) 715

HIGH INTENSITY DISCHARGE LAMPS WITH CERAMIC ENVELOPES

A key technology for the lighting future

Author: P.A. Seinen

Company: Philips Lighting B.V.

Central Development Lighting

Eindhoven, The Netherlands

INTRODUCTION

The most important properties of light sources are the efficacy and the colour quality. Colour quality may be considered as a combination of the colour of the light, the colour stability and the colour rendering. Both colour quality and efficacy play a major role in the selection process of light sources for several application fields. These properties, however, are not of equal importance for each application. Light sources for roads, for example, require high efficacies, whereas the colour quality is of minor importance. Light sources for homes, on the other hand, demand a high colour quality, which limits the selection to lamp families with lower efficacies. Figure 1 shows the efficacy versus the colour quality of a number of todays light sources that are frequently used in various application segments.

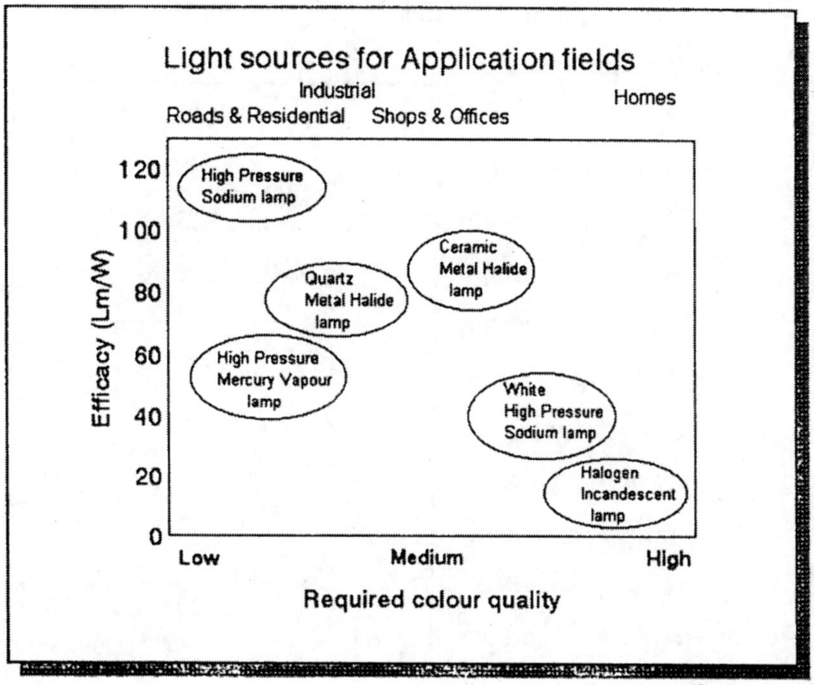

Figure 1

During the last hundred years there has been a continuous drive to improve both efficacy and colour quality of light sources. In this quest for improvement, ceramic envelope materials for High Intensity Discharge lamps have played an important role. Well known examples are High Pressure Sodium lamps, White High Pressure Sodium lamps and Ceramic Metal Halide lamps. In this paper the development of these product families will be discussed.

HIGH PRESSURE SODIUM LAMPS

The first family of High Intensity Discharge lamps with a ceramic envelope was the High Pressure Sodium lamp. Realization of High Pressure Sodium lamps required the use of sodium resistant alumina envelopes. This started the development of a technology new to the lighting industry. The new technology included the manufacturing of translucent alumina tubes, new sealing techniques, sodium resistant sealing frit and electrical feedthrough materials. Once this technology was available, the way to new product families was opened.

WHITE HIGH PRESSURE SODIUM LAMPS

High Pressure Sodium lamps were optimized for maximum efficacy, whereas colour quality was of minor importance. The colour quality can be improved (at the cost of the high efficacy) by further increasing the sodium pressure, as shown in figure 2. Unfortunately, using "ordinary" High Pressure Sodium technology to realize White High Pressure Sodium lamps causes colour instabilities, due to variation of the sodium pressure.

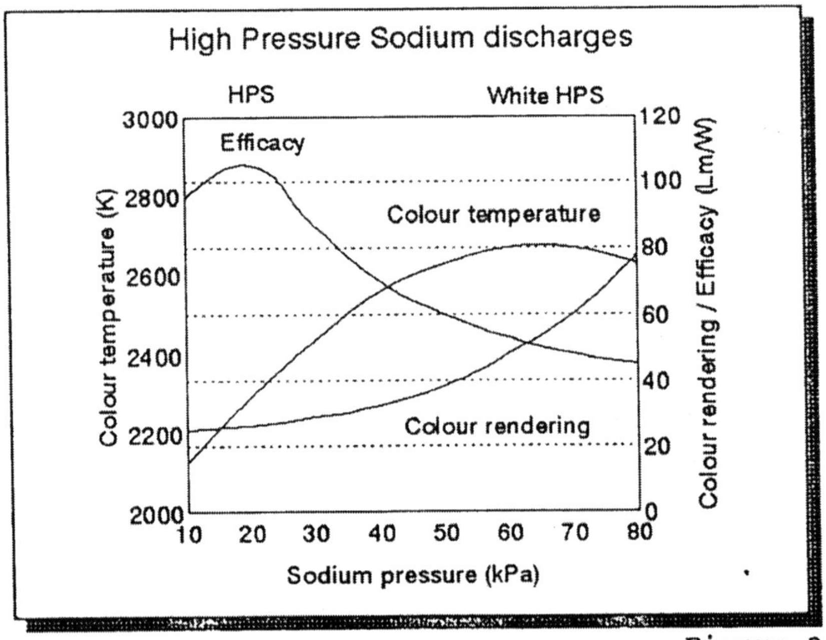

Figure 2

This variation of the sodium pressure is caused by:

* Spread of the tube temperature due to manufacturing spread, and/or blackening of the arc tube wall and/or fluctuations of the mains voltage. These variations were largely eliminated by the use of electronic gear [1] as shown in figure 3.

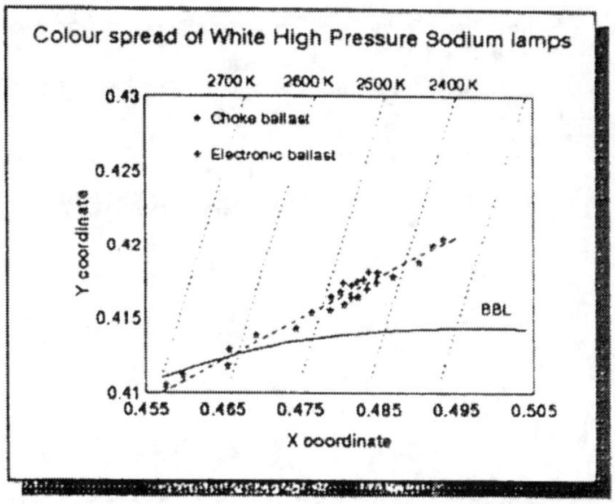

Figure 3

* Decrease of the sodium pressure which is caused by sodium loss processes, due to chemical reactions with various parts of the discharge tube as summarized in figure 4:
  * Reactions with the alumina tube wall, forming sodium aluminates.

    This reaction could be suppressed by lowering the tube wall temperature [2];
  * Reactions with calcium aluminate in the sealing frit.

    This reaction could be suppressed by lowering of the sealing frit temperature by introducing the "protruded plug feedthrough" concept [3];
  * Reactions with calcium aluminate in the tube wall [4].

    This reaction could be eliminated by minimizing the calcium aluminate phase by reducing calcium content of the alumina.

Controlling the variations of the sodium pressure provides White High Pressure Sodium lamps that proved to be an attractive alternative for Incandescent reflector lamps and Halogen lamps in shop lighting

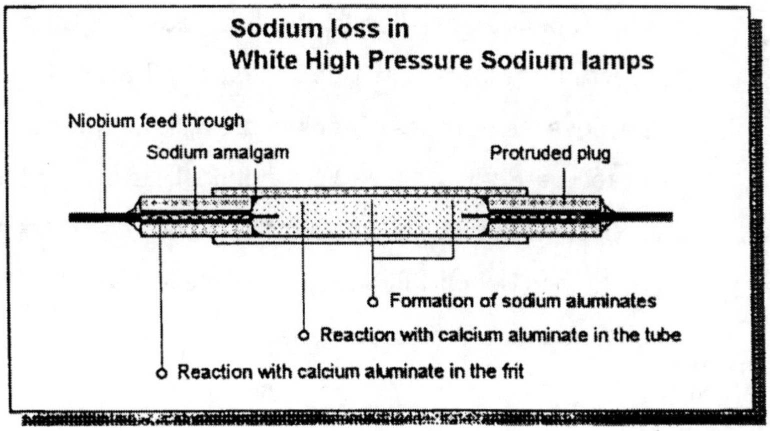

Figure 4

applications. White High Pressure Sodium lamps provide warm-white incandescent-like light with an the appealing rendition of reddish colours. Compared to Incandescent lamps they have a much higher efficacy and a much longer life time.

UNSATURATED HIGH PRESSURE SODIUM LAMPS

Although High Pressure Sodium lamps form an attractive alternative for High Pressure Mercury Vapour lamps, they have a few draw backs.

* The lamps are not retrofit on all High Pressure Mercury Vapour installations in the USA. The strong dependence between the lamp voltage and lamp power, will cause extinguishing of the lamp when the maximum voltage is exceeded, as shown in figure 5.

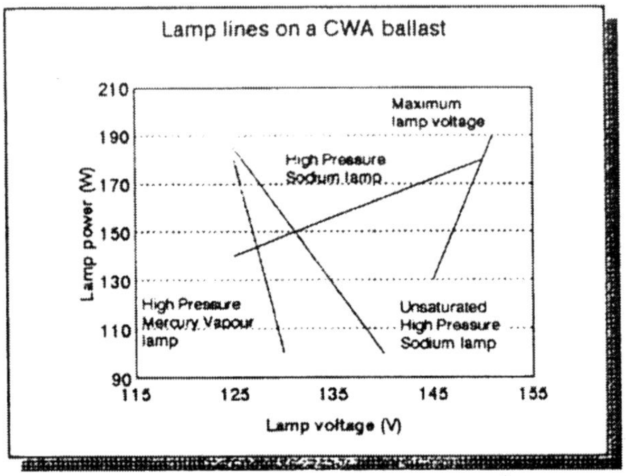

Figure 5

* The lamps tend to cycle near the end of life, which is caused by a rise of the lamp voltage. This rise is due to sodium loss and temperature rise which is caused by a combination of blackening of the tube wall and exhaustion of the emitter [5].

Both draw backs can be eliminated by designing an unsaturated discharge tube concept similar to High Pressure Mercury Vapour lamps.

To realize Unsaturated High Pressure Sodium lamps, the main problem to be solved was rapid depreciation of the luminous flux, due to sodium loss. This sodium loss mechanism is very similar to that of other High Pressure Sodium lamps. However, because the dosages of sodium are extremely small (only 1% of that of Saturated High Pressure Sodium lamps), the utmost care must be taken to prevent even minimal sodium depletion. This requires a rigorous control of all loss processes. Laborious work by Nippon Denchi [6], Iwasaki [7] Sylvania [8] and Philips [9] resulted in revealing and eliminating all loss processes.

The major ones are summarized in figure 6:

- The formation of α sodium aluminate at the end part of the tube wall.

  This reaction could be eliminated by establishing chemical equilibrium by means of a more uniform discharge tube temperature;
- The formation of ß sodium aluminate at the mid part of the tube wall.

  This reaction could be suppressed by lowering its reaction rate by limiting the maximum temperature;
- Reactions with calcium aluminate sealing frit components.

  This reaction could be minimized by eliminating the most reactive calcium aluminate phases by adjusting the frit composition and its processing;
- Reactions with emitter components.

  This reaction could be eliminated by replacing the emitter by a more stable one like yttrium oxide.

Controlling all sodium depletion processes provides Unsaturated High Pressure Sodium lamps which offer a universal solution for retrofitting any High Pressure Mercury Vapour or High Pressure Sodium lamps in existing applications. Unsaturated High Pressure Sodium lamps combine the high efficacy of the normal High Pressure Sodium lamps and the "no cycling at end of life" behaviour of High Pressure Mercury Vapour lamps.

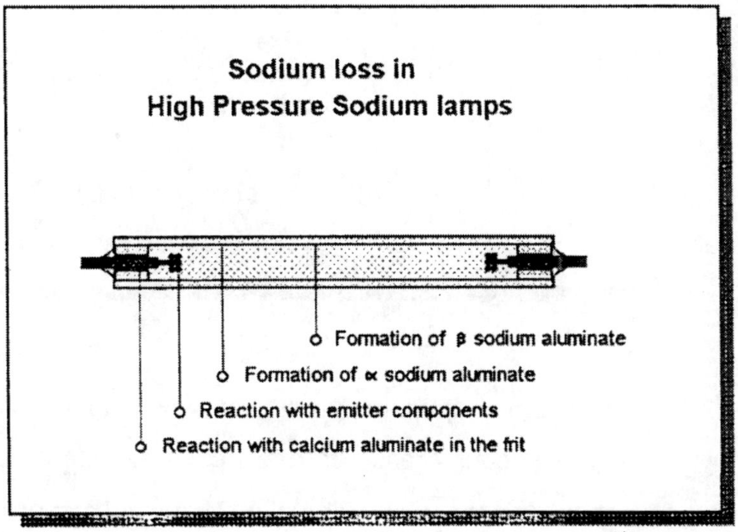

Figure 6

CERAMIC METAL HALIDE LAMPS

Metal halide lamps are commonly made of quartz glass.The lack of resistance of quartz glass towards halides at high temperatures, seriously limits its applicability for these types of lamps. Reactions like sodium diffusion through the quartz wall and reactions of Rare-earth species with the quartz glass, can only be controlled by limiting the wall temperature. Limiting the temperature, however, inevitably limits the attainable colour rendering and control over colour spread, as shown in figure 7. The good

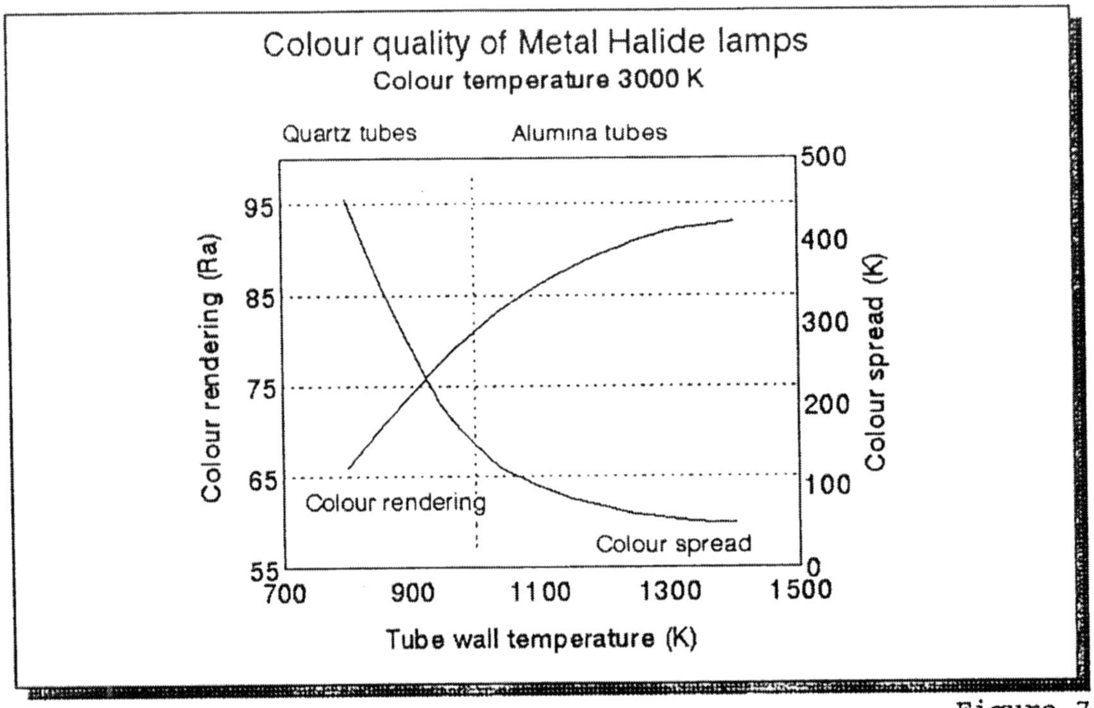

Figure 7

resistance of alumina against corrosive vapours, at high temperatures, enabling high halide vapour pressures and hence excellent colour properties, therefore was an appealing opportunity. The allowable wall temperature for ceramic discharge tubes is significantly higher than for quartz glass tubes, because the reactions described above do not occur in Ceramic Metal Halide lamps. Ceramic Metal Halide lamps, therefore combine the high efficacy and good colour quality with a good colour stability during life time.

Attempts to realize Metal Halide lamps with translucent alumina envelopes have been made since this material was available.

The main problems to be tackled were the relatively short life time and severe colour shifts, which are caused by chemical interactions of metal halides with several discharge tube components:

* Leakage of the discharge tube is caused by corrosion of the calcium-magnesium- aluminate sealing frit.

  This reaction could be suppressed by using the best halide resistant frits based on mixed phases of rare-earth alumino-silicates [10] and lowering the frit temperature by using the "protruded plug construction" (figure 8) as used in the White High Pressure Sodium lamp [11];

* Leakage is also caused by corrosion of the niobium feed through.

  This reaction could be blocked by covering the niobium by a layer of frit [12] as shown in figure 8.

* Colour shift is caused by exchange of the rare-earth species in the halide filling with magnesium and calcium from the frit.

  This reaction could be blocked by using the same rare-earth species for the halide filling and the frit.

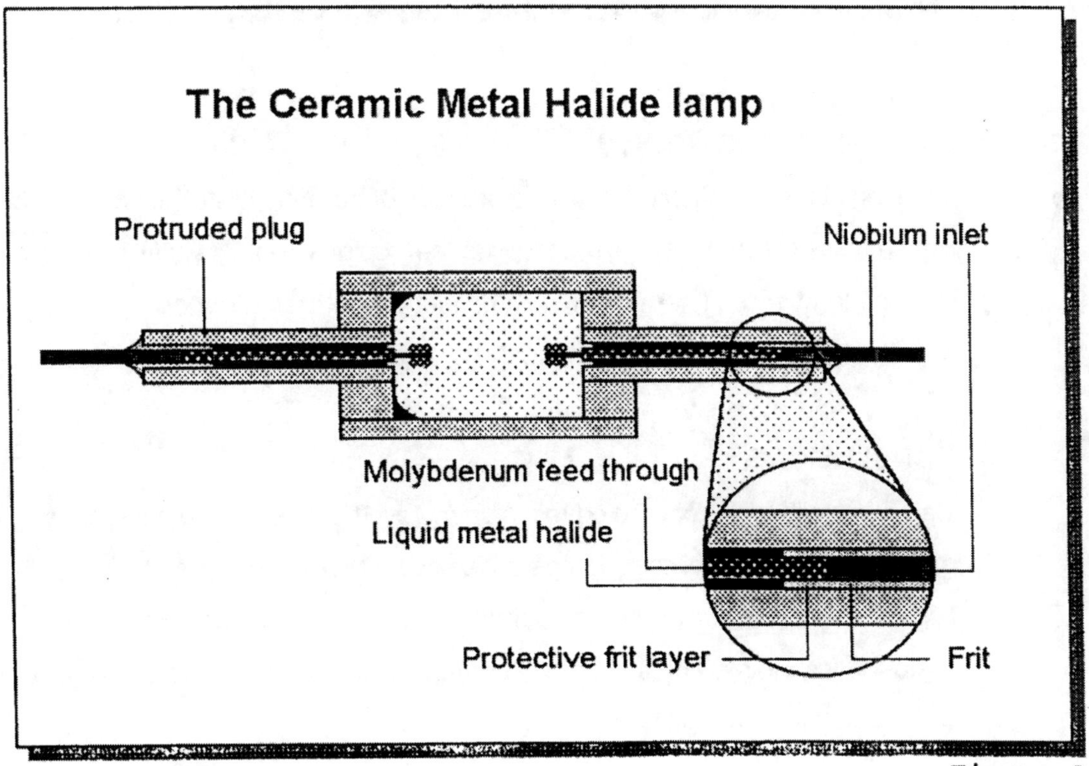

Figure 8

Careful design of the feed through construction and choices for frit materials provided Ceramic Metal Halide lamps which form an excellent alternative for Quartz Metal Halide lamps for shop lighting applications. Ceramic Metal Halide lamps combine a high efficacy and a good colour rendering with a small colour spread and colour shift. Some typical characteristics of present day Ceramic Metal Halide lamps are given in table 1.

Table 1 Typical properties of Ceramic Discharge Metal Halide lamps.

LAMP PROPERTY	TYPICAL VALUE
Lamp power range	35 W - 150 W
Colour temperature	3000 K
Colour rendering (Ra)	> 83
Efficacy	85 lm/W - 90 lm/W
Colour spread: Initial	< ± 150 K
Over life time	< ± 200 K
Average rated life time	> 6000 hours

FUTURE DEVELOPMENTS

Looking at the beneficial chemical properties of ceramic materials especially for Metal Halide types of lamps, it is expected that their colour quality will further be improved in the coming years. Therefore their application will be extended to areas that require excellent colour quality like Studio & Theatres or other critical applications.

CONCLUSIONS

Ceramic envelopes contributed to major improvements of the colour quality and efficacy of High Intensity Discharge lamps as is shown by the success of products like White High Pressure Sodium lamps, Unsaturated High Pressure Sodium lamps and Ceramic Metal Halide lamps. The technology of manufacturing ceramic discharge tubes proved to be a key technology for the lighting future.

References

[1] Carleton S.A., Keyser R.A.J, Journal of the IES 21(no 1): 47-51, 1992.

[2] Akutsu H., Watarai Y., Saito N., Mizumi H., J. Illum. Eng. Soc. 13: 341-349 1984.

[3] Claassens J., Peeters J., Plas R. van den, Tagungsberichte, Licht Band II paper 28 1984.

[4] Hing P., J. Illum. Eng. Soc. 10: 194-203 1981.

[5] Zollweg R., Kussmaul K., Light. Res. and Tech. 15: 179-184 1983

[6] Furukubo H., Takeji Y., Takatsuka K., US patent 4,156,550 1979.

[7] Toshiyuki O., Hiroyoshi Y., Akira K., European patent 0282657 1987.

[8] Ingalls P., Dolan R., Plunk J., Zhu H., Wyner E., Proceedings IES 1994.

[9] Ravi J., Luijks G., Bockstal G. van, Proceedings IES 1992.

[10] Meden Piesslinger G., Klomp J., Oomen J., US patent 4,122,042 1978.

[11] Meulemans C., Janssen M., Amstel A. van, European patent 0215524 1987.

[12] Geven A., Renardus M., Seinen P.A., Stoffels J., Wijenberg C., Dielis H., European patent 0587238 1993.

SHORT ARC METAL HALIDE LAMP WITH NEW INGREDIENTS FOR LCD PROJECTOR

N.Takeuchi, Y.Kitahara and M.Wakamiya
Light R&D Center, Matsushita Electronics Corp.
1-1, Saiwai-cho, Takatsuki-shi, Osaka, 569, JAPAN

H.Omura and M.Tabata
Materials and Devices Laboratory, Matsushita Electric Industrial Co., LTD.
1006, Kadoma, Kadoma-shi, Osaka, 571, JAPAN

Introduction

Recently the increasing demand for LCD projector with brighter screen leads to the necessity for high intensity light source. Metal halide discharge lamps are the most promising light source, however, there arise some problems peculiar to the LCD projector usage. Namely, color characteristics, lifetime and reliability of the discharge lamp would deteriorate with the shortening of the electrode gap to increase the illuminance of the arc discharge. We have studied the radiation characteristics of many metal halide additives. As a result, we found that Gd-Lu iodides have the advantage of a great improvement in color characteristics.

The present paper describes the characteristics of a new metal halide lamp containing GdI_3-LuI_3. Besides, the stress analysis and designing of the arc tubes to increase their reliability are reported.

Experimental

As for the radiation-related problems described above, we examined the radiation characteristics of Gd-Lu iodides. The metal halides were put in the lamps with 4 mm long electrodes gap operated horizontally at 150 watts. Emission properties of the lamps with aparabolic reflector were evaluated by the projected screen characteristics using the standard optical measurement system simulating LCD projector. Spectrum distribution and illuminance at various points on the screen were measured. On a 5.5 hours on-0.5 hours off cycle, lamp lifetime defined by the time at which illuminance at the screen center decrease by fifty percent of its initial brightness was measured.

Stress analysis and designing were conducted for the lamps at 250 watts to improve reliability of the arc tube. The fracture pressure of the arc tube was measured by gradual filling-in of nitrogen gas into the arc tube through exhaust-tube. The crack propagation was recorded with the high speed camera to identify the weekest spot.

Results

1) Emission properties of GdI_3-LuI_3 lamp

Figure 1 shows spectrum distribution at the center of the screen from the Gd-Lu iodides lamp, characterized by strong blue from Gd and a high concentration of energy around 600nm from Lu. The favorable emission for LCD projector usage from the GdI_3-LuI_3 lamp was able to improve the color rendering properties on the screen. The color uniformity on the projected screen was improved in the GdI_3-LuI_3 case in comparison with the DyI_3-NdI_3 case. Also the GdI_3-LuI_3 lamp led to less color change during burning and longer lifetime than the DyI_3-NdI_3 lamp. The results are summarized in table 1.

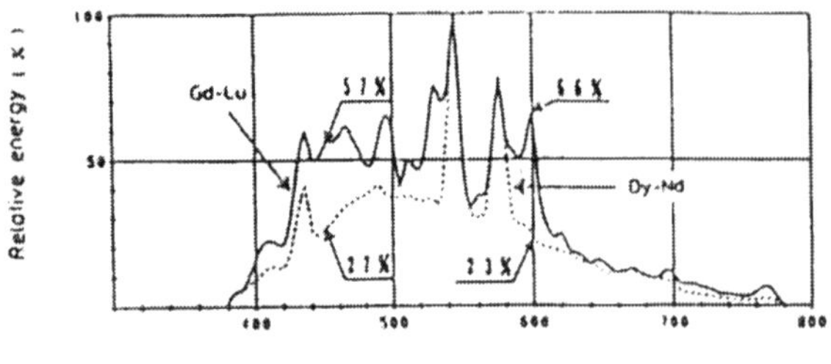

Fig.1 Spectral energy distribution at screen center in the case of a Gd-Lu iodide lamp and a Dy-Nd iodide lamp.

Table1 Characteristics of Gd-Lu iodide lamp and Dy-Nd iodide lamp.

		Dy-Nd	Gd-Lu
Relative intensity of red and blue to green	600 / 545	23(%)	66(%)
	450 / 545	27(%)	57(%)
Difference of color temperature between screen center and corners		1000K	400K
Lifetime		1400h	1600h
Change of color temperature after 1000 hours		1250K	550K

2)Stress analysis and designing

Detailed analysis of the crack initiation with high speed camera revealed that the initial crack was originated mostly around the edge lines of a small opening portion(oblique lines), as shown in Fig.2, which is formed by the pinch-sealed process. We assumed that the fracture strength of the tube is determined by the shape of this opening. The maximum stress σ_{max} at the A and B points is expressed approximately by[1]

$$\sigma_{max} = 2\sigma_0 \times (c/\rho)^{1/2} \qquad (c \gg \rho)$$

where σ_0 is the applied stress, 2c is the opening length and ρ is the radius of curvature, respectively.

In order to increase the fracture strength of the tube, it is necessary to reduce the stress concentration. From the above equation, smaller value of c and/or ρ^{-1} results in the relaxation of the stress concentration. Fracture strength measurement was carried out for tubes with different opening length. Figure 3 shows the arc tube strength depends on the opening length. Based on the results of above analysis, the arc tube was re-designed so as to triple the fracture strength of the tube. Consequently, the reliability of the arc tube was remarkably improved.

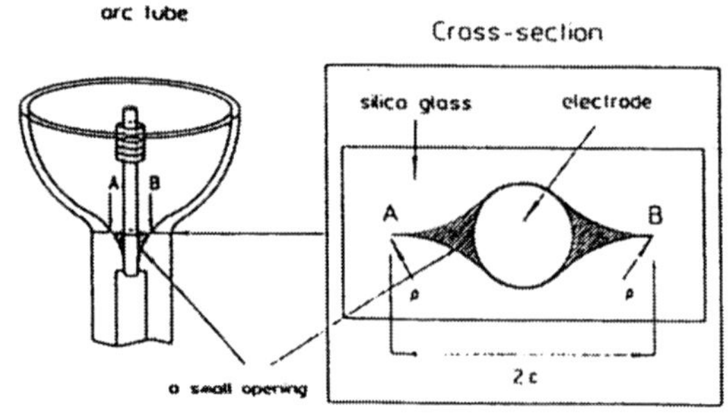

Fig.2 Shape of a small opening.

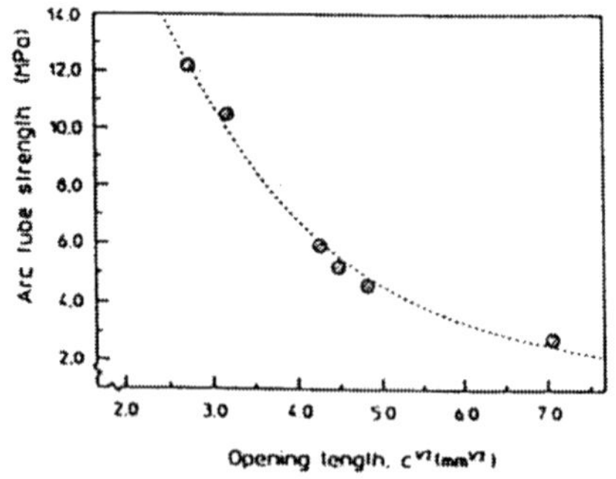

Fig.3 Fracture strength of arc tubes versus opening length.

Reference

1) M.Nishida, ORYOKU SHUCHU, Morikita shuppan, Japan, 1973.

CHARACTERISTICS OF CERAMIC METAL-HALIDE LAMPS AND ITS NEW CONSTRUCTION

M.Ichise,H.Haraguthi,S.Yamazaki
R&DCenter,Iwasaki Electric Corp.
1-20,Fujimi-cho,Gyoda-City,Saitama,361,JAPAN

Metal halide lamp has been steadily estimated for its high efficiency and good color rendering properties,and has come to be widely employed in factories,sports institutions,shops and so on. With respect to the metal halide lamp as a HID(high intensity discherge) light source provided with high efficiency and high color rendering property that has come to be widely utilized in recent years,for the purpose of improving all the more its characteristics ,a lamp was proposed that employs light transmitting polycrystalline alumina tube,from which we have elaborated a proposition of a new light source capable of meeting such requirements,whose description is presented below.

Fig.1 shows the basic construction of the present lamp. An alumina tube with additive MgO is used as its arc tube,and its end portion area are contracted. The upper end portion of the arc tube is so fabricated that a coductive cap made of cermet,formed by adding tungsten to alumina and in which electrode and lead rod are buried,is hermetically sealed to alumina tube with halogen proof glass soler,while regarding its lower end portion an alumina tube and cermet cap are solid-state-sintering during baking process without employing glass solder,thus an attempt for a hermetically integrated structure is made. Moreover,on the inner surface of the cermet cap facing the inside of the arc tube an alumina layer is likewise formed integratedly to prevent exposition of conductive portion in the arc tube. these features have made possible the design of a long life envelope owing to the facts that (1) no corrosion of sealed portions due to excess halogenides takes place, and (2) no aggravation of cermet cap caused by back arc occurs.

The lower table shows p fabricated on experimental basis initial characteristics of the fabricated lamp with a few additives sealed in and tested under conditions such as the input power to the lamp is 150W and the load on envelope wall 37W/cm2. It is seen that the sorts of lamps in which additives Dy-Tl-Li, In-Tl-Li, Sc-Na and Dy-Tl-Na are sealed in have attained such high efficiency and Li additive chracteristics that were unattainable by the lamps employing conventional quartz envelope.

Fig.2 shows the example of the spectral chracteristics of the present lamps(Dy-Tl-Li-I-Hg).

High efficiency and new additives of metal halide lamp is attained by employing alumina tube as is described abave. It is also conceived hereafter that the design of ceramic metal halide lamp will be optimized by selection of additives, and so, which will all the more contribute to the possibility of qualitative improvement.

Sorts of additives	Color temperature(K)	Ra	Luminous flux(lm)	Luminous efficiency
Dy-Tl-Li-I-Hg	4,800	93	13,000	86.7
In-Tl-Li-I-Hg	5,400	87	9,200	61.3
Dy-Tl-Na-I-Hg	4,000	88	14,300	95.3
Sc-Na-I-Hg	3,300	80	16,000	106.7

<Table-1>

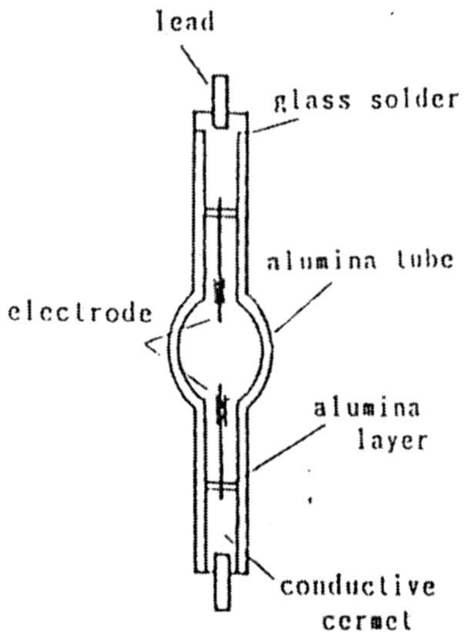

<Fig-1>

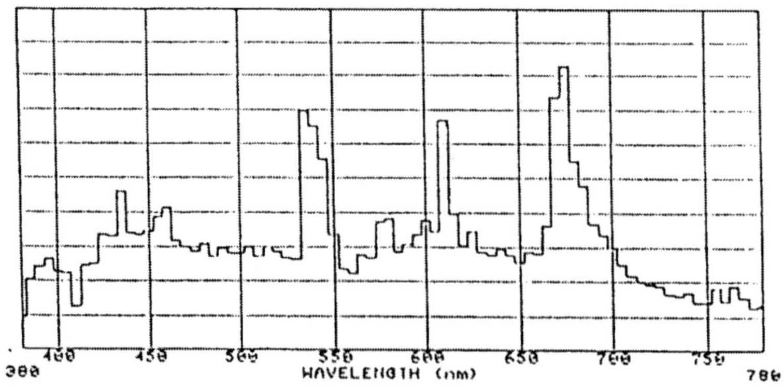

<Fig-2> spectral distoribution of Dy-Tl-Li-I-Hg Lamp

Analysis of the starting condition for Metal Halide Lamp

Makoto Hirabayashi*, Yoshitaka Kemmotsu*, Hirokazu Takahashi**, Mototaka Sone**

* Eye Lighiting Systems Corp. ** Musashi Institute of Technology

1. INTRODUCTION

Recently, HID Lamps are used for the indoor, because of high effeciency, high color rending and long life.

However, some HID Lamp, such as Metal Halide Lamp is not well known about starting mechanism of the lamp operating.

The ballast for M H L has some starting device which can appear at 2 KV~4 KV pulse voltage.

In this study, we found the mechanism of lamp starting process depends on not only pulse voltage but also supply electric power by using Ballast Simulator combined the personal computer and electronic ballast.

2. SYSTEM CONSTRUCTION

In order to measure process of lamp starting energy, we developed Ballast Simulator which has variable supply energy, by using personal computer and electronic ballast showed in figure 1. Electronic ballast is consists of step-down converter and square wave inverter circuit.

The variable lamp supply energy is controled by P.W.M signal which comes from the personal computer. This computer is connected a lamp voltage detector and current transfomer.

The time versus lamp voltage, current and wattege waveform are appeared on its CRT.

Namely, the time versus lamp power consumption in accordance with ballast supply electric power is showed by this system, we are easily able to understand the process of the lamp power consumption .

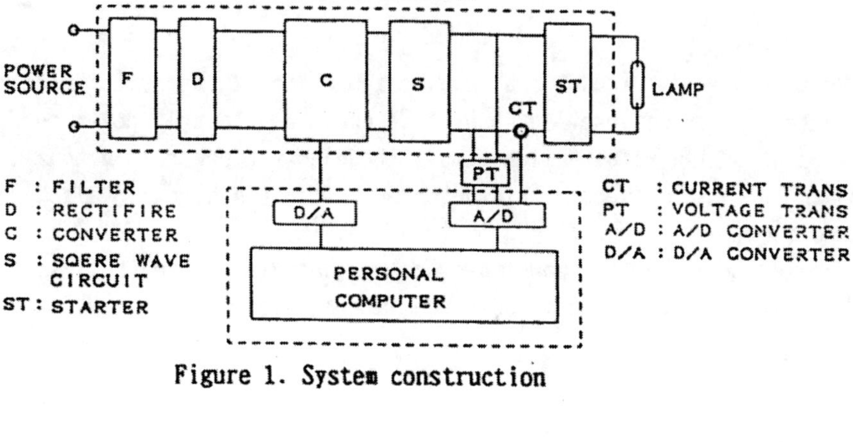

Figure 1. System construction

3. RESULT

Figure 2 shows the time versus lamp voltage, current and wattage waveform by using this system when the 150 W Metal Halide Lamp starts.

We can understand the normal starting condition for the Metal Halide Lamp as this figure shows. Namely, the electric power is gradually consumed in accordance with the electric discharge.

Figure 3 shows each wave form under the falt condition for the Metal Halide Lamp. In this case, the electric power is consumed quickly compared with the normal condition.

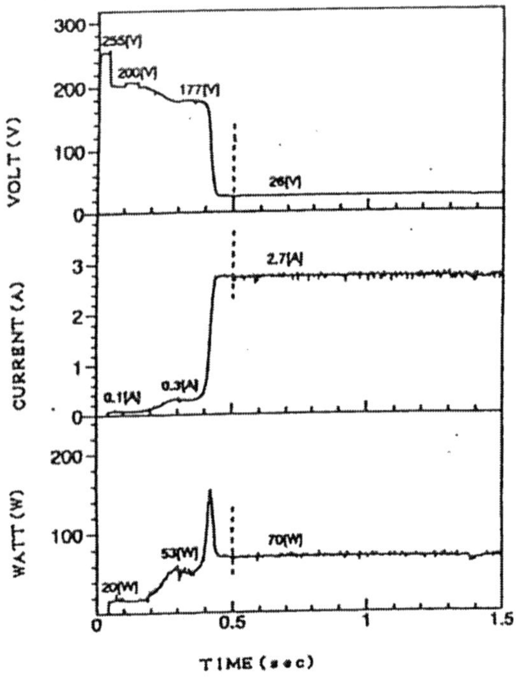

Figure 2. Normal condition wave form of the Metal Halide Lamp

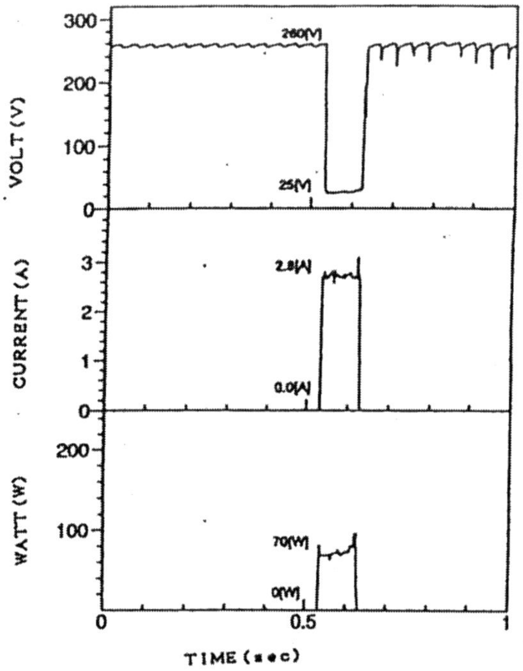

Figure 3. Falt condition wave form of the Metal Halide Lamp

4. CONCLUSION

By this new measurement system, we found the lamp starting mechanism comes from the electric power consumption and it depended on electric power supplied by the ballast.

In case of the Metal Halide Lamp, the electric power is gradually and intentionally consumed in accordance with the growth of the electric discharge under the normal condition.

However, the electric power can not be controled by the ballast, because the electric power is consumed quickly under the falt condition.

ANALYSIS OF THE TIME-DEPENDENT TEMPERATURE DISTRIBUTION OF COMPACT MH LAMPS DURING WARM-UP PHASE

T. Saitoh and Y. Kurimoto
Light R & D Center, Matsushita Electronics Corp.
1-1 Saiwai-cho, Takatsuki, Osaka 569, Japan
Y. Tsutsui
CAE Department, Production Engineering Laboratory, Matsushita Electric Industrial CO., LTD.
2-7 Matsuba-cho, Kadoma, Osaka, 571, Japan

This paper describes a calculation way of time-dependent temperature distribution and then thermal stress in the glass seals of burning metal halide lamps.

Based on the lamp energy balances using finite element analysis together with thermodynamic constants of quartz, tungsten and molybdenum, we calculated the temperature distribution of compact 35W Sc-Na metal halide lamp during warm-up phase.

Figure 1 shows the input lamp wattage in warm-up phase as a function of time. The lamp was operated horizontally. The wattage reaches the maximum (75W) at 2 seconds after ignition, and then decreases gradually to 35W, which is rated wattage. For analysis, the change of the wattage as a function of time was divided into some time intervals, as shown in Table 1.

The thermal losses of the power consumed by the lamp, which is the sum of the electrode loss and non-radiant losses of arc column, were so determined that the calculated temperatures coincided with the results of measurement at some points of the discharge tube surface. Table 1 shows the condition of the time-dependent energy balance, on which the calculation results were in good agreement with the measurement results. The thermal power loss of electrode was assumed to be constant 4.7W independent of time.

Inside the discharge tube, the buoyancy effect creates an upward bowing of the arc. Therefore, based on the results of observation, simplified models were assumed for the arc shape in each time interval between 0 and 6 seconds after ignition, assuming that the arc shape stabilizes in 6 seconds. In Figure 2, the calculated time-dependent temperatures are shown in comparison with the results of measurement.

Time-dependent thermal stress can be derived from thus obtained temperature distribution. For example, Figure 3 shows thermal stress in a cross-section of the glass seal at 10 seconds after ignition.

This analysis is expected to be useful for improving the strength of glass seals in designing new lamps.

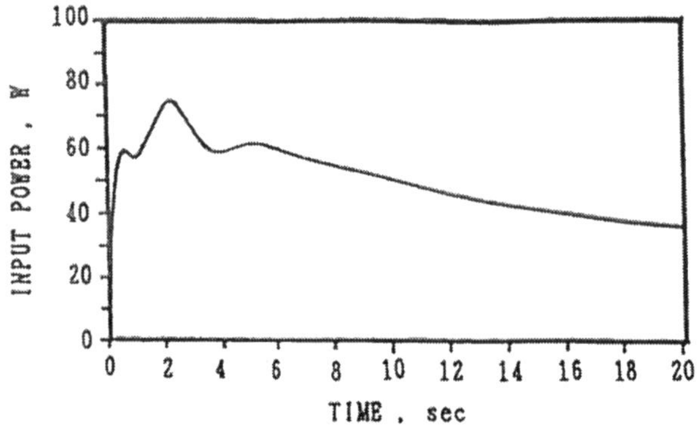

Fig.1 Input lamp wattage in warm-up phase as a function of time

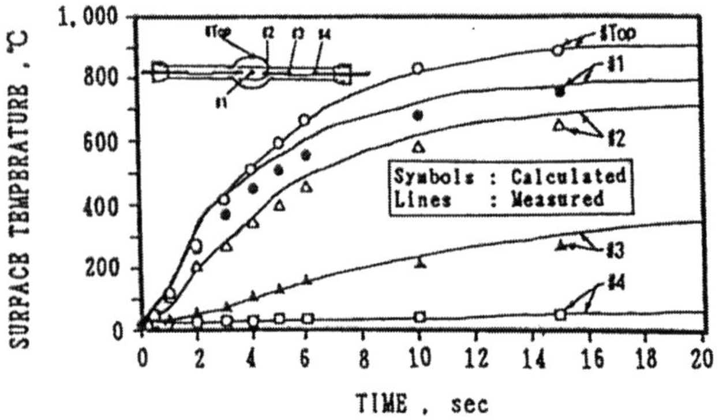

Fig.2 Calculated time-dependent temperature in comparison with the results of measurement

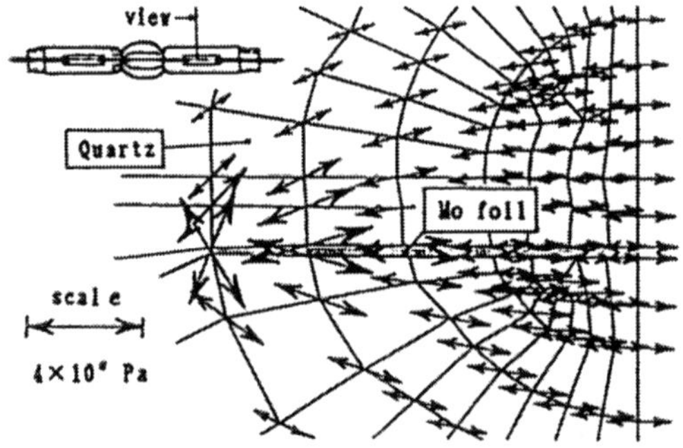

Fig.3 Thermal stress in a cross-section of glass seal at 10 seconds after ignition

Table. 1 Optimized Time-Dependent Energy Balance

Time	[sec]	0-0.5	0.5-1	1-2	2-3	3-4	4-5	5-6	6-10	10-15	15-
Arc	[W]	46	48	44	36	30	27	23	21	16	11
	[%]	83.6	82.8	66.7	49.3	46.2	43.5	40.4	38.2	33.3	31.4
Electrode	[W]	4.7	4.7	4.7	4.7	4.7	4.7	4.7	4.7	4.7	4.7
	[%]	8.5	8.1	7.1	6.4	7.2	7.6	8.2	8.5	9.8	13.4
Power Loss	[W]	50.7	52.7	48.7	40.7	34.7	31.7	27.7	25.7	20.7	15.7
	[%]	92.1	90.9	73.8	55.7	53.4	51.1	48.6	46.7	43.1	44.8
Radiation	[W]	4.3	5.3	17.3	32.3	30.3	30.3	29.3	29.3	27.3	19.3
	[%]	7.9	9.1	26.2	44.3	46.6	48.9	51.4	53.3	56.9	55.2
Input Power	[W]	55	58	66	73	65	62	57	55	48	35
	[%]	100	100	100	100	100	100	100	100	100	100

"LIFE PERFORMANCE IMPROVEMENT of THE SHORT ARC METAL HALIDE LAMP by DC OPERATION"

Tadatoshi Higashi
Ushio inc., Himeji, Japan

1. Introduction

The short arc metal halide lamps have been used as the light sources for liquid crystal projection displays (LCD projector) because they have high luminance and high color rendering property. Since lamps for such purposes must avoid fluctuation of light output, rectangular wave AC power of 250-500 Hz is used as electric input.

Such lamps usually use rare earth metal halides like dysprosium iodide as filling and take very high wall load(-60W/cm2). Their life performance is poor because micro silica crystals are formed on the inner surface of quartz bulb (crystallization), and screen lumen go down below a half of the initial lumen within about 2000 hours, even if the lamp maintains high total lumens yet.

The author have succeeded to remarkably suppress the crystallization of silica by combination of horizontal arc axis operation and DC operation.

2. Suppression of growth of micro crystals by DC operation

(1)Principle

It is estimated that the growth of micro silica crystals comes from attack of free rare earth metal atom to quartz glass. The authors have tried to make use of cataphoreses of DC operation for suppression of the attack of rare earth metal atom to quartz glass.

By DC operation, rare earth metal atoms gathered to high density near the cathode space. The other discharge space becomes to have low density of metal atom and excess of halogen. Halogen of excess easily catch free metal atom. This means the number of rare earth metal atoms reaching the quartz surface as atom decreases remarkably.

(2) Experiment

3 mm arc length lamps were prepared for 125 W, 150 W and 250 W lamps respectively. The discharge bulbs were filled by iodides and bromides of Dy, Nd, Sn, In and Cs suitably. The lamps were fixed to parabolic reflectors respectively and were operated by DC 250W power for more than 3000 hours at the condition that the arc axes were horizontal.

(3) Result

There were no growth of micro silica crystals at 2000 hour operation, and a little growth of micro silica crystals on the upper wall of quartz glass in the near side of cathode at 3000 hour operation.

(4) Light characteristics of DC operation

Fig.1 shows the relation between operation time and screen lumen maintenance of three kind of lamps. **Fig.2** shows the relative luminous intensity distribution of three lamps. Color characteristics on the screens were reasonable.

3. Conclusion

It is proved that the growth of micro silica crystals on the inner quartz surface of the short arc metal halide lamp can be markedly suppressed by horizontal arc axis and direct current operation. Color difference arised in DC operation can be reduced by filling of volatile metal halides adequately.

References
(1) T. Hiramoto : 6th International Sympo. on the Science and Technology of Light Sources. no 67 I (1992)
(2) K. Tomokiyo etal: 6th International Symp. on the Science and Technology of Light Sources. no 73

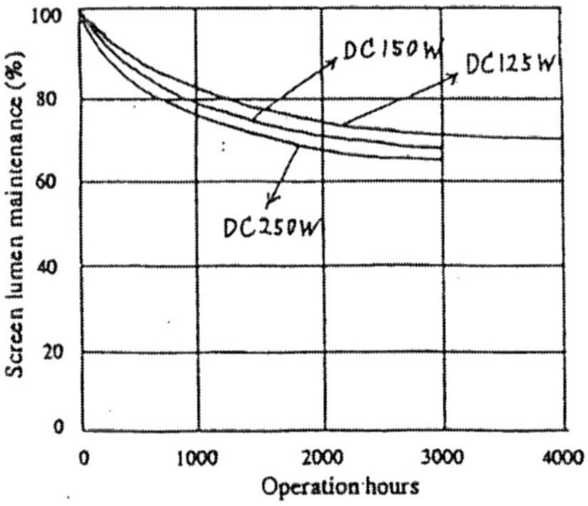

Fig.1

Representative curves of screen lumen maintenance of the125 W ,150W and 250 W lamps,
(3 inche *3 LCD system)

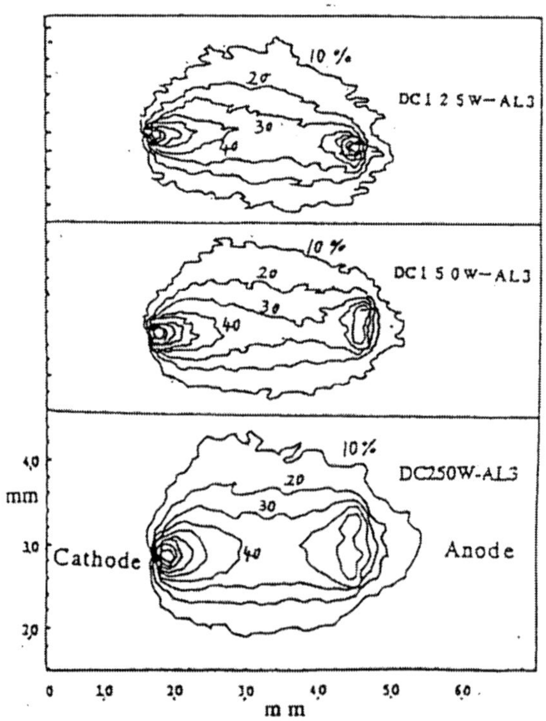

Fig.2 Relative luminous intensity distributions of the 125 W ,150W and 250 W lamp.

INTERACTION BETWEEN ARC-TUBE RADIATION AND BARIUM FILM IN A HIGH PRESSURE SODIUM LAMP

C. CARRETTI, S.P. GIORGI, P. MANINI and A. RENZO
SAES GETTERS S.p.A., Via Gallarate 215, 20151 Milan - ITALY

1. INTRODUCTION

In a high pressure sodium (HPS) lamp the arc tube is mounted within an evacuated outer jacket. This jacket is pumped to a low pressure during processing and then the pressure is further reduced, typically by flashing a barium getter.

Gaseous impurities strongly affect the striking voltage, lumen output and life of this type of lamps: the most dangerous impurity is hydrogen, which both produces a detrimental heat dissipation from the arc tube and penetrates inside the burner increasing the discharge starting voltage. The maximum hydrogen pressure allowed inside the outer jacket is comprised between 10^{-2} and 10^{-3} Pa (1). Some studies, addressed to investigate the residual atmosphere inside an evacuated HPS lamp jacket, indicated that hydrogen is not detectable (2). However, the arc tube radiation should be able to promote H_2 release from the barium film: in fact, on the basis of energetic considerations, ultra-violet components can produce photoelectrons emission, which is the precursor stage of photon induced desorption (3), whereas sodium visible radiation can break barium hydride bonds, providing free hydrogen atoms inside the film.

In this paper, barium films dosed with hydrogen over a wide range of concentrations (from 0.6 to 20 Pa m^3/g) were studied during HPS lamp operation by means of mass spectrometry. The dependence of hydrogen release on the H_2 concentration, film temperature and surface morphology was investigated.

2. EXPERIMENTAL

A properly designed pyrex bulb, containing a sodium arc tube of 400W nominal power was connected to a bench equipped with a quadrupole mass spectrometer, as shown in reference (4).

For each test a ring type barium getter was mounted yielding about 4 mg. Before starting each test, the bench and the bulb were baked under vacuum at 573 K overnight to reduce the background pressure to less than 10^{-8} Pa. The arc tube was also conditioned by switching it on for 2 hours prior the analysis. The barium film temperature was measured by means of a K-type thermocouple directly positioned on it. Hydrogen of 99.99% nominal purity was used for sorption-release studies from the barium film.

The analyses were carried out by measuring the instantaneous flux of gas released from the sample and evaluating the pressure drop across a known conductance.

The background contribution coming from the arc tube and the glass envelope was measured before each flash of the getter.

3. RESULTS

In a first set of tests, each barium film was dosed with hydrogen and then exposed to the arc tube radiation. The H_2 release as a function of time, shown in figure 1 for different concentrations, indicates that hydrogen can be removed from a barium film during lamp operation. The presence of a sharp peak, followed by a slow increase of gas release suggests that two different mechanisms are taking place: a radiation induced H_2 emission followed by a thermally activated hydrogen outgassing. The initial peaks of the curves were integrated as a function of time: regardless the H_2 dose, a release of about 5×10^{18} molecules/m^2 was calculated, this quantity being of the same order of magnitude of a monolayer. This finding can be explained supposing that the interaction between radiation and barium film involves only superficial layers. Therefore, the initial H_2 release appears related to the surface film characteristics. Some studies are in progress to investigate the agreement between these experimental observations and the photon induced desorption models reported in literature.

In a second set of tests, the influence of the barium film morphology on the hydrogen release was investigated after heating fresh films to 600 K, by means of arc tube radiation. As reported in figure 2, barium preliminarily exposed to the lamp burner releases less H_2 with respect to a fresh film and this phenomenon becomes more noticeable for longer irradiation times. On the other hand, increasing the amount of hydrogen dosed on the barium, an increase of the H_2 release from the pre-heated films was measured, as shown in figure 3. The observed phenomena can be explained supposing that the thermal treatment produces a reduction of the surface area of the barium film: this effect, as reported in the literature (5), is caused by the sintering of the grains. Increasing the H_2 doping, metallic barium is converted in barium hydride. The consequent change of the lattice parameters can produce surface fractures, thus increasing the total irradiated area.

To cope with the hydrogen amount released during the lamp operation, non evaporable getters (NEG) such as Zr-Al alloys, which show high pumping speed for hydrogen, could be used. As shown in figure 4, the presence of a NEG reduces the measured hydrogen outgassing, indicating its ability to sorb some H_2 released from the barium film.

4. REFERENCES

(1) A.Barosi and E.Rabusin, Proc. 6th Intern. Vac. Congr. (1974), p.49.

(2) V.D.Meyer, E.F.Wyner and P.L.Fortucci, J.Illum.Eng.Soc., 9, (1980) p.175.

(3) D.H.Baker, R.L.Champion, L.D.Doverspike and Y.Wang, Appl. Phys. Lett. 64 (1994) p.1880.

(4) C.Carretti, P.Manini, A.Renzo and E.Rabusin, Proc. 6th Int. Symp. Sci. Technol. Light Sources LS6 (1992) p.53.

(5) H.J.R.Perdijk, Vacuum, 10 (1960) p.204.

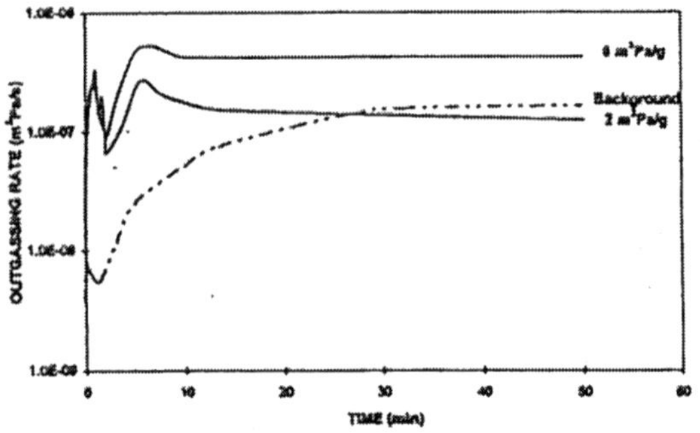

Fig. 1 - Hydrogen outgassing evolution during barium film irradiation at different H_2 concentration

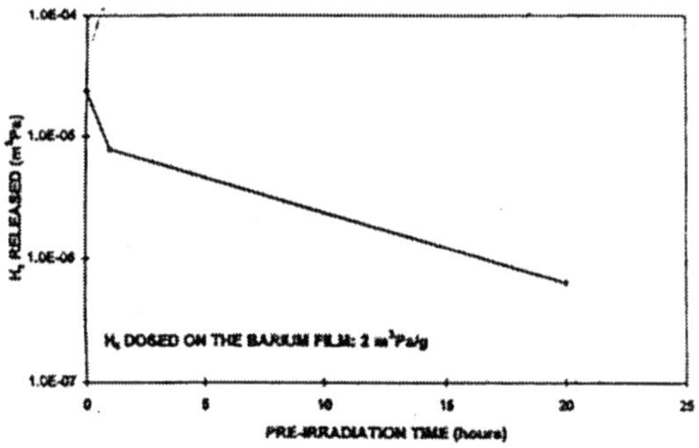

Fig. 2 - Total amount of H_2 released at the arc tube switching on as a function of preliminary irradiation time

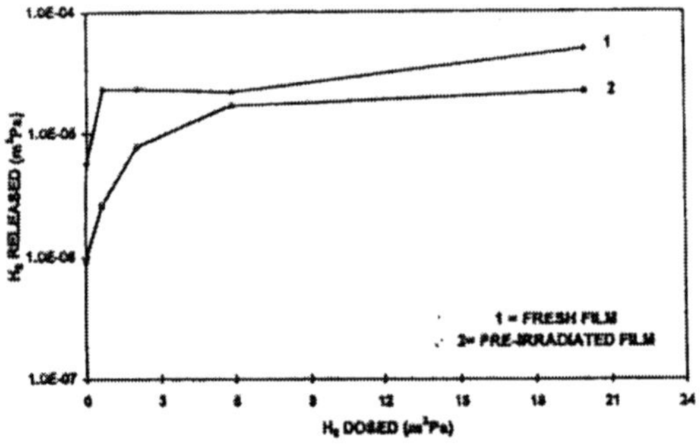

Fig. 3 - Total amount of H_2 released at the arc tube switching on as a function of concentration

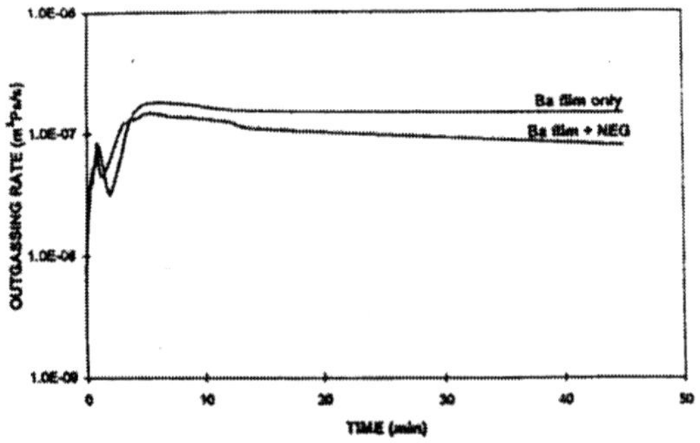

Fig. 4 - Comparison between H_2 outgassing rate in presence of a barium film and with an additional Zr-Al alloy

FACTORS INFLUENCING THE MINIMUM ARC SUSTAINING CURRENT

W. YAN, F.P. DAWSON
Department of Electrical and Computer Engineering,
University of Toronto, Ontario, Canada, M5S 1A4

Pulsed high pressure arc lamps are often operated with a small sustaining current[1] during the interpulse period, so as to: minimize the arc expansion time; to increase the lamps operating lifetime; to enable operation at higher repetition frequencies[2],[3]. It is common in practice to select a sustaining current level to coincide with the minimum electric field on the arc E-I characteristic curve. One of the difficulties in designing power sources to provide the sustaining current is to predict the appropriate level for the sustaining current. This paper shows that the radius of the arc tube, the amount of convection and radiation, and the assumptions regarding the native of the radiation model strongly influence the choice of the minimum sustaining current whereas the pressure and boundary temperature have only a slight effect on the minimum sustaining current level. The extended Ellenbass-Heller equations were solved using Patankar's well-known simple algorithm, subject to prescribed boundary conditions. Fig. 1 shows that the pressure has little impact on the minimum sustaining current. Fig. 2 shows that the amount of radiation has an impact on the minimum sustaining current level. With radiation, a smaller minimum sustaining current is obtained. Fig. 3a and Fig. 3b show that a change in the thermal conductivity in the outer regions of the plasma column have an influence on the minimum sustaining current level. These simulations were made in an effort to roughly simulate the effect of convection on minimum sustaining current. In Fig. 3a, the thermal conductivity values, which range from 3.8×10^{-4} W/cm°K at 1000 °K to 0.002 W/cm°K at 7000 K, were increased to a constant value of 0.004 W/cm°K for all temperatures less than 7000K. In Fig. 3b, thermal conductivity was increased yet another order of magnitude to 0.04 W/cm°K for all temperatures less than 7000°K. It is seen that when the thermal conductivity is at the value of 0.004 W/cm°K, the change in the value of minimum current is indistinguishable since temperature profiles are only slightly affected by this change. When the value of thermal conductivity increases to 0.04 W/cm°K, the minimum sustaining current is increased. Fig. 4 shows that the boundary temperature has no influence on the minimum sustaining current level.Fig. 5 shows that the use of an optically thick model has a significant effect on the minimum sustaining current level.

Reference

1.ILC Technical Bulletin 2, A Guide to Flashlamps for Pulsed Solid State Lasers, ILC Technology, Inc., 1983.

2.Ronald L. Stephens and W.F.Hug, "Simmering lamps live longer", Laser Focus, No.8, pp.38-40, July, 1972.

3.J.Jethwa and F.P.Schafer, "A reliable high average power dye laser", Applied Physics, No.4, pp.299-302, 1974.

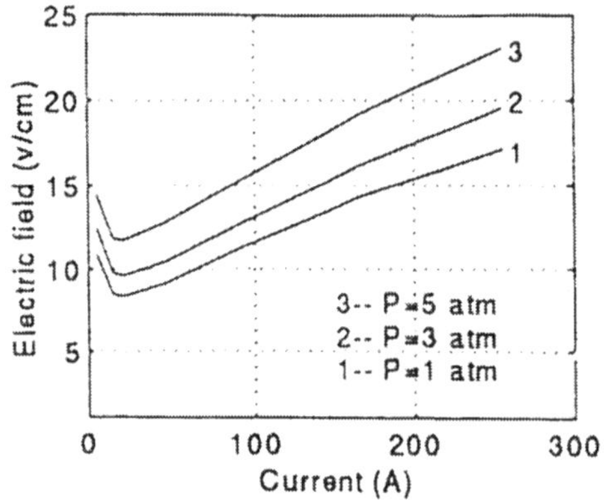

Fig. 1 The effect of pressure on the minimum sustaining current

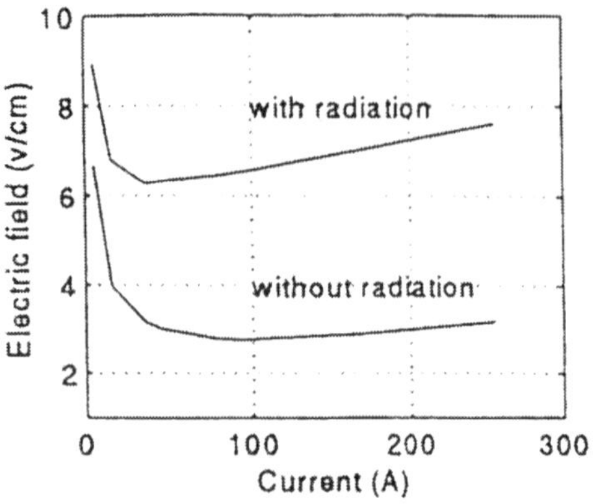

Fig. 2 The influence of the radiation on the minimum sustaining current

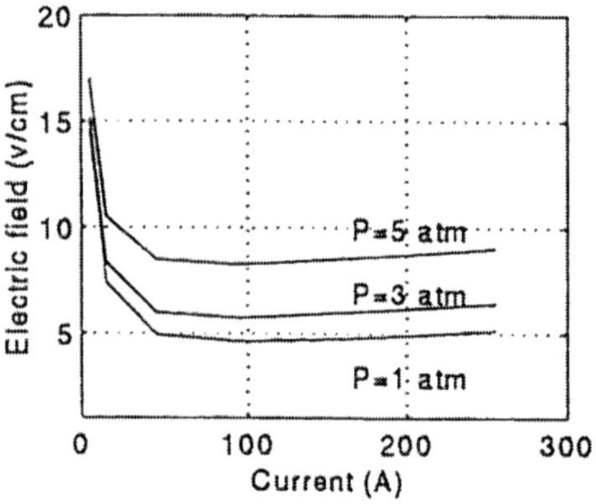

Fig. 3(a) Thermal conductivity increased to 0.004 W/cm°K for all temperature less 7000 °K

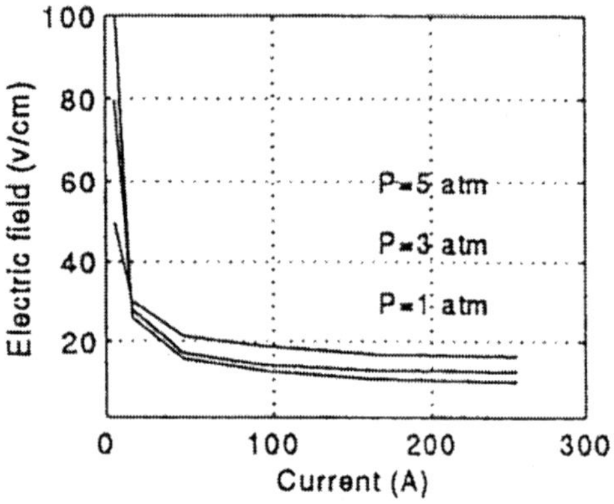

Fig. 3(b) Thermal conductivity increased to 0.04 W/cm°K for all temperature less 7000 °K

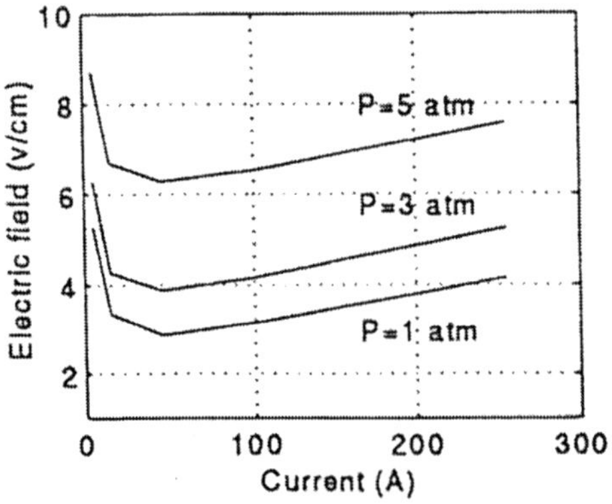

Fig. 4 Boundary temperature is increased from 360k to 1500k.

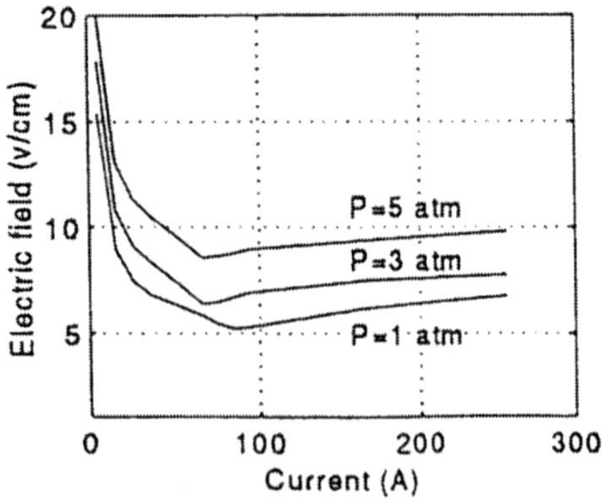

Fig. 5 The influence of the optically thick model on the minimum sustaining current

PROGRESS IN SULFUR LAMP TECHNOLOGY

B.P. Turner, M.G. Ury, D.A. MacLennan and Y. Leng
Fusion Lighting, Inc.
15700 Crabbs Branch Way
Rockville, Maryland USA 20855

At LS:6, Fusion Systems Corp. announced the discovery of an unique, electrodeless lamp utilizing Group VI elements (e.g. sulfur).[1] Athough this technology had shown several attractive features such as high efficacy, continuous spectrum, long life, good color rendering and stability, and an enviromentally benign fill, the sulfur lamp had been regarded more as a laboratory curiosity than a viable alternative for general lighting applications.

In the past year, the sulfur lamp has started to make the transition from the laboratory to real life applications. The use of 3.1 kW sulfur lamps in conjuction with light pipes to illuminate the east bay of the US National Air and Space Museum and the US Department of Energy's Forestall building have been technically, economically and asthetically sucessful. Despite the sucess of these demonstration applications, it is apparent that the complexity of the 3.1 kW lamp is a significant barrier to wide spread use.

The most common characteristics of the 3.1 kW lamp that have been cited as objectionable are the need for forced air cooling and the large lumen output of the lamp. The obvious way to circumvent these "problems" is to make a lamp which had significantly lower power loading. However, it was expected that the high microwave efficacy would fall with lower power loading. Nonetheless, a lamp has been built which has lower power loading, no forced air cooling which has higher intrisic microwave efficiency and higher wall plug efficiency than the 3.1 kW lamp. The Solar 1000™ lamp system, employs a 40 mm (OD) lamp envelope and a single magnetron source which couples 825 W of microwave power into the lamp to produce approximately 134,000 lumens. Table I gives the energy balance of this new lamp system. A prototype lamp has been evaluated by the US National Institute of Standards and Technology. NIST has certified that this lamp operates at 97 lumens/ wall plug watt.

TABLE I: Energy Balance

Component	Power (W)	Subcomponent	Power (W)
Bulb	825	visible radiation	500
		uv radiation	85
		ir radiation	5
		"heat"	235
Lamp Chassis	495	Magnetron losses	390
		Blower/rotator/misc.	63
		filament transformer	32
Power Supply	75	Transformer losses	75
TOTAL	1385		

This table dramatically points out the advantages and disadvantages of this lamp technology. The instrinsic efficiency of the bulb is quite high, exceeding 160 lumens/ microwave watt. The system efficiency falls drastically because of the relatively inefficient conversion of electric power into microwave power. However, compared to other sources with near 100 LPW performance, the light beam is relatively "cool." That is, there is a low amount of infrared radiaton from the plasma or from the lamp envelope. Well over half of the "waste" heat is generated outside the bulb and hence outside the distribution optics. This can be in a significant advantage for applications in which heat cause problems. This table also implies that higher power systems will have better efficiency because of a constant "overhead" of energy consuming items such as filament transformers and blowers.

The spectrum of the lamp is shown in Figure 1. The CCT is approximately 5400 °K with a CRI of 79.

Figure 1: Sulfur spectral power distribution

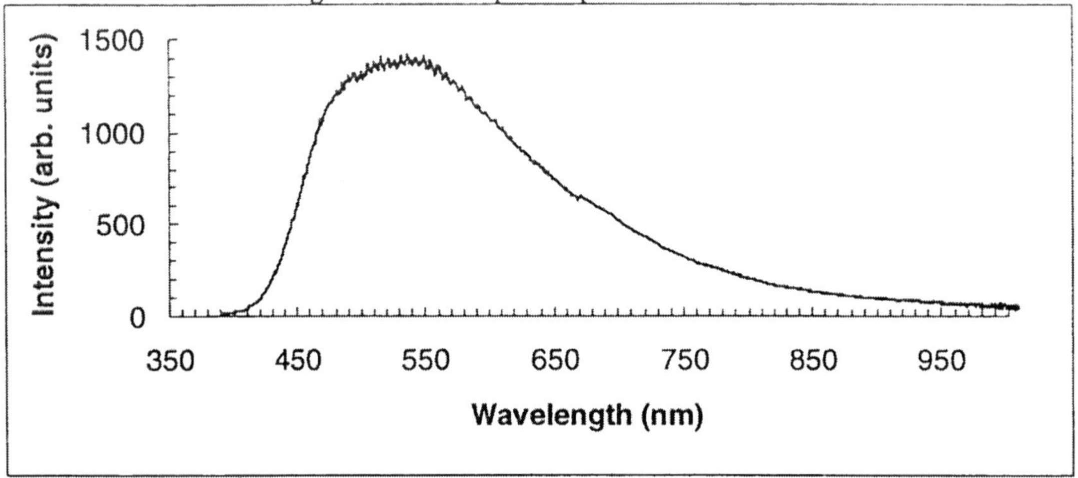

Turner et. al.[2] have shown that the characteristic spectra of the sulfur and other group VI molecules can be explained by considering the vibrational temperature of the plasma, the transition strengths of the vibrational manifold (Frank-Condon factors), the molecular density and the radius of the lamp. That is, the optical depth (density radius product) and the temperature of the gas determine the spectra. By going to larger radius than the 3.1 kW lamp, the sulfur density was significantly reduced. This reduction in density minimizes the amount of "sludge" (high order sulfur chains) and reduces quenching collisions. Reducing both of these quanities more than overcome the efficacy penalty associated with a reduction in power loading.

The Fusion Lighting team has used an Elebaas type of analysis to measure the slope efficiency of this particular embodiement of the suflur lamp to be 220 lumens/microwave watt. This slope efficacys shows that there still exists room for improvement in efficacy; with improvements in instrinsic efficacy and with improvements of microwave driving sources, the sulfur lamp technology hold the promise of being not only the most efficacious source of high quality light known, but a pratical source as well.

---

[1] J.T. Dolan, M.G. Ury and C.H. Wood, "A novel high efficacy microwave powered light source." Presented at LS:6, Budapest. 1992

[2] B.P. Turner, M.G. Ury and D.A. MacLennan, "Microwave excited sulfur lamp" 47th Gaseous Eletronics Conference. Gaithersbury, Maryland 18-21 October, 1994.

LOW POWER MICROWAVE DISCHARGE IN SULFUR LAMPS

A.N. Didenko, G.A. Lyakhov, K.F. Shipilov, E.A. Vinogradov
Department of Physical and Technical Problems of
Energetic RAS, General Physics Institute RAS.
Moscow, Russia

Microwave power discharge in gas sulfur + argon mixtures has been used recently to develop the new practical electrodeless light source of high efficiency [1, 2]. Such sources have solar-like spectrum and high ecological quality. To advance the application field, it is necessary to search the possibilities to lower of microwave power guided to excite a discharge in quartz bulbs containing sulfur + argon mixtures of different concentrations. Spherical bulbs at the end of waveguide were energized by X-band-magnetron producing 6 W of microwave power (Wavelength 3 cm) [3].

It is known that breakdown electrical field changes from 60 to 100 V/cm when argon pressure grows from 1 to 7 torr. In our arrangement we have had 35 V/cm at the end of waveguide. In such underthreshould conditions it is necessary to use the external spark 30 kV-discharger producing argon plasma with short time red lighting. Nevertheless, microwave power 3 W guided to sulfur + argon filled bulbs was sufficient to maintain the stable long-time discharge with white sulfur lighting. The temperature of outer surface of bulb did not exceed 150°C.

It is very important to determine the optimal frequency range for discharge exiting. Three length scales must be estimated for this purpose: the free path length, the electron oscillation amplitude and diffusion length (which is proportional to the bulb dimension). These estimates show that optimal frequency range is near 100 – 1000 MHz if pressure is equal to 1 – 10 torr. It allows us to excite the argon + sulfur discharge without external ionisator at frequency about 300 MHz.

Thus, we realized the white light source with low power microwave pumping and without external cooling. These results allow us to conclude the possibility to use electrodeless discharge for constructing high quality light sources of wide application range.

1. J.T. Dolan, M.J. Ury, C.H. Wood
   The 6th International Symposium on the Science and Technology of Light Sources, 1992, p. 301
2. D.O. Wharmby. IEE Proc. 1993, A140 (6), p. 465
3. A.N. Didenko, E.A. Vinogradov, G.A. Lyakhov, K.F. Shipilov
   Rep. of Russian Academy of Sciences (in press)

PULSATING OPERATION OF ELECTRODELESS HID LAMP

Keiichi Shimizu Ichiro Yokozeki Kouzo Uemura Akira Ito Akihiro Inouye
Toshiba Lighting and Technology Corporation
1-201-1 Funakoshi-cho Yokosuka-City Kanagawa Pref. Japan 237

INTRODUCTION

Inductively coupled electrodeless lamp systems have been realized because the high frequency power supply is easily constructed due to the recent progress of semiconductor devices. The conventional construction of the circuit is: commercial AC is rectified and smoothed into DC, which is inverted to the high frequency voltage and supplied to the lamp(1). When applying this method to high power systems (such as hundreds of wattage), we have a problem that the power dissipation and the cost of DC supply are so high, it is difficult for general lighting to realize. We found it appropriate for the electrodeless lamp systems that inverting rectified but not smoothed commercial AC into high frequency, and supplying high frequency voltage to the lamp.

CIRCUIT CONFIGURATION

Figure 1 shows the circuit configuration. (1) is a commercial AC supply. (1) is rectified by a rectifier (2), and supplied to a high frequency inverter (3). Inverted high frequency voltage goes through a matching circuit (4), then is supplied to an excitation coil (5) and operates a lamp (6). We call this system as the pulsating operation system.

Figure 2 shows the waveforms of some part of the circuit. The intensity of magnetic field captured by a pickup coil is similar to the form of the rectified, pulsated voltage. However input current is distorted a little, this circuit satisfies the harmonic current limit for the lighting equipment.

FORM OF THE ARC

Figure 3 shows the distribution of the arc intensity. Each curved line corresponds to the input supply phase of 0,30,60,90,120,150 degrees respectively, on the basis of zero cross point of input voltage. This figure shows the tendency that when the arc intensity is high, the diameter of the arc ring widens; and when the arc intensity is low, the diameter shrinks.

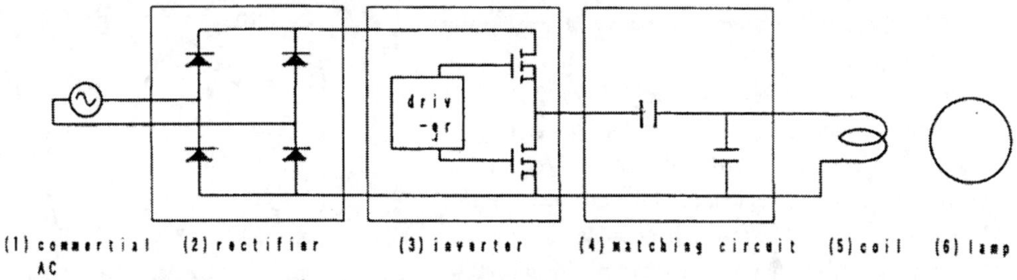

Fig.1 circuit configuration

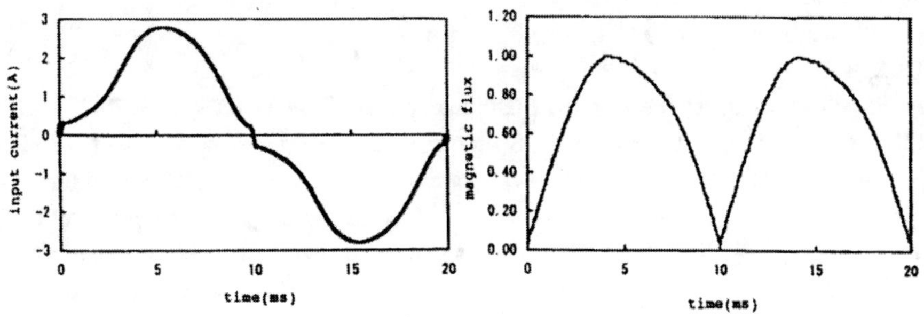

Fig.2 electrical waveforms

Assume that dividing this distribution curve into two portions, left and right, along the line at horizontal position =0mm. We considered the center of the gravity of each portion as the center of the arc. We calculated these points, the center of the gravity of each portion, and defined that the distance between these two points be the diameter of the arc ring. Figure 4 shows the change of the diameter of the arc ring across time. It widens and shrinks in accordance with the input supply, and reaches the lowest point about 20 degrees behind the zero cross of input supply.

SYSTEM EFFICIENCY

We measured the input wattage and luminous output when supplying commercial AC, or supplying smoothed DC, respectively, to the system shown in Figure 1, and compared the system efficiency. In this system, a ripple of light output is larger than that of common (electroded) HID . This causes the decrease in luminous efficiency, but this decrease nearly equals to the loss causing when using an AC/DC converter.

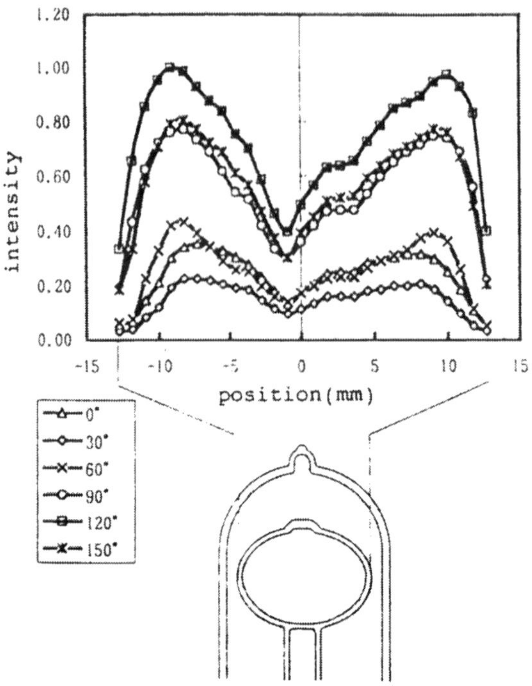

Fig.3 intensity distribution

CONCLUSION

The pulsating operation system, which supplies the lamp with high frequency voltage inverted from rectified but not smoothed commercial AC, is appropriate method to realize the electrodeless lighting system.

1. Low harmonic distortion and high power factor without the circuit that corrects the waveform of input current.
2. High efficiency with no power loss in waveform-correcting-circuit.
3. The electrodeless lighting system with uncomplicated circuit configuration and low cost.

Inductively coupled electrodeless discharge has the function of changing the length of the arc according to the supply voltage. We found it possible to maintain the arc stable when supplying the lamp with pulsating operation system.

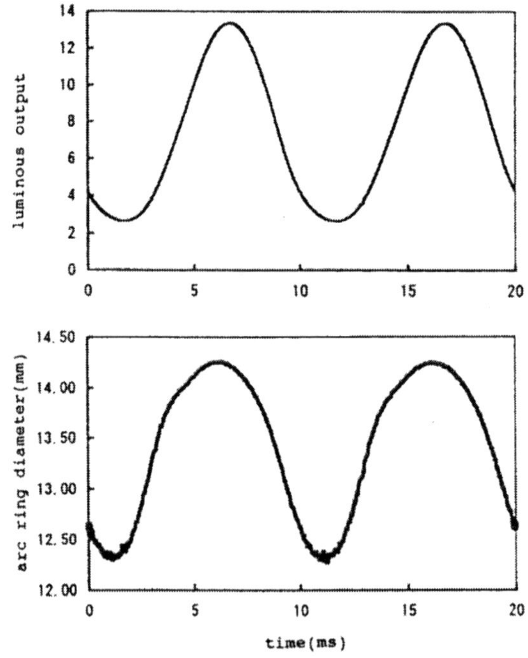

Fig.4 arc ring diameter/luminous output

REFERENCE

(1) C.N.Stewart et al, Inductively coupled HID lighting system LS:7 N0.78-L(1992)

(2) M.E.Duffy et al, Diagnostics and model of an inductive HID Hg Discharge LS:6 No.80-P(1992)

(3) K.Uemura et al, Development of exciter-coil for electrodeless metal halide lamp system Proceedings of annual conference of IEIJ No. 36(1995)

(4) I.Yokozeki et al, Study of an electrodeless lamp system using phase locked loop control Proceedings of annual conference of IEIJ No.38(1995)

EXTRAPOLATION OF RADIAL TEMPERATURE PROFILES IN HIGH-PRESSURE MERCURY LAMPS

B. Freisinger, G. Hartel, H. Schöpp
Institute of Low-Temperature Plasma Physics
Robert-Blum-Straße 8-10
17489 Greifswald, Germany

Motivation

High-pressure mercury lamps (HPML) have a large field of applications in public lighting. To determine the qualities of the lamps and to calculate the radiation transport the knowledge of some basic parameters is necessary. One basic scientific problem is to reach a well adapted radial temperature profile. An attempt was made to fit measured temperatures in the axis up to the wall of the envelope using electrical conductivity.

Methods

To determine the axis temperatures in HPML different methods and their variations are described in the literature. Some spectroscopical methods use the measured emissivity of optically thin lines or self-reversed emission of optically thick lines. In this work three spectroscopical methods were selected. On the one hand the absolute line intensity of the mercury lines 576.96 nm/579.066 nm were measured. These lines are optically thin [1,2]. For a LTE plasma one can get the radial temperature profile after ABEL-inversion up to the reduced radius of about 0.33. The pressure in the discharge tube was calculated self-consistently. These temperatures are verified by the procedures of BARTELS [3] for the self-reversed optically thick line 546.1 nm and by the ratio of the emission coefficients of two optically thin lines (407.781 nm/690.716 nm). The axis temperatures agree well within the experimental error.

Experimental set-up

Plasmas are generated in cylindrical quartz lamps. Two different types of lamps were examined. One type had an inner diameter of 12 mm at an electrode spacing of 10 mm, the other one had a diameter of 17 mm and a spacing of 27.2 mm. The lamps were filled with 25 mg respectively 32 mg of mercury and contained some kPa Ar as starting gas. They are operated in the AC discharge mode (250W) with a commercial available inductive ballast. The radiation of the lamp passes a mirror system to the entrance slit of a 0.5 m spectrograph with gratings of 600 1/mm and 1800 1/mm. The signals are recorded by means of a gated OMA system. For an absolute radiation calibration a xenon radiation standard is used [4].

Extrapolation of temperature profiles

The determination of the plasma temperature by absolute line intensity is in most cases only possible in the inner parts of the arc. The attempt is made to get further information about the outer parts of the arc using the electrical conductivity profile in the cross section Q

$$\frac{I}{E} = \int_Q \sigma \, dq$$

with the electrical current I, the field strength E and the conductivity σ. For this purpose the temperature between the inner part and the wall is interpolated by a polynomial of 3rd degree. The polynomial coefficients are determined in such a way that the interpolation is continuous and has the same derivative in the connection point as the measured temperature profile. Fixed points of the interpolation are the wall temperature and smooth crossing to the measured profile. The electrical field strength E is a free parameter and can be changed considering the measured voltage and electrode distance. The mercury pressure is calculated by integrating the mass density over the discharge vessel assuming temperature distributions behind the electrodes and comparing the result with the whole mass. Figure 1 shows the results of the interpolation for the second type of lamps with a current of 5.8 A. An agreement with experimental results can be achieved.

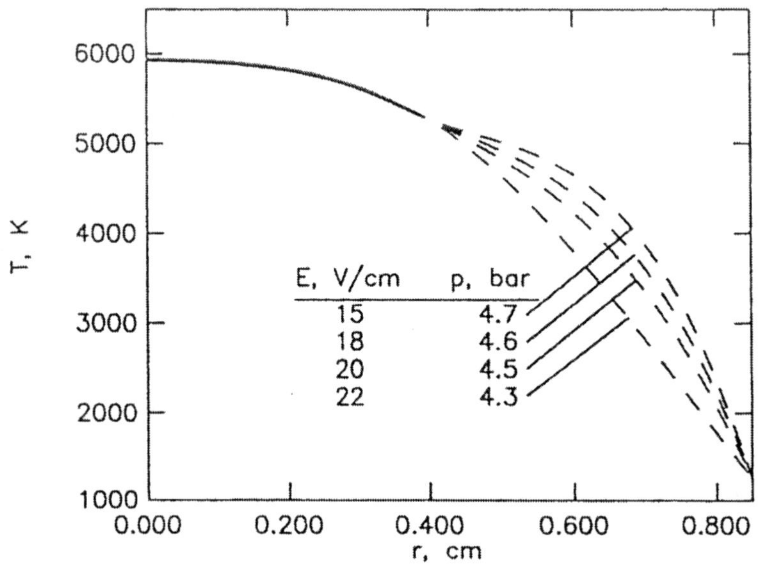

Figure 1: Interpolation of the temperature profile between the inner region of the arc with temperatures from the Hg 577/579 nm intensity (full line) and the wall using the electrical conductivity profile (broken lines). E and p are the assumed electrical field strength and the deduced mercury pressures.

Acknowledgements

This work was supported by the Federal Ministry of Education and Research via project 13 N 6190.

References

[1] Karabourniotis D Drakakis E Zacharopoulos B (1992) *J Phys D Appl Phys* **25** 188-194

[2] Dakin J T Gilliard R P (1986) *J Appl Phys* **60** 1281-1288

[3] Bartels H (1950) *Z. Physik* **127** 243-273

[4] Guenther K and Radtke R (1975) *J Physics E* **8** 371-376

Utilization of HID Illumination in Medical Diagnostics and Treatment

Charles N. Stewart, Douglas M. Rutan, William H. Lagerway
Welch Allyn, 4619 Jordan Road, Skaneateles Falls, NY 13153, USA

Medical diagnosticians rely on suitable illumination, including adequate illumination levels and faithful color rendition. Recent advances in CCD technology and high efficiency lighting have permitted non-invasive diagnostics and minimally invasive surgery employing fiber optic lighting coupled with distal mounted CCD imagers. These technologies result in both less expensive procedures and better medical outcomes.

We present the results of a design of a lighting system that was developed to provide optimum illumination for medical and industrial endoscopes. Through Monte Carlo ray tracing techniques[1] based on experimental measurements of luminous intensity distribution through 4π steradians we have successfully modelled the lamp's light output. By optimizing the lamp, reflector, and the optical system, we have been able to develop a system that provides the physician with 5500K illumination with very good color rendition and optical system efficacies well above halogen and many HID alternatives. This high efficacy results in smaller system size and lower weight, which allows battery operation where necessary, as well as consuming less electrical power.

A key factor in the efficiency of illumination systems for endoscopes is the collection efficiency of the lamp and fiber optic system. This efficiency is controlled by several factors: inherent efficiency of the lamp in lumens per watt (lpw); source size (*i.e.,* arc gap or filament size); luminance of the source, which is related to the lamp efficiency and the source size; magnification of the reflector; numerical aperture of the fiber optic bundle; and size of the fiber optic bundle.

Fiber optic bundles used for medical diagnostics must be as small as possible, consistent with adequate illumination; generally the bundle size is in the range of 1.0 to 5.0 mm. The arc tube used in this study has an arc gap of 1.2 mm; it is desired, therefore, that the optical system magnification be no more than four. The arc tube output, of course, should be as large as possible, compatible with a short arc gap and a long lifetime. Fortuitously, the design of an elliptical reflector with a magnification of four results in a numerical aperture of approximately 0.7, which provides wide angle illumination to the working area.

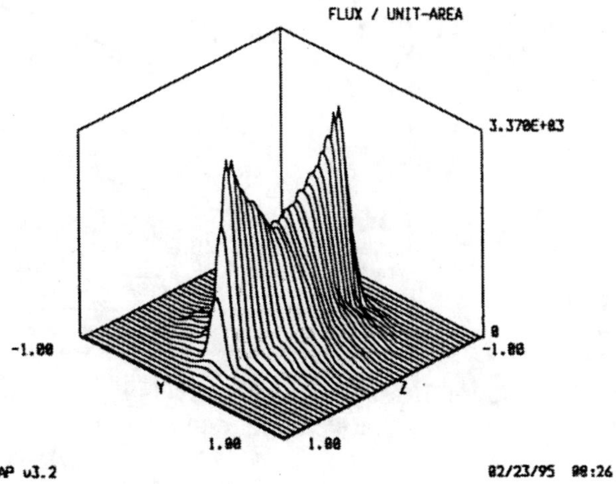

Figure 1. Model results for luminance distribution from the 1.2 mm arc.

For this work we measured the luminous intensity distribution of the arc tube as a function of angle for three burning orientations. We were then able to construct a model of the arc that gave an excellent representation of the projected arc intensity, as shown in Figure 1.

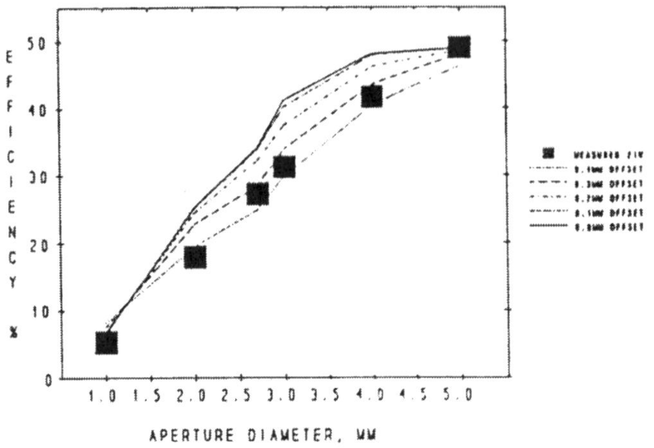

Figure 2. Comparison of model results with measurement for insertion efficiency into apertures.

This model was used to predict the quantity of light projected into an aperture at the focal length (f_2) of the elliptical reflector. The agreement between model and measurement was excellent when the axial position of the arc tube was treated as an adjustable parameter. Even when the axial position was fixed at the first focus of the reflector (f_1) the agreement was very good. Model and measurement are shown in Figure 2.

Combining the several parts of the optical system and comparing to other lighting system alternatives, as shown in Figure 3, we find that the combination of the arc tube's high efficacy (60 lpw), the collection efficiency of the reflector (> 60%), and the low magnification of the reflector (4.0) gives an extremely efficacious lighting system. System efficacy is approximately 30 lpw for 5.0 mm bundles; comparable systems range from 5 to 15 lpw. The system performs particularly well in comparison to halogen lamps because of their lower lamp efficacy, larger source size, and higher magnification.

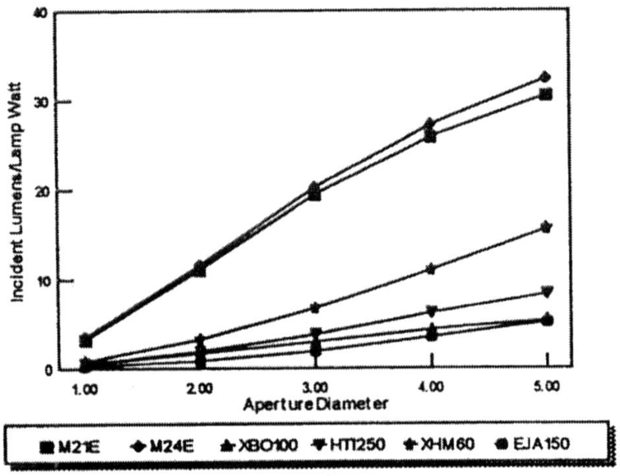

Figure 3. Measured efficacy as a function of aperture diameter for several sources.

In summary, we have found that a combination of measurements and optical modelling shows that a compact arc source can be designed into an optical system to give system efficacies of 30 lpw into 5.0 mm fiber optic bundles while maintaining excellent color rendition. Additionally, the low input wattage of the system allows light weight, portable systems to be designed for the medical practitioner or the industrial technician - a capability not previously achievable in high brightness systems.

1. Advanced Systems Analysis Program (ASAP), Breault Research, Tucson, AZ.

ON THE CALCULATION OF THE MERCURY DISTRIBUTION IN VERTICALLY OPERATING HIGH PRESSURE MERCURY DISCHARGE LAMPS

K. CHARRADA AND G. ZISSIS

Centre de physique des Plasmas et de leurs Applications de Toulouse
118 rte de Narbonne, F-31062 Toulouse, France

INTRODUCTION

At the beginning of the transitory phase in a high pressure mercury lamp device, the discharge is established in a Hg-Ar low pressure mixture (a few torr of Ar and a few mTorr of Hg corresponding to the Hg saturated vapour pressure at room temperature). During warm up, the gas heating causes evaporation of the mercury, so increasing its vapour pressure until it reaches a few atmospheres in the steady state, where a thermal equilibrium is attained [1]. The matter here is how the mercury amount, initially introduced as a liquid in the burner, will be distributed over the three main discharge zones, namely the regions lying behind both the lower and the upper electrodes, and the inter-electrode volume. Obviously, discharge practically occurs in the inter-electrode region. Thus, out of the total mercury amount introduced in the burner only the active quantity present in the third region should be considered. The remaining mercury amount is regarded as a loss.

In this paper, we try to clarify this point and discuss to what extent the mercury distribution in these characteristic zones is related to the electrical arc current, the arc tube radius and finally the total mercury amount initially introduced. Our discussion is based upon calculations obtained with a numerical approach. This simulation of arc properties is performed with a two-dimensional fluid code using a finite element scheme.

MODEL

We assume that the burner presents axial symmetry and that cylindrical electrodes are mounted. We also consider that our arc is in Local Thermodynamic Equilibrium (L.T.E.). Under such assumptions, all discharge properties can be deduced from the temperature distribution. Indeed, the required material functions namely thermal and electrical conductivities, specific heat of arc plasma as well as net emission coefficient are assumed to be fully described by the local temperature only. It is also assumed that plasma flow is laminar. We neglect the viscous dissipation and we suppose that the electric field is purely axial. We assume that the contribution of magnetic forces is negligible.

Under these conditions, the plasma is governed by usual conservation equations (mass, radial momentum, axial momentum and energy). These equations are completed with boundary conditions, details of which can be found in reference 2.

RESULTS AND DISCUSSION

The mercury amount distribution in the tube is closely related to temperature profile in the plasma. So, in order to calculate the mercury distribution we must first determine the temperature distribution in the discharge. Note that the hot gas coming out of the arc channel, heats the region behind the upper electrode. Thus, the mercury amount in this region should be less than that behind the lower one. The mercury amount accumulated in the regions behind each electrode as a function of the square of arc tube diameter is given in figure 1. The corresponding pressure is 1 atmosphere and the current value is maintained at 3 A. The arc tube has a total length of 7 cm with an inter-electrode spacing of 5 cm. The 1 cm long electrodes have a radius of 0.1 cm for all cases.

Figure 1 shows that an increase in arc tube diameter leads to an increase of the mercury amount accumulated in these regions. So, a small radius leads to a reduction of this mercury loss. However, the arc tube radius should be high enough so as to keep the thermal loss quite low.

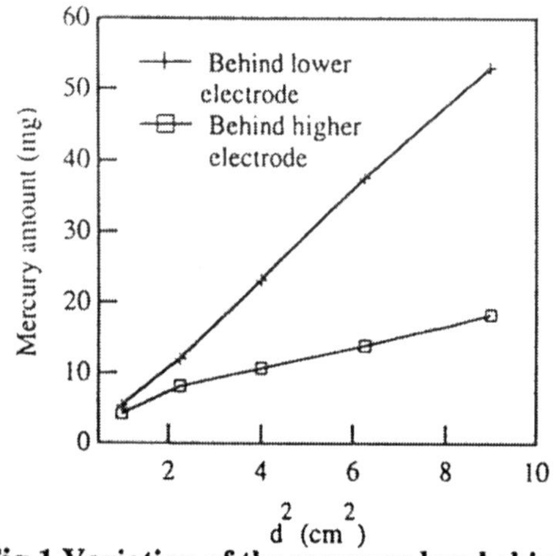

Fig.1 Variation of the mercury loss behind electrodes as function of the tube diameter

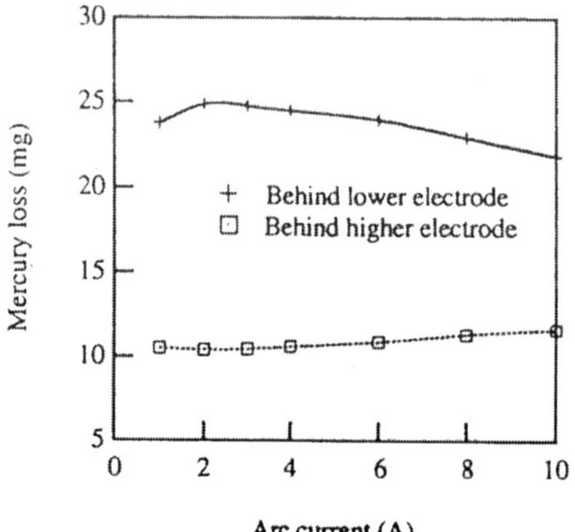

Fig.3 Variation of the mercury loss behind electrodes as function of the arc current

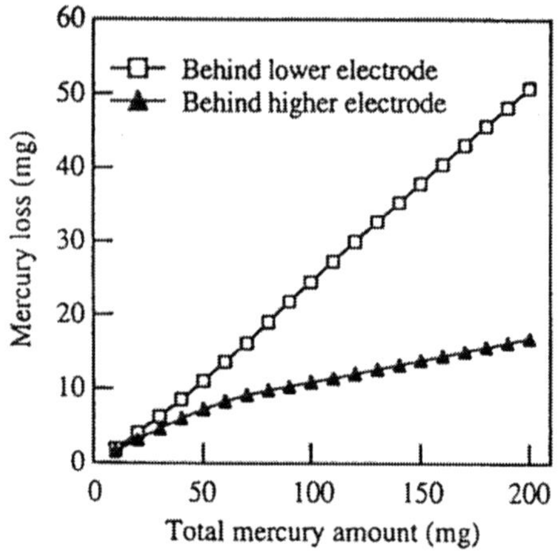

Fig.2 Variation of the mercury loss behind electrodes as function of the total mercury amount

In figure 2 we show the variations of the mercury amount behind the electrodes as function of the total mercury amount introduced in the burner at current of 3.2 A. Here, the arc tube has a total length of 9.2 cm and a radius of 0.925 cm. These results show that this evolution is quite similar to the situation where the tube diameter varies at constant pressure. Nevertheless, dependence of these quantities upon arc current is particular (cf. figure 3). We observe that there follows only a small variation of the mercury quantities accumulated behind the electrodes when current varies. This is due to the fact that the temperature of these regions does not vary enough with arc current.

CONCLUSION

In this paper, a particular attention is paid to the region behind the electrodes in order to determine the mercury distribution in the plasma. A study of this distribution variation as function of arc current, arc tube diameter and finally total mercury amount is discussed here.

REFERENCES

[1] M. Stambouli, K. Charrada and J.J. Damelincourt, "Thermalization of the high pressure mercury lamp positive column during the warm-up phase", IEEE Transactions on plasma science, vol. 43 No. 2, 1995

[2] K. Charrada, "Etude hydrodynamique d'un plasma de décharge mercure haute pression. Analyse de l'influence des transports de masse. généralisation à un système à deux fluides", Thèse de l'Univ. Paul Sabatier 1995

THE IGNITION CHARACTERISTICS OF ELECTRODE-LESS HID LAMP

Shin Ukegawa, Shigeaki Wada, Atsunori Okada, Miki Kotani
Matsushita Electric Works, Ltd., 1048, Kadoma, Osaka 571, Japan

In the previous works[1][2][3], low-pressure inductive RF discharge has two discharge modes.
The first discharge that is capacitive mode is initiated between adjacent turns of the coil to which voltage is applied. When the coil current is large enough to induce an azimuthal RF field, it can maintain the ionization process. The other is the second discharge mode, inductive mode.
In the case of low-pressure, the first discharge is glow-like mode with low visible output. But in the case of high-pressure (Xe pressure >6.65kPa), the first discharge is string-like mode. And between them (around 7kPa of Xe pressure), there is also another mode between the glow-like and string-like modes simultaneously in the lamp. We are interested in the first discharge mode of high-pressure Xe, shifting process of the mode from string-like discharge to arc-like discharge and the characteristics of hot-restart of Xe-metal halide lamp.

When the first discharge is glow-like discharge (around 100Pa), there is a minimum ignition point. And also the minimum voltage of mode-shifting-point from the first discharge to the second discharge is around 100kPa.[4]
In the case of higher pressure, we cannot expect any lower mode-shift-voltage. But we found the decrease of mode-shift-voltage around 30kPa.[Fig. 1] And we can see the string-like discharge in the glow-like discharge at that time. Fig2 shows the sketch of discharge-mode-shift. The arc tube is sphere shape and the diameter is 30mm. The arc tube is filled with NaI-TlI-InI and 39.9kPa of Xe.
The arc tube is excited by a 36mm diameter, 3~5 turn induction coil. The coil is driven by a 13.56MHz power source. The first discharge is string-like-discharge(not glow-like). As increasing the coil current, the string-like-discharge moves and lengthen along the induced electrical field(azimuthal direction). When the string-like-discharge connect itself and a short path is made in the lamp, instantly the string-like-mode shifts to the second-discharge-mode(arc-like-mode).

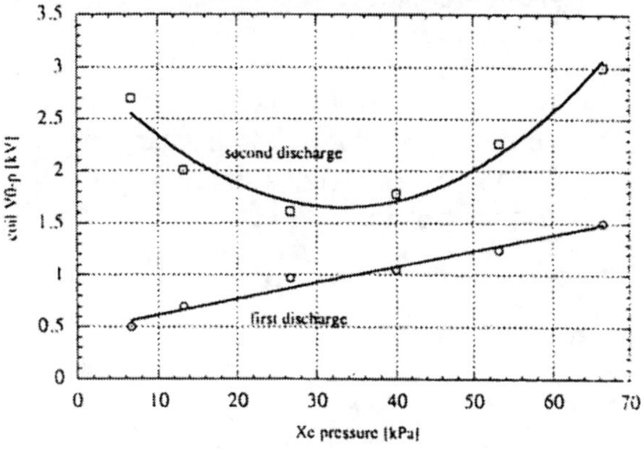

Fig. 1 : The starting voltage vs. Xe pressure

(We confirmed by using 1/4000sec high speed video tape recorder system).
So the mechanism of the shift of the discharge-mode at high Xe pressure is different from that of low Xe pressure, and it reduces the mode-shift voltage.

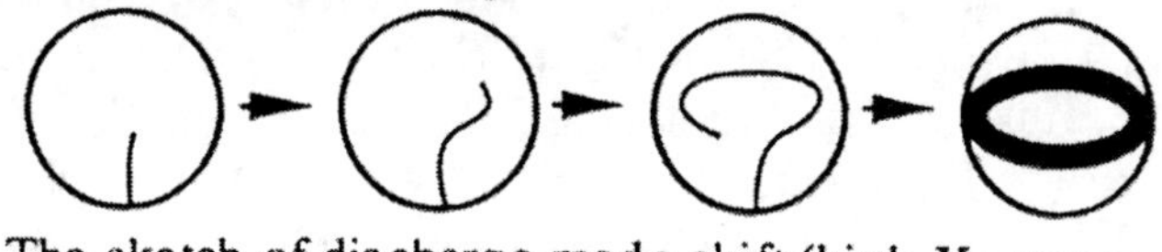

Fig.2 : The sketch of discharge-mode-shift (high Xe pressure)

Fig.3 shows the starting voltage characteristic. We measured the voltage of the induction coil at the timing of starting the first-discharge and shifting the discharge-mode. The parameter is numbers of the coil-turn(3,4,5turns). The first discharge voltage does not increase as increasing coil turns, but the second discharge voltage does. It means that when the first discharge is string-like, the discharge is maintained by capacitive coupling between adjacent turns of the coil. And the mode-shift is influenced by inductive coupling. (same as at low pressure)[5]

Fig.4 shows the characteristics of hot-restart of Xe-metal halide.
We measured the induction coil voltage at shifting the discharge-mode, the temperature of the lamp and observed the shape of the first discharge.
As decreasing the temperature of the lamp, the string-like-discharge by metal halide is made first, the glow-like-discharge by metal halide is made second, mixing discharge of the glow-like-discharge by metal halide and the string-like-discharge by Xe(+XeI*excimer) third, and Xe(+XeI*) string-like-discharge is finally made.
We found that when the first discharge is glow-like-discharge by metal halide, the voltage of the mode-shift was higher than other mode.
This phenomenon is similar to the initial ignition characteristic as varying Xe pressure.
So the shape of the first discharge is supposed to be related to the mode-shifting voltage.

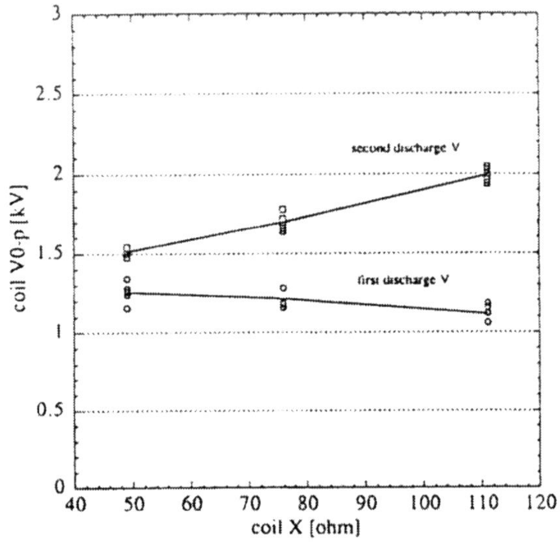

Fig.3 : The starting voltage vs. numbers of the coil-turn

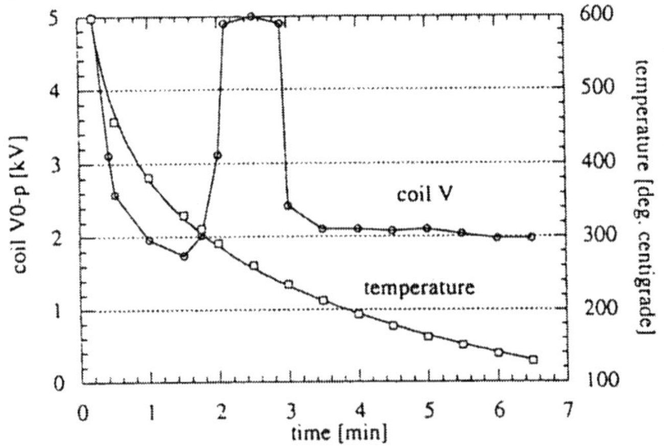

Fig.4 : The characteristics of hot-restart

References
[1]M.Shinomiya et al.,J. of the IES,20(no.1):44 (1991)
[2] J.Schlejen,paper 76:I,LS 6,Budapest(1992)
[3] R.B. Piejak et al.,Plasma Sources Sci. Technol.1(1992)
[4] T.Takamoto et al.,J.I.E.E.J.,Vol.91,No.4(1971)
[5] H.U. Eckert,High Temperature Science 6,99-134(1974)

NOVEL MICROWAVE POWERED HIGH INTENSITY DISCHARGE LAMPS

Walter P. Lapatovich

OSRAM SYLVANIA INC.
60 Boston Street
Salem, MA 01970

INTRODUCTION

Low wattage compact high intensity discharge (HID) lamps having good lumen output and superior color rendering and long life have been sought for years. The applications for such devices are limited only by imagination and run the gamut from task lighting, roadway illumination,[1] fiber optic illumination[2], and projector[3] applications. Electroded lamps of small size or low wattage are certainly available but tend to suffer from limited lifetimes of a few thousand hours. Additionally, fabrication issues such as electrode alignment and press pockets produce lamps with great lamp-to-lamp variability especially with regard to color stability. In small lamps any amount of blackening from the electrodes during start-up or over life is unacceptable since this affects the lumen maintenance dramatically.

Electrodeless high intensity discharge (EHID) lamps offer a number of advantages over present electroded technology. Electrodeless lamps can be small with high and uniform luminance eliminating the hot spots directly in front of the electrodes in conventional HID lamps. They exhibit excellent lumen and color maintenance, are flicker-free, efficacious, and demonstrate long life. The lifetime is ultimately limited by interaction of the chemical species with the arc tube material, vitreous silica also called fused quartz. The lamps can be made in various geometries including cylinders, ellipsoids and spheres. In addition, the lamps can utilize more reactive chemistries for achieving special spectral outputs for specific applications.[4] The trade off may be reduced life if the fill chemistry is incompatible with the arc tube material. In general, a tungsten electrode is usually the most reactive component in lamps. An EHID lamp containing sulfur[5] produces quasicontinuum emission with high CRI, but the fill cannot be used in electroded lamps because of tungsten sulfide formation.

EHID lamps should be thought of in terms of a system. The system consists of the power unit (or ballast in analogy to electroded lamps), the applicator which delivers power to the lamp, and the lamp itself. The ballast or power source converts the mains power into a form usable by the EHID lamp - usually high frequency. This could be 13.56 MHz for inductively coupled EHID lamps such as the GE Solenoidal Electric Field (SEF) lamp,[6] or microwave frequencies. Microwave frequencies of interest usually lie in one of the allowed Industrial Scientific and Medical (ISM) bands.[7] The band of choice is centered around 2.45 GHz with a total bandwidth of 0.1 GHz. This is a world-wide free radiation band in which microwave ovens operate, and accordingly sources of high power microwaves are readily available. For research purposes, lamps can be operated at other frequencies and the ISM band centered around 915 MHz is a good choice. However, from a practical standpoint operation outside the ISM bands causes interference with communications networks and is not permitted. In certain circumstances operation of low pressure electrodeless lamps at 2.65 MHz is permitted in Europe.[8]

POWER SOURCES

Various sources for high frequency power are available. The discussion which follows will be limited to microwave frequencies where OSRAM SYLVANIA INC. has concentrated its efforts. Electron tube devices which are affordable are limited to magnetrons. TWT's, klystrons and other crossed field devices such as the gyrotron, gryo-klystrons or peniotron are either too expensive or generate inappropriate power levels.[9] Good magnetrons will operate with a prime power to microwave conversion efficiency of about 80% at high output levels of several hundred watts to kilowatts RF. At lower power level which might be appropriate for many of the applications mentioned earlier, the efficiency is less, more like 40-60%. At this level of conversion efficiency, solid state devices become very attractive.

Figure 1. is a photograph of a GaAs power module designed and built by OSRAM SYLVANIA INC. using metal semiconductor field effect transistors (MESFETS). The device operates at 2.45 GHz with a 40% dc to microwave conversion efficiency. The output is a single SMA connector. Refinements in device technology and integration of the various subunits (e.g. the oscillator, preamplifier, and final amplifier stages) onto a single substrate have reduced size and weight and insertion loss at the interconnects. Advances in solid state electronics, such as the development of SiC, GaN and doped diamond semiconductors will make available high frequency, high power supplies which operate at higher efficiency at lower cost in the future.[10]

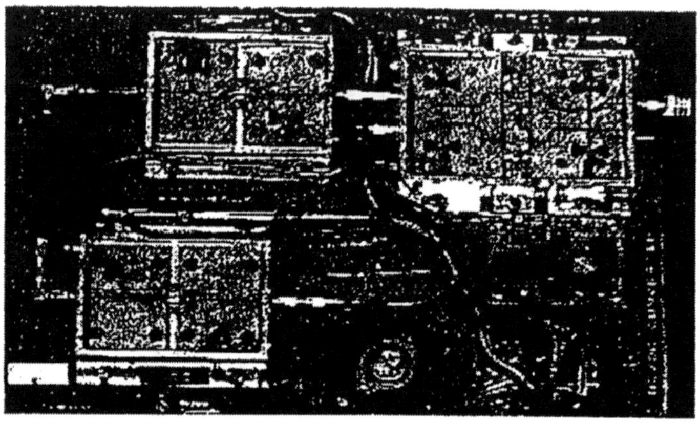

FIGURE 1. GaAs MESFET power source for 35W at 2.45 GHz. The oscillator, driver or preamplifier and final amplifier are mounted on individual carriers for ease of testing. The integrated unit is more compact.

LAMP CONFIGURATIONS AND POWER APPLICATORS

EHID lamps can be configured in many geometries as shown in Figure 2. It is the application which dictates the geometry, for example if a long thin arc is required to mimic a filament then a thin tube with large aspect ratio (length/diameter) is preferred. For applications requiring a point-like source a sphere is more appropriate (see Figure 2) The key challenge is to design the electromagnetic applicator to energize these lamps and still provide for useful light output. Typical cylindrical lamps range in size from inner diameters of 1 mm to 10 mm with outer diameters from 2 mm to 15 mm and lengths from 3 mm to 20 mm. Power densities are on the order of 1 to 1.5 kW/cm^3.

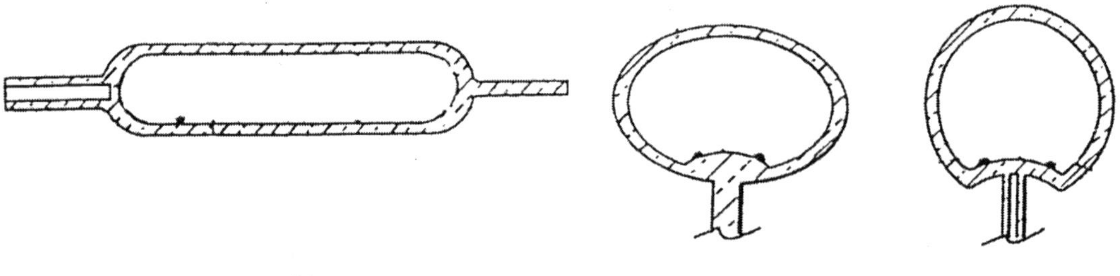

(a) (b) (c)

FIGURE 2. In (a) a cylindrical lamp is shown appropriate for a filament application. In (b) and (c) the ellipsoidal and spherical lamps are shown. Typical dimensions are on the order of a few millimeters.

How the power is coupled to the lamp will affect the optical transparency of the applicator. Techniques for coupling to lamps have included resonant cavities, waveguides, termination fixtures[11], induction coils, capacitor plates, etc. The basic physics for coupling has been reviewed in the literature and essentially falls into two categories.[12] Inductive coupling via an oscillatory magnetic field is principally used at lower frequencies of a few hundred kilohertz to a few hundred megahertz. Capacitive coupling via the electric field component is generally used from a few hundred megahertz to higher frequencies.

Power Level	Termination Fixtures	Microstripline Applicators
400+ W	1292 g	156 g
	1527 cm^3	314 cm^3
100-150 W	390 g	21 g
	339 cm^3	70 cm^3
25 - 50 W	177 g	19 g
	196 cm^3	46 cm^3

TABLE 1. The advantages of the compact microstripline applicators are compared to termination fixtures. Size and weight are reduced. The printed circuit applicators are also low loss and manufacturable.

A power applicator is shown in Figure 3. The applicator consists of a planar transmission media (PTFE fiberglass circuit board) with printed microstrip transmission lines and discrete tuning capacitor chips. The transmission line encompassing the

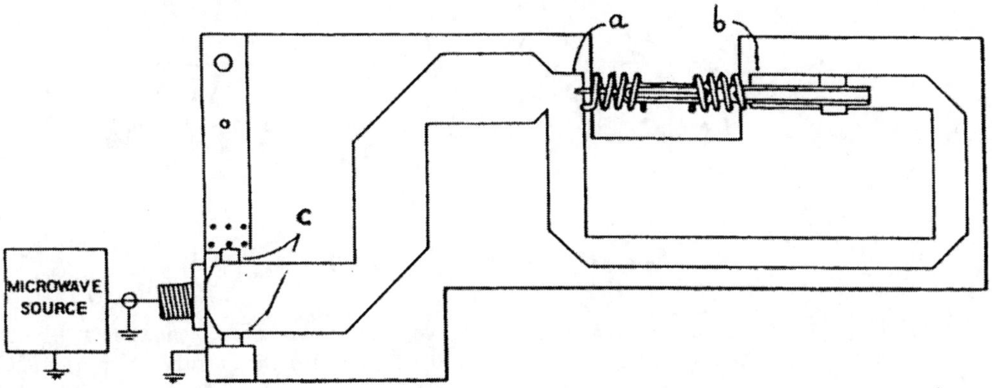

FIGURE 3. Power applicator for 25 W EHID lamp at 915 MHz. The substrate is 0.060" thick PTFE fiberglass with 1 oz. copper traces. The helical field applicators are wound from 0.020" (0.5mm) diameter nickel wire.

lamp forms a half wave balun between points (a) and (b) in the figure.[13] This is fed by a lower impedance line (hence wider width) connected to an SMA connector. Chip capacitors (c) are used to fine tune the impedance match. The field applicators are the parts (helices in the figure) attached at the ends of the balun which transform the wave with electric field components perpendicular to the plane of the board to a wave with field components parallel to the board and parallel to the lamp axis. In this case, the field is similar to that in an electroded lamp, i.e. directed between the electrodes along the lamp axis. The electric field magnitude of the various field components in the vicinity of the helix is calculated using a sheath helix model.[14] The axial field is dominant. The radial and azimuthal components manifest themselves during start up via the formation of filamentary discharges in close proximity to the coil ends as shown in Figure 4.

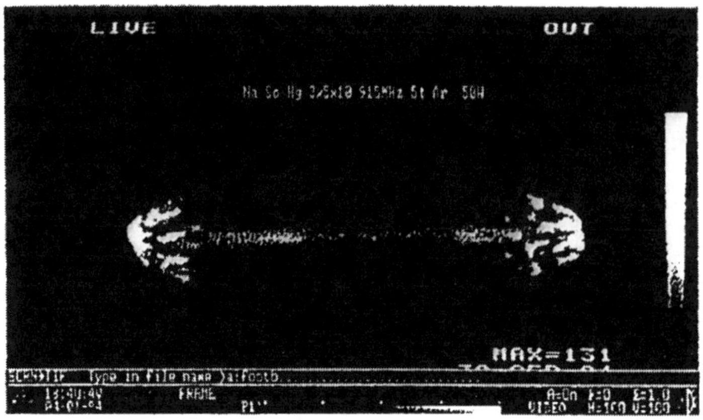

FIGURE 4. An EHID lamp on startup. The discharge is primarily low pressure(subatmospheric) and filamentary "fingers" are seen in the vicinity of the helical applicators.

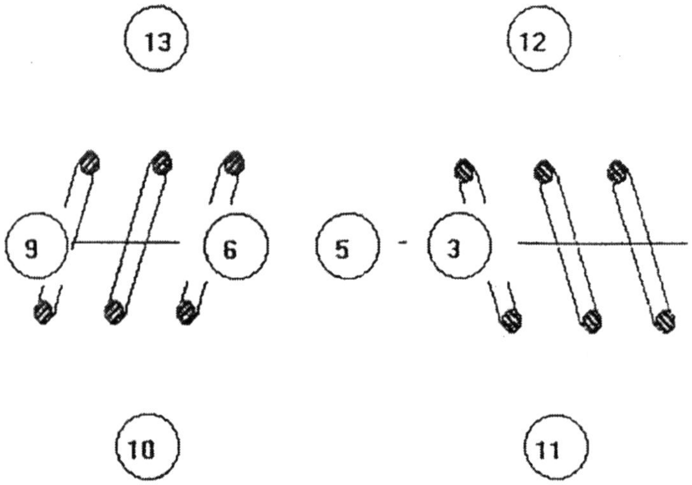

FIGURE 5. Field strength measurements made in the vicinity of a helical applicator with a Luxtron probe. The helix was wound from 0.040" (1mm) diameter nickel wire and had 3 turns with 8 mm ID, 10 mm ID, with 3.1 mm pitch. At position 3-41 kV/m, 5-30 kV/m, 9-20 kV/m, 10-6.2 kV/m, and 13-7.2 kV/m were measured with an uncertainty of about 0.5 kV/m.

The balun automatically ensures a phase shift of 180° between the two helix feed points. Consequently the electric field will add vectorally in the gap between the helices. This gives rise to a strong axial directed field powering the lamp in a push-pull configuration. The power deposition is adequately described in the literature.[15,16,17] The electric field strength was measured in a large diameter applicator using a Luxtron microwave field probe. These results are shown in Figure 5.

LAMP BEHAVIOR

Lamps are generally filled with mercury, argon gas, and vaporizable salt. They are ignited by a UV starting aid typically. The ensuing low pressure discharge slowly absorbs microwave power. The media is weakly ionized, essentially uniform and reflects most of the microwave power. This is seen in Figure 6. which shows the real (resistive) parts of an EHID discharge impedance as a function of applied power. Each measurement was taken at a stable operating point except for the rapid transition zone. This occurs at a power level of 2-3 watts where sufficient power is delivered to raise the wall temperature to the boiling point of mercury. At this point, the discharge becomes increasingly resistive due to electron-mercury scattering and power is more effectively transferred to the gas.

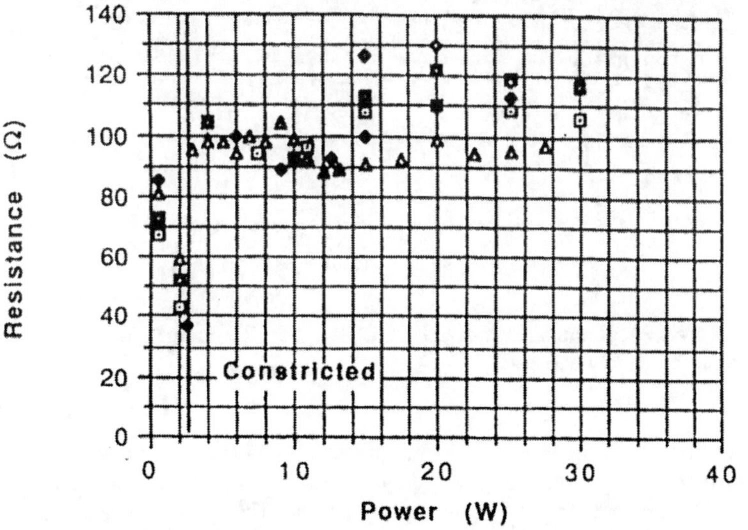

FIGURE 6. The resistive part of the lamp and helical field applicator as measured for various power levels. Data for lamps with various mercury doses are shown and account for the spread in the data. The discontinuity near 2-3 watts corresponds to arc constriction, rapid tube warming and vaporization of the mercury fill.

The time from ignition to steady state operation is reduced in EHID lamps compared to convertional electroded HID lamps. EHID lamp warm-up times are limited only by their thermal capacity and ability to match the available microwave power source impedance. If the reflection coefficient is kept low fully developed arcs can be achieved in five seconds without overpowering the lamp. This is in marked contrast to stabilization time of tens of seconds to minutes for electroded lamps. Electroded lamps routinely receive 4 to 6 times the steady state current on warm-up to force the lamp to stabilize faster. Needless to say, this overcurrent stresses the electrodes causing sputtering and blackening the arc tube eventually. A comparison of the wall temperatures for a 35W EHID lamp and a 35W electroded D1 type (automotive headlamp) lamp[18] with high pressure xenon is shown in Figure 7. The EHID lamp mass is reduced by the absence of the glass to metal seals and no heat is lost via conduction through the electrode inleads.

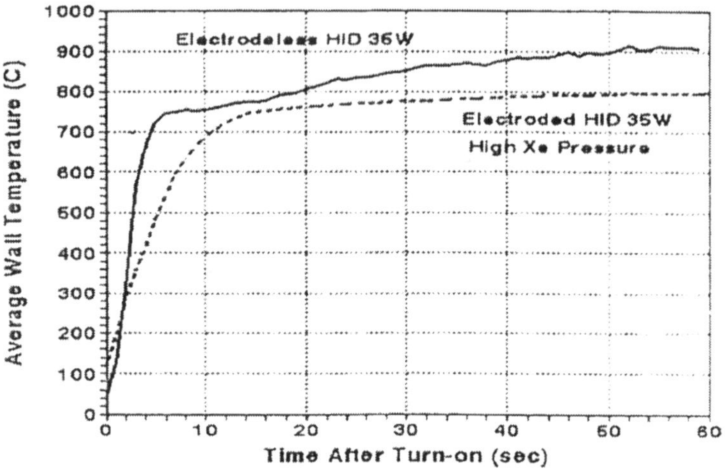

FIGURE 7. Warm up comparison of 35 W EHID and HID lamps. The temperature is monitored using an infrared imaging camera.

During the lamp warm up process, distinct phases are observed visually which correspond to low pressure mercury inert gas emission, mercury increasing pressure with arc constriction, vaporization of the salt with color change, and finally full operation. The spectral output changes from ignition to fully developed arc in much the same way as in an electroded arc lamp, viz. mercury emission, mercury with some additive and fully operation with spectrally broadened lines. This is seen readily in Figure 8.

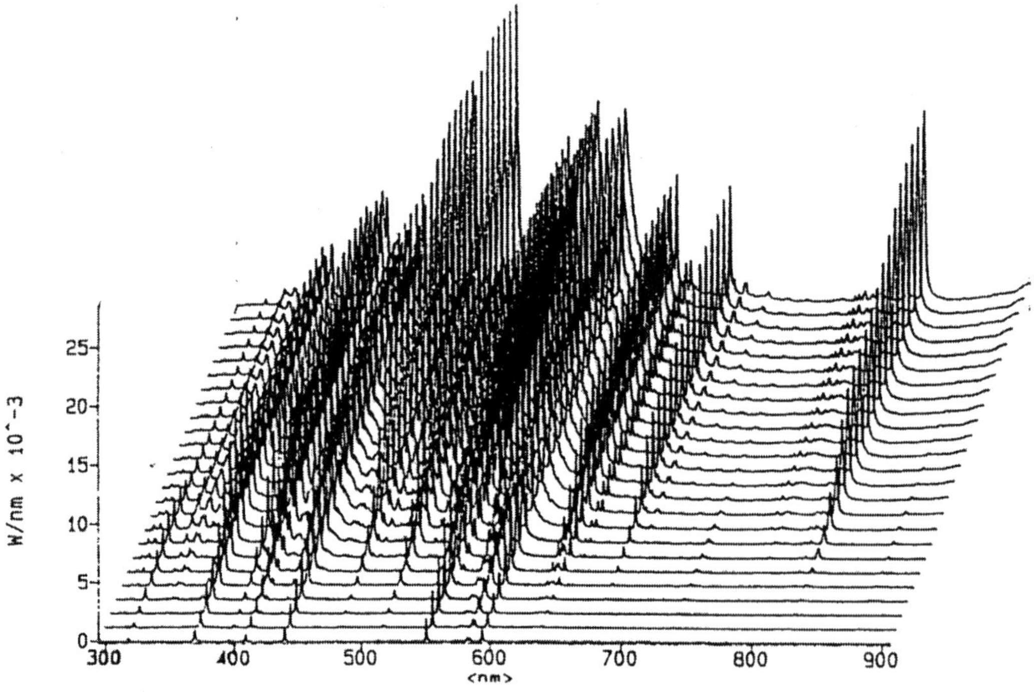

FIGURE 8. The temporal development of spectral emission from a 50W EHID lamp containing sodium scandium iodide and mercury. The time steps are approximately 0.52 s intervals.

In steady state operation the condensate forms a well defined pool in the cold zone which for horizontally burning lamps is the lower lateral surface as shown in Figure 9.

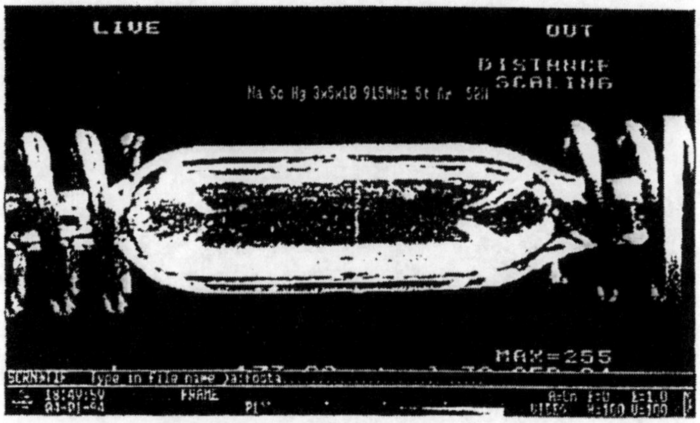

FIGURE 9. A 50 W EHID lamp mounted in applicator. The condensate forms a pool on the lower lateral surface during horizontal burning. The extent of the pool depends on the power input to the lamp, the salt dose, and the thermal distribution during operation.

A steady state spectrum of a 50 W Na-Sc-iodide lamp[19] containing 1.5 mg of Hg and 5 torr of argon gas is shown in Figure 10.

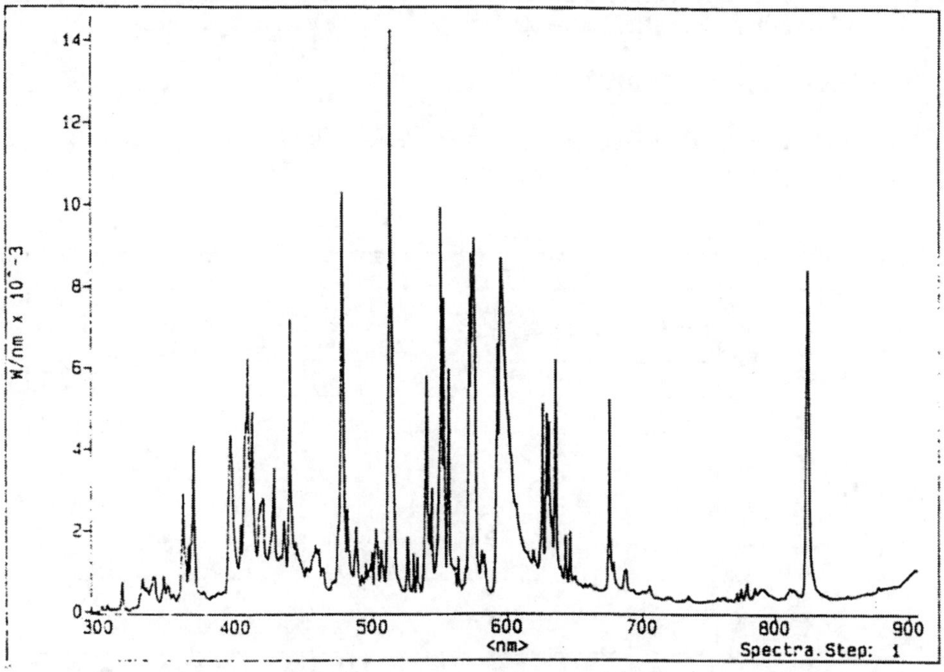

FIGURE 10. Spectral power distribution from 50W EHID arc containing Na-Sc iodide, Hg and Ar.

The scandium and sodium lines dominate the spectrum. The mercury lines are reduced in amplitude from those observed in comparably loaded electroded lamps. This is attributed to a slightly flatter temperature distribution in the microwave powered arc. The modified spectral distribution is the reason for the efficacious (100 lm/ $W_{microwave}$) operation of the lamp and also the higher CRI ~85 compared to 65 for electroded lamps containing this same chemistry. Other chemistries can also be used.[20]

The actual thermal distribution for a 915MHz 50W Na-Sc lamp mounted as shown in Figure 9 is shown in Figure 11. The image is recorded with an Inframetrics model 760 thermal imaging system with an 8-12 micron passband. In this spectral region the silica is opaque with an emittance near 0.98. The image is related to the exterior thermal distribution of the lamp.

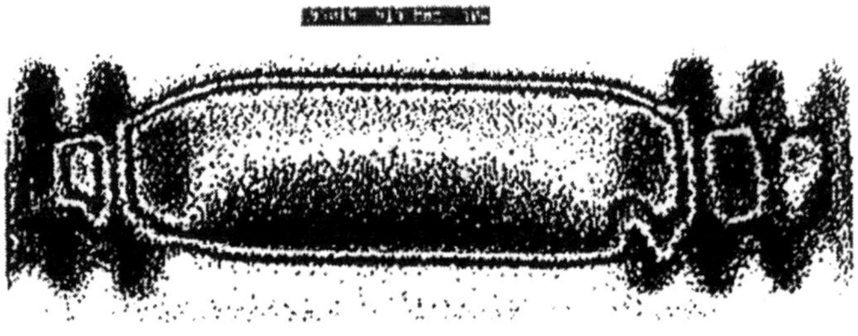

FIGURE 11. Thermal image of horizontally burning EHID lamp at 50W. The cold zone is the lower lateral surface of the lamp. The hot spots appear near the field applicators in the ends of the lamp.

The arc is bowed as expected from a horizontal burner with a maximum temperature of 831°C on the upper lateral surface. The cold spot is approximately 750°C. The ends near the applicator are hottest with temperatures not exceeding 921°C. The lamp burns in ambient air and requires no cooling or spinning. Such lamps have been operated for extended periods of time with little degradation. Figure 12. shows the relative lumen maintenance for a family of EHID lamps over burn time. The life limiting mechanism appears to be silicate formation with the chemical fill species.

SUMMARY AND CONCLUSIONS

A novel technique for energizing small electrodeless high intensity discharge (EHID) lamps is described. These lamps are bright, compact devices suitable for a number of applications

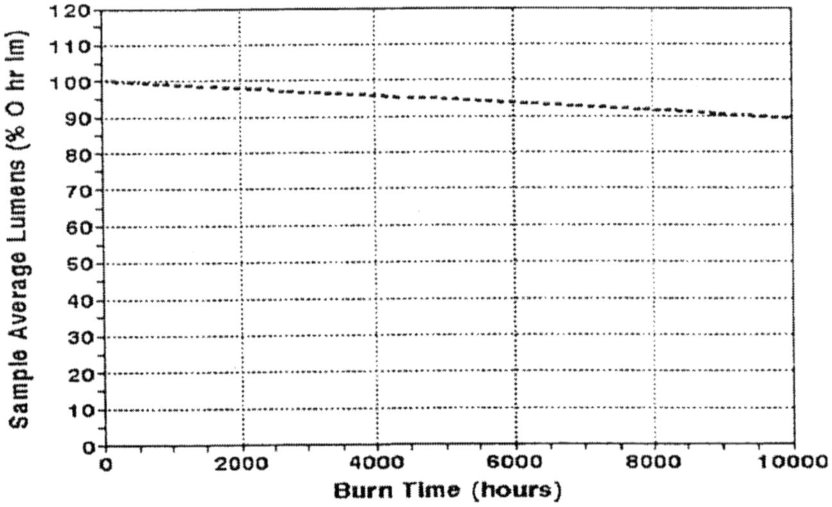

FIGURE 12. Good lumen maintenance is a benefit of EHID lamps. No blackening occurs in the absence of electrodes. The ordinate is expressed as a percentage of the 35W lamp zero hour average lumen output. The dotted line represents a least squares fit to the data. An estimated uncertainty is about 6%.Z

especially where mating to a optic is required. The lamps can operate at low power levels from solid state ballasts.

The lamps warm up quickly and stabilize rapidly due to the low thermal mass of the lamp envelope. EHID lamps outlive conventional electroded lamps by more than a factor of two in terms of failures and far surpass the conventional lamp in lumen maintenance.

The overall system efficacy is important. This is a product of the mains to microwave converter efficiency, the applicator efficiency, and the lamp efficacy. With microstripline technology the applicator losses can be kept small. With current high frequency technology an efficiency of 50% for conversion of prime power to microwave is achievable. However to fully utilize the EHID advantages higher ballast conversion efficiency is required.

REFERENCES

---

[1] K. Jennings and H. Krause, Paper 950913, SAE Publication SP-1081, Soc. of Auto. Engineers, Warrendale, PA, pp. 29-33 (1995); see also J. M. Davenport and R. L Hansler, U. S. Patent No. 5,222,793, issued June 1993, and U. S. Patent No. 4,811,172 issued Mar. 1989; and R. Nuemann et. al. U. S. Patent No. 5,213,406 issued May 1993.

[2] J. F. Caminada, ed., International Lighting Review, Vol. 3 (1993); see also P. Timson and B. Gregson, Fiber Optic Lighting and Sensing Technology, Eurotec Fibre Optics Ltd., Doncaster, U.K. (1994).

[3] G. E. English and H. L. Rothwell, Journal of IES, pp.131-144, Oct. (1993); also G. English, H. Rothwell and M. Beschle, U. S. Patent No. 4,766,348 issued Aug. 1988, and W. Keeffe, H. Rothwell and G. English, U. S. Patent No. 4,308,483 issued Dec. 1981.

[4] W. P. Lapatovich, W. M. Keeffe, R. W. Libermann, and J. Maya, U. S. Patent No. 4,672,267, issued June 1987.

[5] J. T. Dolan, M. G. Ury and C. H. Wood, Paper 74:L, 6th International Symposium on the Science and Technology of Light Sources, Budapest, Hungary, 30 Aug.-3 Sept. (1992).

[6] J. M. Anderson, P. D. Johnson, C. E. Jones and T. H. Rautenberg, Final Report LBL-19035, DOE Contract No. DE-AC03-76SF00098, Lighting Systems Research Group, Lawrence Berkley Laboratory, Berkeley, Ca. (1985); also M. E. Duffy, J. T. Dakin, G. E. Duffy and M. M. Secen, Paper 80:P, 6th International Symposium on the Science and Technology of Light Sources, Budapest, Hungary, 30 Aug.-3 Sept. (1992).

[7] J. M. Osepchuk, IEEE Transactions on Microwave Theory and Techniques, Vol. MTT-32, No. 9, p. 1220 (1984).

[8] J. Schlejen, Paper 76:I, 6th International Symposium on the Science and Technology of Light Sources, Budapest, Hungary, 30 Aug.-3 Sept. (1992).

[9] K. Yokoo, Y. Nakazato, and S. Ono, IEEE Transactions on Electron Devices, Vol. 38, No. 12, pp. 2677-2681 (1991).

[10] F. A. Olsen, Microwave Journal, Vol. 38, No. 2, pp 24-46 (1995); see also T. B. Ramachandran, ibid, pp. 91-107 (1990).

[11] W. H. McNeil, J. M. Lech, P. O. Haugsjaa and R. J. Regan, U. S. Patent No. 4,178,534, issued Dec. 1979; also P. O. Haugsjaa, R. J. Regan, and W. H. McNeil, U. S. Patent No. 3,999,927 issued Nov. 1976.

[12] D. O. Wharmby, Paper 36:I, 5th International Symposium on the Science and Technology of Light Sources, York, England, 10-14 Sept., 1989.

[13] S. J. Butler, W. P. Lapatovich, and J. R. Bochinski, U. S. Patent No. 5,144,206, issued Sept. 1992.

[14] R. M. Bevensee, *Electromagnetic Slow Wave Systems*, John Wiley & Sons, New York (1964).

[15] M. Moison and J. Pelletier, eds., *Microwave Excited Plasmas*, Elsevier, Amsterdam (1992).

[16] J. F. Waymouth, *Microwave Discharges: Fundamentals and Applications*, Proc. of NATO Advanced Research Workshop May 11-15, 1992, Vimeiro, Portugal, Plenum (1993).

[17] S. Offermans, J. Appl. Phys. Vol. 67 No. 1, pp. 115-123 (1989).

[18] U. Alder, ed., *Automotive Handbook*, pp.668-689, Robert Bosch GmbH, Stuttgart, Germany (1993).

[19] W. P. Lapatovich, S. J. Butler, and R. K. Smith, U. S. Patent No. 5,339,008, issued August 1994.

[20] J. Kramer and W. P. Lapatovich, U. S. Patent No. 4,801,846, issued January 1989.

RECENT PROGRESS OF EXCIMER RADIATION -RESEARCH, DEVELOPMENT AND APPLICATION-

Minoru Obara
Keio University, Department of Electrical Engineering,
3-14-1 Hiyoshi, Kohoku-ku, Yokohama-shi, 223 Japan

Introduction

Recently, high-power UV and VUV light sources using excimer medium have been developed and their potential applications have experimentally been evaluated. As a high-peak-power pulsed *coherent* UV and VUV light source, discharge-pumped rare-gas-halide excimer lasers have been developed for practical use such as LSI lithography, pulsed laser deposition of high-performance electronic materials, laser angioplasty, and photorefractive keratectomy. As a high-power *incoherent* UV and VUV light source, excimer lamps have recently been developed. The excimer lamps can efficiently generate UV and VUV photons as a spontaneous emission, which could be pumped by electrodeless discharge such as a microwave discharge and a dielectric barrier discharge so far. These excimer lamps opened up versatile, novel applications. Here, we will review the recent progress in research, development and applications of excimer lamps.

Physics of Excimer Lamp

Excimer lamp is to efficiently extract excimer photons as a spontaneous emission from excimer medium. The spontaneous emission lifetime of the excimer is approximately proportional to λ^2. **Figure 1** shows the spontaneous emission lifetime of some excimers. It is seen that their lifetimes are shorter than several 10 nsec. Therefore, it is technically easy to efficiently extract photons with less collisional deactivation of excimer states under the low pressure operational conditions. Moreover, low excitation rate pumping (100 W/cm^3 or less) is acceptable while the rate of several MW/cm^3 is required for the efficient excimer laser oscillation. Due to low pressure operation with low excitation rate pumping, long pulse excimer emission and cw excimer emission are achievable only in excimer lamp, and microwave discharge and dielectric barrier discharge can be used. The excimer pumping mechanisms for excimer lamp are mainly due to neutral reaction channels, while excimer laser pumping is dominated by ionic reaction channels [1] . **Table 1** shows the excimer emission spectra, which are potentially obtainable by excimer lamp. The shortest excimer wavelength is of NeF excimer emission at 108 nm. The spectral width of excimer lamp is extremely wider than that of the corresponding excimer laser, but it is still within several 10 nm, which is dependent upon the operating pressure [2] .

The necessary conditions for a light source to make UV photochemistry applicable in an industrial process are (1) high efficiency of the light source, (2) a narrow spectrum of radiation, and (3) low photon cost. An UV photon below 250 nm has a photon energy higher than 5 eV, while chemical bonds like C-C, C-H, etc. have 3-5 eV bonding energy. Radiation of an UV photon can directly break these chemical bonds at room temperature and then initiate chemical reactions. High-power KrF (248 nm) and ArF (193 nm) excimer lasers are coherent light sources, and mercury lamps and xenon flashes are incoherent light sources in the UV spectral region below 250 nm. However, no light source which meets all the necessary conditions has been developed recently.

Microwave-Discharge-Pumped Excimer Lamp

Just after the advent of the laser, electrodeless discharges such as microwave and RF discharges were used to pump gas lasers due mainly to less contamination in the laser gases than discharges with electrodes. Therefore, He-Ne laser, Ar ion laser, CO_2 laser, etc. were pumped by microwave and RF discharges.

At dawn of the excimer laser, spectroscopists used neon-sign / Tesla transformer discharge and RF discharge of the low pressure excimer laser gas mixtures due mainly to no-contaminant discharge. However, it was reported in 1989 that high-efficiency operation of KrF(*B-X*) excimer lamp was excited by microwave discharge [3] .

Among the rare gas halides, the KrF excimer lamp is expected to be the most efficient source based on a KrF(*B-X*) fluorescence measurement with the KrF laser [3] and a simulation analysis [4] . However, for KrF lasers, neither long pulse operation greater than 1 msec nor multikilohertz operation has been achieved, which are crucial for increasing the average laser output power. Microwave discharge excitation should be a simple and practical excitation technique for a quasi-continuous excimer light source which can eliminate the discharge instabilities inherent in a pulsed excimer discharge because of the electrodeless discharge. These result from a premature termination of the spatially uniform avalanche discharge and reactions between the lamp medium and the electrodes. Earlier works of microwave excitation of XeCl and XeF excimer lasers were reported by Mendelsohn *et al.* [5] ,Christensen and Waynant [6] ,respectively. Recently, magnetron tubes are widely used for microwave oven and became inexpensive due to consumer electronics devices. If this tube is used as a driver, a very cheap UV light source will be realized.

In 1989 Kumagai and Obara [3] reported that a new UV light source: a quasi-continuous KrF(*B-X*) excimer lamp pumped by a microwave discharge was developed for the first time. The highest average KrF fluorescence power was 53W obtained with an intrinsic efficiency of 8.3%, when the deposited average microwave power was 678W with a peak power of 987W. The KrF peak output power and the power efficiency were 120W and 12.1%, respectively. **Figure 2** shows plots of the average KrF fluorescence power and the intrinsic efficiency which is defined as the average KrF fluorescence power divided by the average microwave deposited power, as a function of the average microwave deposited power for the 100 Torr He/Ne=1/1 diluted gas mixture.

The experimental setup which they used was composed of a microwave oscillator with a magnetron tube, directional couplers, a three stub tuner, and a resonant cavity with a discharge sphere. When a dc voltage of approximately 4kV is applied to the magnetron tube, it oscillates, radiating microwaves at a frequency of 2.45GHz. These microwaves propagate through the directional couplers, the three stub tuner, and a coupling hole into a cylindrical cavity made of a metallic mesh and create a strong electromagnetic field in the TE_{111} mode. This produces a discharge in a mixture of F_2 and Kr diluted with He or Ne and UV light is emitted within the narrow KrF spectrum. The inside diameter of the spherical quartz discharge tube is set at 3 cm to prevent discharge nonuniformity due to electromagnetic field inhomogeneities and skin depth effects. **Figure 3** shows typical time histories of the microwave power deposited into the gas mixture and the KrF fluorescence. Both a stable discharge and the KrF fluorescence continued over 8msec with a 100Hz repetition frequency. The quasi-dc voltage applied to the magnetron tube periodically has a rest time of approximately 1 msec to prevent anode current runaway in the magnetron tube. The complete cw operation of the excimer lamp will easily be realizable if the complete cw magnetron tube is just used. KrF fluorescence pulses of 5-9-μsec duration with 55-kHz operation frequency were also produced from a microwave-discharge-pumped KrF mixture [7] . A high average KrF power of 43 W was obtained with an intrinsic efficiency of 12.5% when the average microwave power was deposited into the 25-Torr mixture of F_2/Kr/He/Ne=2/1/48.5/48.5(%). A peak KrF optical power of 215 W was also obtained with a power efficiency of 15.9%.

A quasi-continuous ArF(193 nm) fluorescence of over 8 msec pulse with 100 Hz operation frequency was also generated in a microwave-discharge-pumped ArF lamp [8] . The ArF fluorescence power was 29 W obtained with an intrinsic efficiency of 4.4% at a deposited power of 655 W. In addition, a peak ArF power of 53 W was achieved with a power efficiency of 5.5%.

The KrF excimer lamp was improved by Nakamura, Kannari and Obara [9] using an Ar buffer gas in place of the He/Ne buffer. This lamp emitted 193 nm radiation additionally. Dual wavelength emission is useful for applications that need both the 248 and 193 nm.

The XeCl excimer lamp(308 nm) was demonstrated by Kitamura, Mitsuka and Sato [10] with a maximum continuous wave power of 80 W at an excitation power of 10 kW. The lamp house is shown schematically in **Fig. 4.** The irradiation intensity was 0.2 W/cm^2 in the focused area of 20 x 1cm^2 with a lamp lifetime of 160 hours, and at an excitation power of 1.5 kW. With a 35-kW and 1.2μsec pulsed excitation the obtained optical pulses were 1 mJ for the XeCl excimer(308 nm) and 0.28 mJ for the KrCl lamp (222 nm).

A high-efficiency vacuum ultraviolet F_2 excimer lamp (157 nm: *D'*-*A'*) was theoretically investigated by Hatakeyama, Kannari and Obara [11] and developed by Kumagai and Toyoda [12] . A total F_2 fluorescence power of 18.9 W was experimentally achieved with an intrinsic efficiency of 6.2 W and at a deposited microwave power of 310 W.

Dielectric Barrier Discharge-Pumped Excimer Lamp

Dielectric barrier discharges are one kind of capacitively coupled electrodeless discharges. Theses discharges were successfully used to drive ozonizers and often to pump gas lasers. The discharge power density of the barrier discharge is not so high but moderate. This power density level is also suited to efficiently pump CO_2 lasers.

The UV and VUV excimer lamps pumped by dielectric-barrier discharge were developed by Kogelsschatz et al. [13-17] . Neiger has recently reviewed the the dielectric barrier discharges [18] . Due to the capacitive electrical coupling the choice of discharge geometries is very flexible. Therefore, the lamp geometry is able to fit any application. **Figure 5** shows a schematic of different silent discharge configurations for generating incoherent excimer radiation [19] .The cylindrical UV source radiating to the inside can be used to irradiate fluids passing through the inner tube or to treat fibers or wires. Planar UV sources radiating one or both sides have practical applications for the irradiation of large flat substrate [19] . Recently, this research has been concentrated into rare gas excimer radiation such as Xe_2 (172 nm), Kr_2 (146 nm), and Ar_2 (126 nm). So far, the discharge characteristics [13,16,17] , discharge frequency effect [18] , rare gas excimer formation kinetics and spectral characteristics [12,13,16,20] were investigated. The high efficiency characteristics of the dielectric barrier discharge pumped Xe_2 lamp were demonstrated [21] , as shown in **Fig. 6.** The typical relative efficiency is 10:2.5:1 approximately for Xe_2 (172 nm), Kr_2 (146 nm), and

Ar_2 (126 nm), respectively [22].

Recently, wide aperture excimer lamps were investigated and developed [20,23,24]. The Xe_2 excimer lamp with an aperture size of 230 x 230 mm^2 generated an optical output of 9 W, corresponding to the intensity of 17 mW/cm^2. A largest one is 580 x 680 mm^2 and can generate an output of 60 W (16 mW/cm^2) [23].

The windowless VUV excimer lamps have recently been developed [25]. A dielectric barrier discharge between parallel dielectric tubes, as shown in **Fig. 7**, is used to pump the rare gas excimers as Xe_2 (172 nm), Kr_2 (146 nm), and Ar_2 (126 nm). The VUV optical window materials such as CaF_2, , MgF_2, LiF etc. are very expensive and may not be available in the large size. Moreover, these materials are easily subject to VUV radiation damage, which results in deteriorating VUV transmission. The deposition rate of palladium from spin on palladium acetate films could be enhanced drastically by a factor of 40 compared to the rates achieved with a VUV excimer light source with a cylindrical quartz envelope. The effect of excited rare gases and rare gas ions was not evaluated.

These dielectric barrier discharge pumped excimer lamps are now commercially available. For example, a 2-W Xe_2 excimer lamp (172 nm, FWHM=14 nm) has an operational lifetime of 1000 hours. The corresponding irradiation intensity reaches 10 mW/cm^2 [26].

Tarasenko *et al.* of High Current Electronics Institute, Tomsk, Russia have extensively been working on the high-power excimer flashlamps [27]. These are pumped by various types of discharge (longitudinal discharge, dielectric barrier discharge, and continuous glow discharge). Excimer lamps with a radiating surface area of 570, 840, and 1300 cm^2 have been created. The average radiation power achieved for KrCl (222 nm) is up to 4 W and for XeCl (308 nm) and XeF (350 nm) is about 10 W. The possibility of an inductive energy storage with a semiconductor opening switch being used for a longitudinal discharge pumped excimer lamp has been demonstrated.

Applications of Excimer Lamp Radiation

The photochemical processing is a low-temperature and selective processing, which can minimize undesirable effects. One of the major limitations of photochemical processing is the lack of sufficiently intense UV and VUV light sources. After the advent of the excimer lamps, many photochemical applications were demonstrated [19,28].
Direct a-Si:H photodeposition was demonstrated with a large area dielectric-

barrier discharge lamps of Xe_2, Kr_2, and Ar_2 [29,30] . The deposition rate of 5 nm / min was achieved for a-Si:H from Si_2H_6 [20] . Direct photodepositions of SiO_2 [31] and Si_3N_4 films [32] were also achieved using a Xe_2 excimer lamp. Palladium acetate films were decomposed by Xe_2 radiation from a new windowless lamp in order to form thin palladium (Pd) layers that can be used as activators for electrodeless metal plating processes [25] .

Surface modification was carried out with rare gas excimer UV/VUV light sources [28] . Modification of polyethylene terephthalate (PET) surface was demonstrated with a Xe_2 excimer lamp [33] .

Surface cleaning of silica glasses was successfully demonstrated by phot-oxidation with 172 nm photon [34] . The great potential of the VUV photolysis (172 nm) of water was demonstrated as a new method for the oxidative degradation of organic pollutants in water [35] . As a model substance 4-chlorophenol was rapidly degraded independent of the dissolved oxygen concentration.

Summary

The high-efficiency, high-power UV/VUV excimer lamps will make selective photochemical processing a reality mainly for the semiconductor and related industries in the near future. The high-power output and operational lifetime seem to be trade-off. Therefore, the use of gas-flow system with gas purifier is one of the clever ways to compromise both features. This modified (even complexed) system will still be more valuable rather than discharge-pumped excimer lasers in terms of photon cost. In the VUV region of the spectrum rare gas excimer lamps can open up new material processings. VUV emission from pulsed discharge-pumped rare gas supersonic jets was investigated, but was not described here [36] .

References

[1] M. Obara: "Laser, Rare Gas-Halide," Encyclopedia of Physical Science and Technology, Vol.7, p.192 (1987) Academic Press.

[2] Ch.K.Rhodes ed.: "Excimer Laser," (1984) Springer-Verlag.

[3] H.Kumagai, M.Obara: Appl. Phys. Lett. 54, p.2619 (1989).

[4] H.Kumagai, T.Hatakeyama, I.Nakamura, M.Obara: Trans. IEEE of Japan 111-C, p.129 (1991) (in Japanese).

[5] A.J.Mendelsohn,R.Normandin,S.E.Harris J.F.Young: Appl. Phys. Lett. 38, p.603 (1981).

[6] C.P.Christensen, R.W.Waynant: Appl. Phys. Lett.41, p.9 (1982).

[7] H.Kumagai, M.Obara: Japn. J. Appl.Phys. 28, p.L2228 (1989).

[8] H. Kumagai, M. Obara: Appl. Phys. Lett., 55, p.1583 (1989).

[9] I.Nakamura, F.Kannari, M.Obara: Appl. Phys. Lett. 57, p.2057 (1990).
[10] M.Kitamura,K.Mitsuka H.Sato: Appl. Surface Science 79/80, p.507 (1994).
[11] T.Hatakeyama, F.Kannari, M.Obara: Appl. Phys. Lett. 59, p.387 (1991).
[12] H.Kumagai, K.Toyoda: Appl. Phys. Lett. 59, p.2811 (1991).
[13] B.Eliasson, U.Kogelschatz: Appl. Phys. B 46, p.299 (1988).
[14] U.Kogelschatz: Pure and Appl. Chem. 62, p.1667 (1990).
[15] B.Gellert, U.Kogelschatz: Appl. Phys. B 52, p.14 (19991)
[16] B.Eliasson, U.Kogelschatz: IEEE Trans. Plasma Science 19, p.309 (1991).
[17] B.Eliasson, U.Kogelschatz: IEEE Trans. Plasma Science 19, p.1063 (1991).
[18] M.Neiger: Proc. of The 6th International Symposium on the Science and Technology of Light Sources, Budapest, p.75 (1992).
[19] U.Kogelschatz: Appl. Surface Science 54, p.410 (1992).
[20] F.Kessler, G.H.Bauer: Appl. Surface Science 54, p.430 (1992).
[21] H.Matsuno, T.Igarashi, T.Hiramoto, Y.Ohnishi, N.Hishinuma, F.Takemoto, K.Kasagi: Properties of Excimer Lamp Pumped by Dielectric Barrier Discharges, LS-93-20, p.1 (The Illuminating Engineering Institute of Japan (1993).
[22] H.Sugahara, H.Matsuno, Y.Ohnishi, T.Igarashi, T.Hirano, T.Asahina: Paper 30p-ZE-3, Extended Abstracts of The 41st Spring Meeting, 1994, The Japan Society of Applied Physics and Related Societies, p.32 (1994).
[23] T.Igarashi: The 42nd Spring Meeting of The Japan Society of Applied Physics and Related Societies, (1995).
[24] P.Bergonzo, P.Patel, I.W.Boyd, U.Kogelschatz: Appl. Surface Science 54, p.424 (1992).
[25] H.Esrom, U.Kogelschatz: Appl. Surface Science 54, p.440 (1992).
[26] Ushio Inc.: Catalog, 1994.
[27] V.F.Tarasenko: private communication.
[28] H.Esrom, U.Kogelschatz: Thin Solid Films 218, p.231 (1992).
[29] F.Kessler, H.-D.Mohring, G.H.Bauer: Mat. Res. Soc. Symp. Proc. 192, p.559 (1990).
[30] C.Manfredotti, F.Fizzotti, M.Boero, G.Piatti: Appl. Surface Science 69, p.127 (1993).
[31] P.Bergozo, U.Kogelschatz, I.W.Boyd: Appl. Surface Science 69, p.393 (1993).
[32] P.Bergonzo, I.W.Boyd: Appl. Phys. Lett. 63, p.1757 (1993).
[33] T.Igarashi, H.Matsuno, S.Nozawa, T.Matsushima, T.Hirano, M.Murahara: 27a-HE-2, Extended Abstract of The 54th Autumn Meeting, The Japan Society of Appl. Phys., p.1 (1993).
[34] S.Iso, T.Igarashi, S.Nozawa, M.Dan, H.Matsuno, T.Hiramoto: 30p-ZE-2, Extended Abstract of The 41st Spring Meeting, The Japan Society of Appl. Phys. (1994).
[35] L.Jakob, T.M.Hashem, S.Buerki, N.M.Guindy, A.M.Baun: J. Photochem. Photobiol. A: Chem. 75, p.97 (1993).

[36] S.K.Searles, J.E.Tucker, B.L.Wexler, M.F.Masters: IEEE J. Quant. Electron. 30, p.2141 (1994) and references therein.

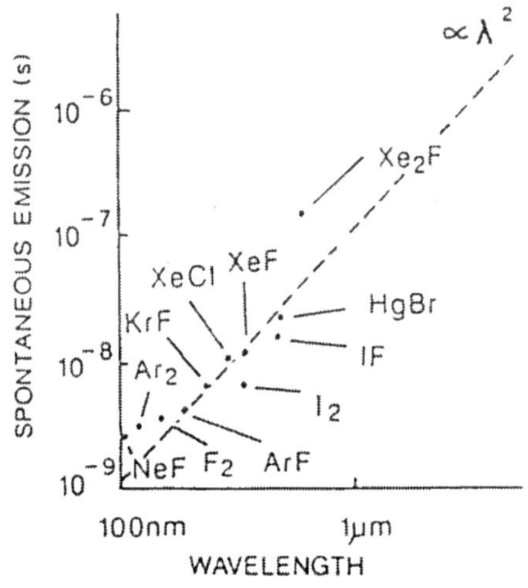

Fig. 1 The spontaneous emission lifetime of some excimers.

Table 1 The excimer emission spectra, which are potentially obtainable by excimer lamp [2] .

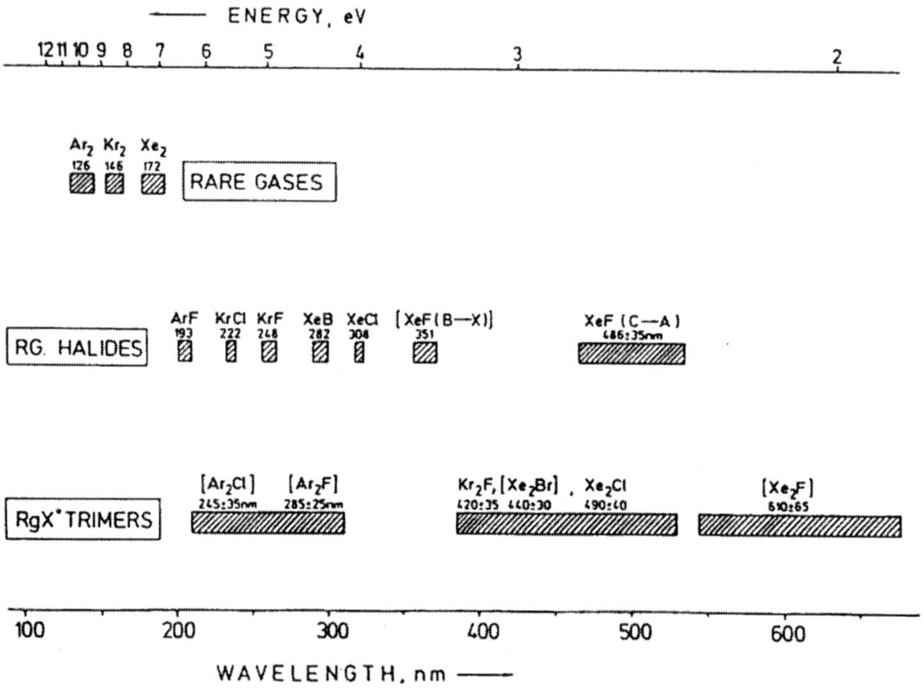

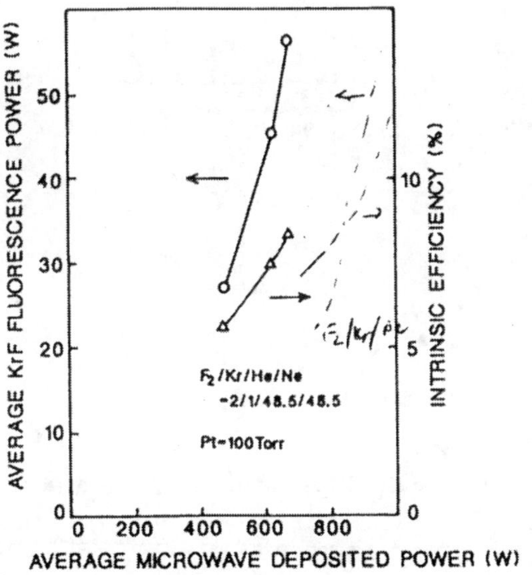

Fig. 2 Plots of the average KrF fluorescence power and the intrinsic efficiency [3].

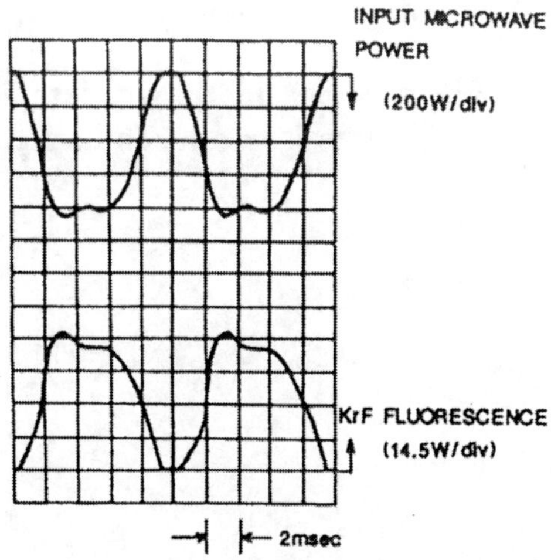

Fig. 3 Typical time histories of the microwave power deposited into the gas mixture and the KrF fluorescence [3].

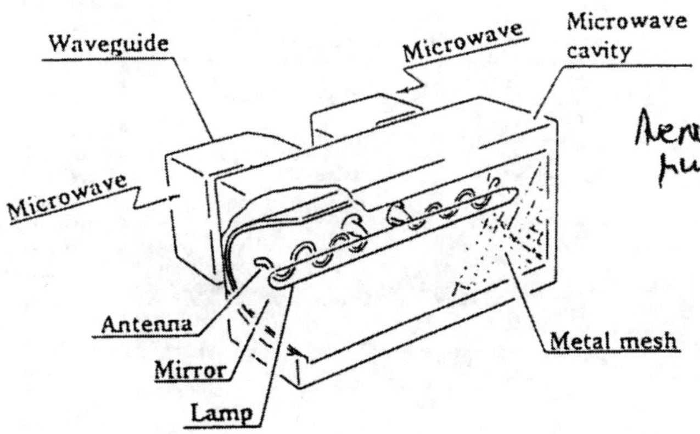

Fig. 4 Schematic of XeCl excimer lamp house pumped by microwave discharge [10].

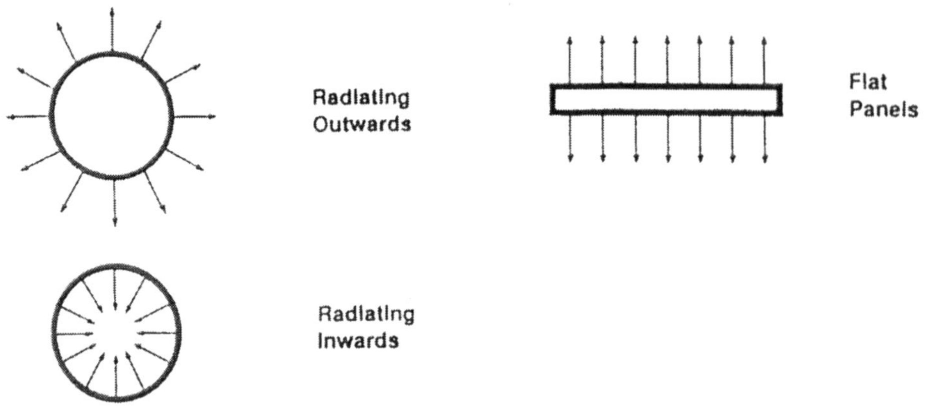

Fig. 5 Schematic of different silent discharge configurations for generating incoherent excimer radiation [19] .

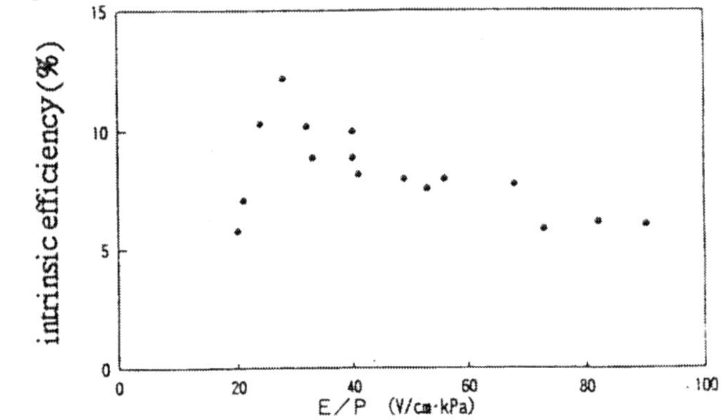

Fig. 6 The high efficiency characteristics of the dielectric barrier discharge pumped Xe_2 lamp [21] .

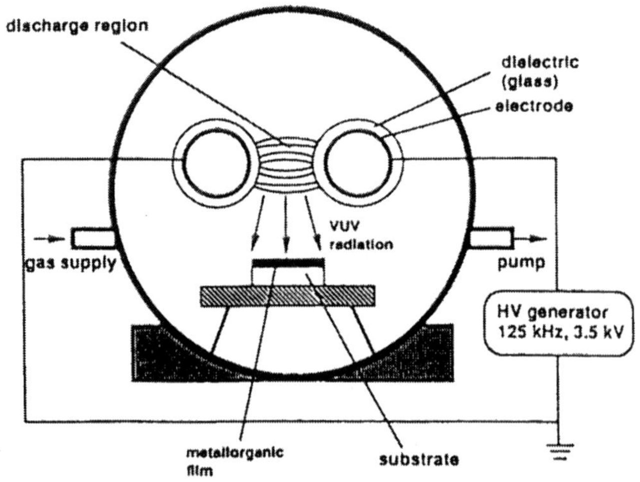

Fig. 7 Schematic of the windowless VUV excimer lamps [25] .

A Flat Fluorescent Lamp with Xe Dielectric Barrier Discharges

Takahiro Urakabe, Shigeki Harada, Takeo Saikatsu and Masao Karino
Mitsubishi Electric Corporation, 5-1-1, Ofuna, Kamakura, Kanagawa 247, Japan

We have studied the new type of flat fluorescent lamps for LCD backlighting, which utilizes Xe(xenon) dielectric barrier discharges.[1] The lamp is characterized with excellent luminance uniformity and good luminous efficacy.

Fig. 1 shows the schematic view of the flat fluorescent lamp. The lamp has two dielectric plates, which are coated with phosphors. A transparent electrode is formed on the outside of the front side dielectric plate and a metal electrode is formed on the outside of the rear side dielectric plate. Xe or Ne-Xe(neon-xenon) mixtures are filled in the space between the two dielectric plates.

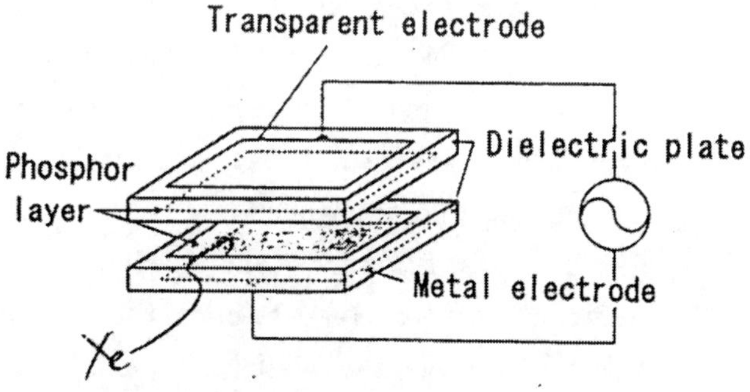

Fig. 1 Schematic view of the flat fluorescent lamp.

When AC voltage is supplied between these electrodes, discharges are generated in the gas discharge space. And then Xe is excited and VUV(Vacuum ultraviolet) rays are radiated. VUV rays are transformed into visible rays by the phosphor.

Fig. 2 shows the cross section of the practical flat fluorescent lamp. Two soda-lime glass plates are sealed with a lead glass, having a gap width of 0.5~2.0mm. A metal electrode is formed on the inside of the rear side glass plate and is covered with a dielectric layer in order to lower the input voltage.

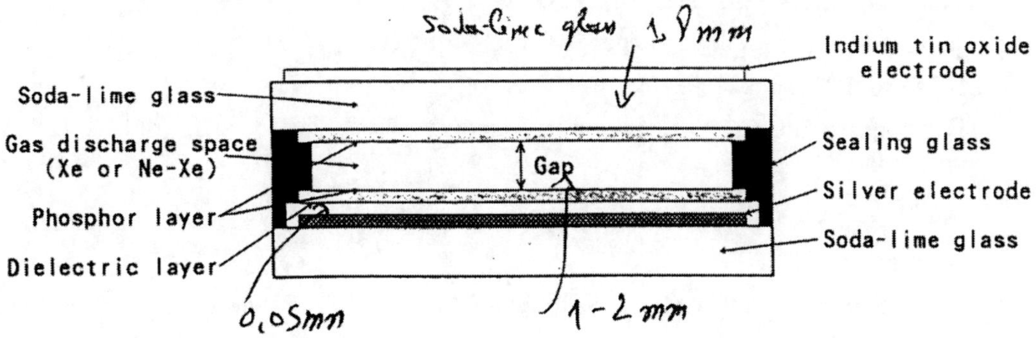

Fig. 2 Cross section of the practical flat flourescent lamp.

This lamp is featured, as described below.

(1) High luminance and high luminous efficacy.

(2) Excellent luminance uniformity.

(3) Stable luminous characteristics in every lamp size.

(4) Good operation in low temperature.

The feature (2), (3) and (4) are ascribed to the structure of the lamp. Discharges do not concentrate on part of the electrode because it is covered with a dielectric material. Accordingly light is uniformly radiated. The lamp has good luminous characteristics in every lamp size, because the gap width is kept at a certain value irrespective of size. And due to the absence of mercury, the lamp has good characteristics in operating under a low temperature.

We have examined the effect of such factors as gas pressure, discharge gap, phosphor and operating waveform to obtain high luminance and luminous efficacy. Fig. 3 shows luminance and luminous efficacy when the lamp is operated by 20kHz sine wave and 18kHz pulse one. The size of the lamp is 72x54mm, the discharge gap width is 1.2mm. And the thickness of the soda-lime glass is 1.8mm and the gas pressure(Xe) is 22.6kPa. High luminous efficacy of 27lm/W has been obtained by the pulse operation.

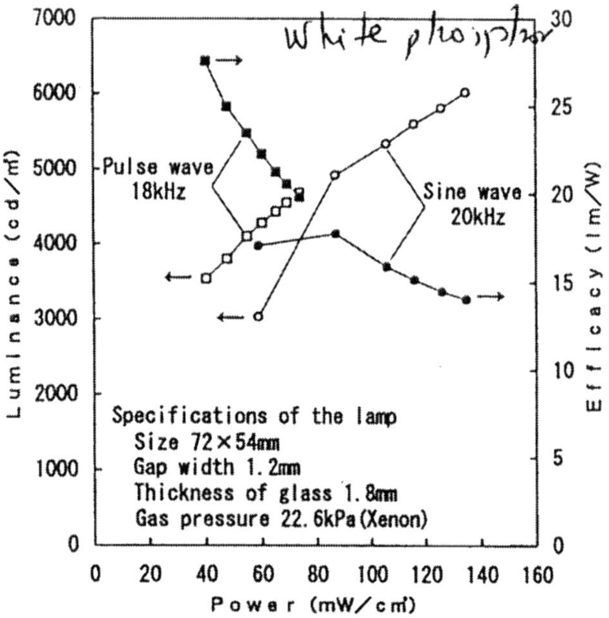

Fig. 3 Luminance and efficacy of the flat fluorescent lamp operated by two different waveforms.

The flat fluorescent lamp described above has some excellent characteristics and is expected to apply to LCD backlighting or other lightings in future.

Reference

[1] S. Harada, T. Urakabe, K. Sano and T. Saikatu: "A Study of electrode Structure in Surface-discharge Panels", Proc. Annual Conf. of Illum. Engng. Inst. Japan (1993), p. 46

DEVELOPMENT OF Ar, Kr EXCIMER LAMPS USING DIELECTRIC BARRIER DISCHARGE

H.Sugahara,Y.Ohnishi,H.Matsuno,T.Igarashi and T.Hiramoto
USHIO INC.,1194 Sazuchi,Bessho-cho,Himeji-city,Hyogo-pref.,Japan

Introduction

Recentry VUVexcimer lamps using dielectric barrier discharge (silent discharge) have been developed [1] . These VUV excimer sources are used for surface cleaning, surface modification, photochemical processes and so on [2] [3] [4] .But these lamps are not easy to handling because of those gas flow structure. Then we developed Ar,Kr head-on type excmer lamps using dielectric barrier discharge. These consist of a sealed-off structure and it is easy to use.

Design

Figure 1 shows a schematic structure of the excimer lamp. This lamp consists of a double-walled coaxial quartz tube and have an MgF_2 window on one side of the tube. The electrodes made from metal foil are set up on the outer surface of the outside wall and on the inner surface of the inside wall respectively. An annular part is the discharge space. Sinusoidal voltages are applied between the two electrodes with the frequency of about 20 kHz. Pure Ar or Kr gas is enclosed in it for 126 nm or 146 nm radiation respectively. The MgF_2 window is attached to the central part of an annular metal cap made from a nikel plate with frit, and the metal cap is indium-metalizingly attached to the quartz glass. The lamp is operated inside a lamp jacket cooled with water in order to protect the sealing part and to prevent the window degradation. A product **pd** of a gas filling pressure **p** and a discharge gap length **d** for both Ar and Kr excimer lamps is about 40 kPa・cm. An input power is 20 W.

Performance

Figure 2 shows spectra of Ar, Kr ,Xe and 30 W-type deuterium (D_2) lamp. These are measured on 20 W input power for the excimer lamps. Relative radiative intensities were measured with a VUV spectrometer at a place of 200 mm apart from the lamp surfaces. Peak wavelength for Ar and Kr excimer spectra are respectively 126 nm, 146 nm and FWHM are respectively 10 nm, 13 nm. The spectra of Ar and Kr excimer lamps are so intensive as high as 4 and 7 times respectively compared with that of the D_2 lamp in the same spectral region. Radiative intensity was measured with a fluorescence detector with sodium salicylate. The sensitivity of this detector has been calibrated using a D_2 lamp calibrated in Prof. Nakagawa's laboratory of Saitama University.

The intensities of the Ar and Kr excimer radiation are 2 and 5 mW/cm^2 at a head of the lamps respectively. Life of the Ar and Kr excimer lamps are respectively 500 and 700 hours provided that the lamp life is defined as the operating time untill reducing by half of the initial radiation output.

Conclusion

We have developed Ar,Kr head-on type excimer lamps using dielectric barrier discharge that can be used in not only laboratories but also in industrial fields.

Refereces

[1] B. Gellert and U. Kogelschatz: Appl. Phys., B52,14 (1991)

[2] P. Bergonzo, P. Patel, I.W. Boyd and U. Kogelschatz.:Appl. Surf. Sci., 54,424 (1992)

[3] H. Esrom and U. Kogelschatz: Appl. Surf. Sci., 54, 440 (1992)

[4] F. Kessler, H. D. Mohring and G. H. Bauer: Met. Res. Soc. Symp. Proc.,192 ,559 (1990)

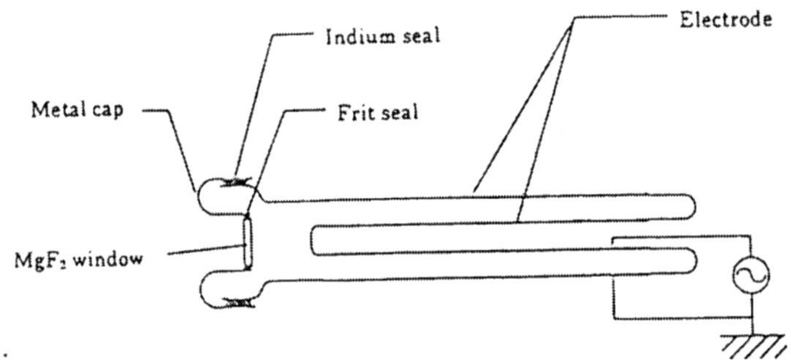

Figure 1 Schematic structure of excimer lamp

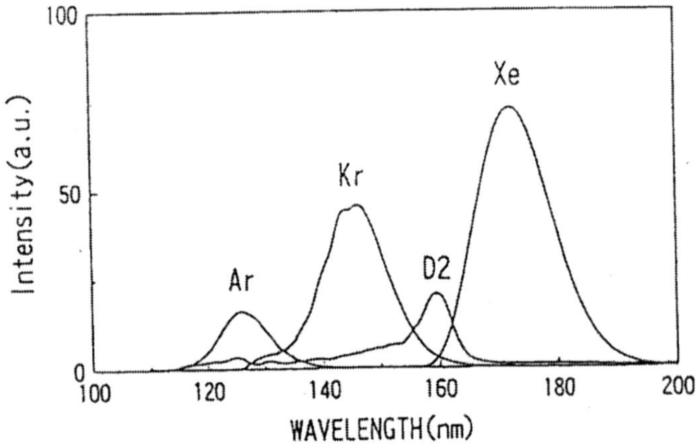

Figure2 Spectra of Ar,Kr,Xe head-on type excimer lamps and a deuterium lamp measured at a distance of 200mm

MEASUREMENT OF SINGLE - EXCIMER - FILAMENTS IN A BARRIER DISCHARGE AT VARIOUS PRESSURES AND DIFFERENT MIXTURES

S. Kaiser, H.- P. Popp
Lichttechnisches Institut, Universität Karlsruhe (TH), Kaiserstraße 12, D-76131 Karlsruhe, Germany

E. Arnold, K.- J. Dietz, A. Hofmann
Heraeus Noblelight GmbH, Heraeusstraße 12-14, D-63450 Hanau, Germany

There is a wide field of applications for incoherent, quasi-monochromatic and selective UV-radiation such as UV-curing, surface modification and photochemistry. Excimer lamps combine all these specific properties. Excimer are molecules which exist only in excited states and immediatelly dissociate after returning to the ground state. If at least one dielectrical layer is placed between the electrodes this type of discharge is called barrier discharge.There have been several studies about UV excimer radiation from dielectric barrier discharge [1], [2] in the past.
Subject of this work are high resolution measurements of single $KrCl^*$ and $XeCl^*$ filaments in barrier discharges at different filling pressures and mixture ratios. Presupposing that the interaction between filaments is negligible it is possible to conclude from the radiation characteristic of a single filament to that of lots of filaments occuring in a commercial excimer lamp.The results are the base to increase the UV efficiency by optimizing the operating conditions for excimer lamps. The UV-radiation was analysed at a frequency of 235 kHz. The thickness of the quartz layers was 1.5 mm. The distance between the quartz layers was 5.5 mm. Electrodes with 8 mm diameter were used.

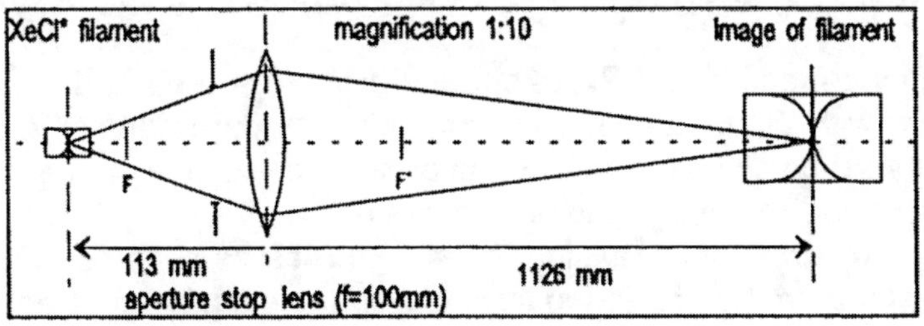

Figure 1: optical system

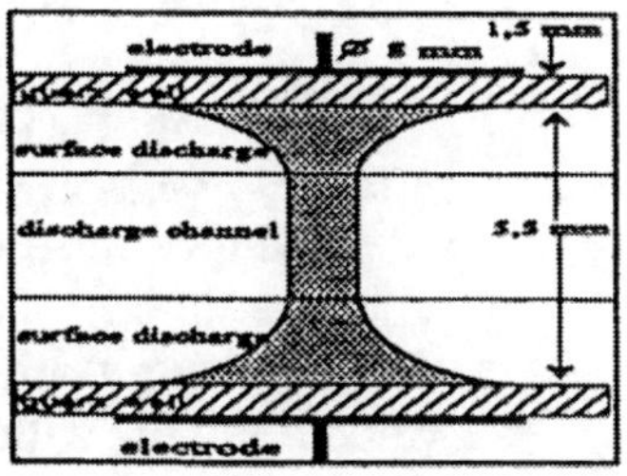

Figure 2: shape of filament

The radiance of filaments in $KrCl^*$ and $XeCl^*$ was measured side-on. The filaments were magnified ten times to obtain a good resolution (Fig.1). They were projected on an pupil (80μm). It was possible to measure the whole depth of the filaments up to 10 mm. The radiation of the filaments was measured by moving the lens and detected by a photomultiplier tube. According to Fig. 2 the filaments are divided into different areas, the surface discharges and the discharge channel.
As the distribution of radiance was symmetrical to the middle of the gap as a consequence of integrating the radiance over the different polarities only one half of

the filament had to be measured (Fig.4 and Fig.5).
The filling pressure was varried from 120 mbar up to 750 mbar. Measurements were made at a chlorine concentration from 0.4 up to 3.0 compared to standard mixtures of excimer lasers.

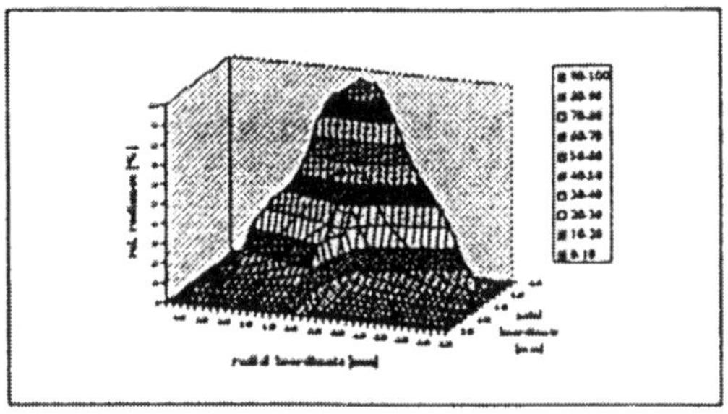

Figure 3: KrCl* filament, p = 170 mbar

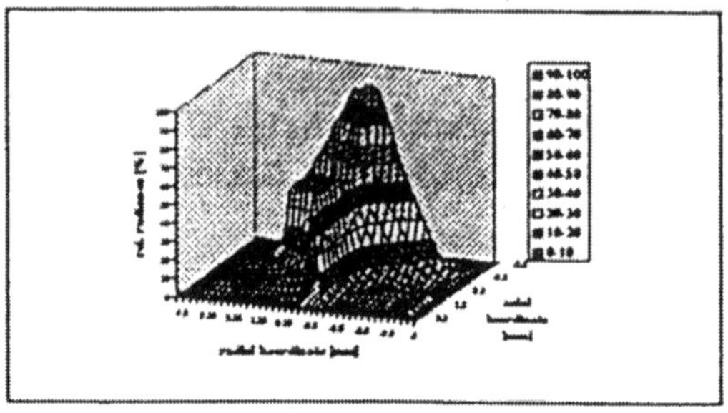

Figure 4: KrCl* filament, p = 350 mbar

Fig.3 and Fig.4 show two filaments in KrCl* burned at different pressures but with identical gas mixtures at a frequency of 235 kHz. It can be seen that the relative radiance of the channel increases by increasing the filling pressure.

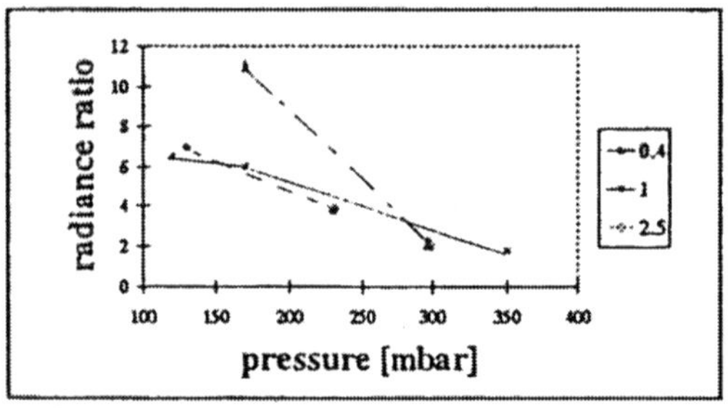

Figure 5: radiance versus pressure

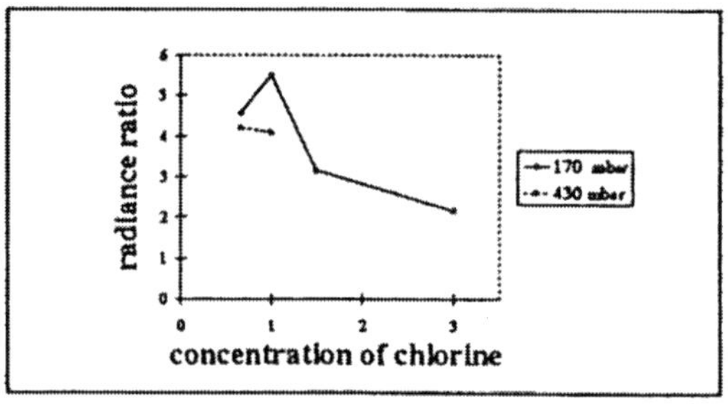

Figure 6: radiance versus mixture ratio

Fig. 5 shows the radiance ratio between the surface discharge area and the channel area versus pressure with the concentration of chlorine as parameter. Fig.6 shows the radiance ratio of surface discharge area to channel area versus concentration of chlorine. In both cases the radiance ratio is decreasing.
The measurements show that by increasing the filling pressure or the concentration of chlorine the ratio of the emitted radiance from the surface discharge area and the discarge channel area is decreasing in the examined parameter range.The diameter of the surface discharge is getting smaller, too. KrCl* and XeCl* filaments almost show the same behaviour.
Although the radiance of the channel area increases by increasing the filling pressure or the pressure of chlorine most excimer radiance is emitted from the surface area. The amount is about 65% up to 85% of the total radiance.

References:

1. B. Eliasson and u. Kogelschatz, "UV Excimer radiation from dielectric barrier discharges", Proc. 40th Anual Gaseous Electronics Conf. GEC 87, 174, Atlanta, USA
2. V. Schorpp," Die dielektrisch behinderte Edelgas-Halogenid-Excimer-Entladung: Eine neuartige UV-Strahlungsquelle", Dissertation, Karlsruhe Universty, 1991

DYNAMICS OF THE DISCHARGE PUMPING IN AN ArF EXCIMER LASER

Shinji Nagai*, Hiroaki Sano*, Hideo Furuhashi**, Yoshiyuki Uchida**, Jun Yamada***, Akihiro Kono*, and Toshio Goto*
*Department of Information Electronics, Nagoya University,
Furo-cho, Chikusa-ku Nagoya, 464-01, JAPAN
(Phone +81-52-789-4421 Fax +81-52-789-3164)
**Department of Information Network Engineering,
***Department of Electronics, Aichi Institute of Technology,
Yachigusa, Yakusa-cho, Toyota, 470-03, JAPAN

Discharge-pumped excimer lasers have been developed and improved through studies of the excitation mechanism. ArF excimer lasers have the shortest wavelength among discharge-pumped excimer lasers and attract the attention of many researchers. However, corresponding work for ArF excimer lasers is quite limited. The buffer gases used in excimer lasers have effects upon the laser performance and play an important role in the ArF plasma. The purpose of the present work is to investigate experimentally and theoretically the effect of buffer gases on the excitation mechanism of the laser.

In this study, a capacitor transfer type excimer laser with automatic spark UV preionization controlled by a thyratron switch was used. The length of the electrodes is 60 cm and the cross-section of the discharge space is 0.8 cm (width) × 1.8 cm (height). The gas mixtures of F_2:Ar:M = 0.2:10.0:89.8 (M is the He/Ne diluent) was used for the measurements of the Ne* and He* densities. The total gas pressure was 2.5 atm and the charging voltage of the storage capacitors was 30 kV in all measurements. To clarify the effect of the buffer gas, the number densities of the excited atoms of the buffer gas were measured in ArF discharge plasma. They were measured by a laser absorption probing using cw ring dye laser (Coherent Model 7500). From the absorption coefficient profile obtained by measuring transmitted laser intensities at different wavelengths, the particle number densities of excited atoms were calculated[1]. The Ne* (3s $^3P_2$) and He* (2p ^3P) atom densities were measured in the Ne and He diluent discharges, respectively. The absorption lines used were 3s $^3P_2$ → 3p $^3P_1$ (line center at 588.19 nm) for Ne* and 2p ^3P → 3d ^3D (line center at 587.56 nm) for He*. The discharge voltage and ArF* spontaneous emission intensity (193 nm) were also measured. The peak densities of Ne* and He* atoms are 1.7×10^{12} cm^{-3} and 2.1×10^{12} cm^{-3}, respectively. Although the pulse duration of the ArF* spontaneous emission for the Ne diluent discharge was somewhat longer than that for the He diluent discharge, the peak intensities are almost the same. Therefore, it seems that no distinct difference in the laser performance was observed as a result of the use of Ne and He buffer gases. In the case of XeCl excimer lasers, higher laser output power was obtained by using Ne diluent than by using He diluent[2,3]. To clarify the reason, the temporal behaviors of the fractional concentration of the He* (2p ^3P) in the total He atom density (denoted by [He*]/[He] below) were measured for the different Ne fractional concentrations in He/Ne diluent. The gas mixture ratio was F_2:Ar:M = 0.2:10.0:89.8 and the Ne fractional concentration in He/Ne diluent was varied. Figure 1 shows the results. Since the excited He* atoms are mainly formed by direct electron impact excitation, [He*]/[He] gives information about the power used for the excitation of atoms. Furthermore, it is closely connected with the formation of excited Ar* atoms and Ar^+ ions, which have important roles in the formation of ArF* excimer molecules. The increase in [He*]/[He] becomes rapid with Ne fraction. This fact shows that the excitation rate of He increases, that is, the power used for the excitation of atoms increases with Ne fraction in this stage. However, the rise time and the pulse width of the [He*]/[He] temporal profile decrease with Ne fraction. This is because the discharge voltage drops in a shorter time when Ne buffer gas is used; it is due to the low breakdown voltage.

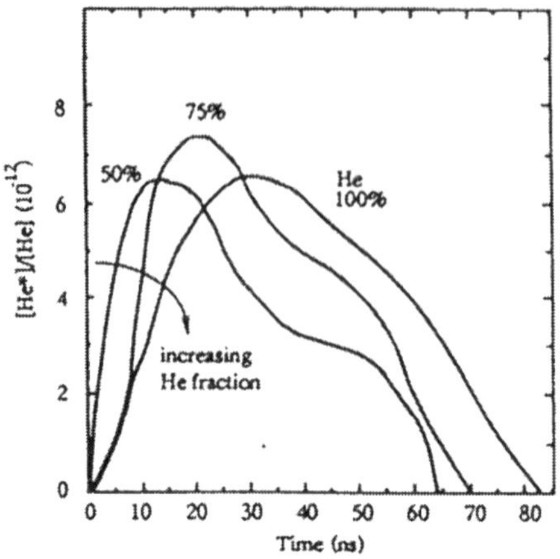

FIG. 1. Temporal behavior of the [He*]/[He] density ratio on the Ne fractional concentration in He/Ne diluent.

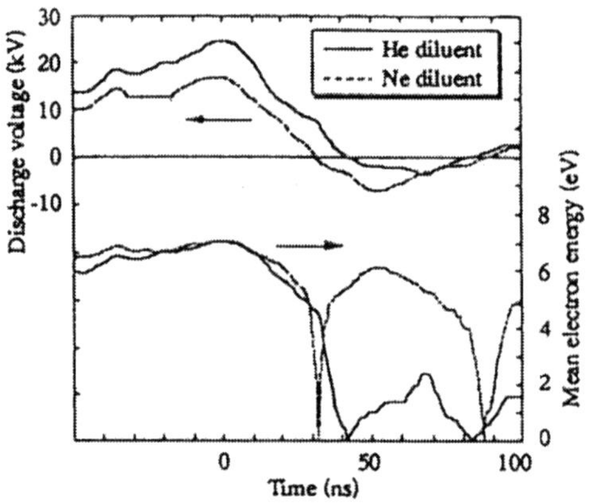

FIG. 2. Temporal behavior of the mean electron energy. The discharge voltages are also shown.

Subsequently, a computer modeling was performed to consider about the buffer gas effect from the point of view of the electron energy distribution. They were obtained by solving the steady state Boltzmann equation using the same conditions with the above experiments. And the temporal behavior of the mean electron energy was calculated using the electric field obtained in the experiments. Figure 2 shows the result. The mean electron energy for the Ne and He diluent are same in the stage after the start of the discharge.

From the results, we may conclude as follows. At the beginning of the discharge, the mean electron energy are same in comparison with Ne and He diluent. Therefore, it is considered that the electron densities in the Ne diluent are larger than in the He diluent from the difference of the excitation intensity between Ne diluent and He diluent. The power used for the excitation of atoms in He diluent is smaller than that in Ne diluent. However, by using the He diluent, the total energy used for excitation of atoms increases owing to the long excitation period.

[Reference]

1) A. C. G. Mitchell and M. W. Zemansky, *Resonance Radiation and Excited Atoms*, Cambridge University Press (1934).
2) R. C. Sze, IEEE J. Quantum Electron., QE-15, 1338 (1979).
3) R. C. Sze, J. Appl. Phys., 50, 4596 (1979).

MULTIFREQUENCY LASER EMISSION FOR ILLUMINATION GENERATED BY FOUR-WAVE RAMAN MIXING

Hiroyuki Kawano and Totaro Imasaka
Department of Chemical Science and Technology, Faculty of Engineering, Kyushu University, Hakozaki, Fukuoka 812, Japan

Introduction

The first laser was invented in 1960, and since then a large number of scientists developed many lasers and found various nonlinear optical phenomena. A laser has unique properties, i.e. monochromaticity, directivity, coherence, high intensity, and capability of generating ultrashort pulses. Especially, monochromaticity and directivity are important natures for application to illumination. However, there has been an inevitable problem in a laser, i.e. a narrow emission region. Many kinds of lasers have been manufactured so far, but each of them has its own limited range. For example, an argon ion laser and a krypton ion laser are widely used as light sources for illumination, but the emission wavelengths are, unfortunately, limited. Moreover, the wavelengths are fixed, and the intensity distributions are not so flat. A dye laser and an optical parametric oscillator recently developed are tunable, but these lasers also have limited tunable ranges, and additionally a multifrequency laser, which is useful for color illumination, has not yet been developed. In 1989, a new laser technology, i.e. two-color stimulated Raman scattering, was reported by our group. We succeeded in generation of multifrequency laser emission, which covers the entire visible region simultaneously. It is called "Rainbow Stars", because it contains all the rainbow colors and twinkles like stars. This emission consists of many Stokes and anti-Stokes rotational emissions separated by 587 cm^{-1} and vibrational emissions separated by 4155 cm^{-1}, which are produced by introduction of a two-color beam separated by 587 cm^{-1} into molecular hydrogen pressurized to 10 atm typically. This frequency separation corresponds to the rotational energy of ortho-hydrogen between J=1 and J=3 levels in the vibrational and electronic ground state.

Theoretical Bases

Generation of multifrequency laser emission is based on four-wave Raman mixing. The intensity generated is described by eq. (1).

$$I_{FWM} \propto \omega_S^2 I_P^2 l^2 \tag{1}$$

The intensity, i.e. the conversion efficiency, depends on the frequency and the intensity of the pump laser, and also the interaction length of the laser beam and the Raman medium.

Results and Discussion

Figure 1 shows the dependence of the interaction length on the efficiency. The interaction length is multiplied by the reflection number in the multi-pass Raman cell. The efficiency is substantially improved by focusing the beam five times in hydrogen. The efficiency is also frequency dependent. For example, the efficiency in generation of a rotational line decreases with increasing the laser wavelength, but the decrease of the efficiency is more serious in stimulated Raman scattering producing a vibrational line. Thus rotational emissions are more pronounced by pumping hydrogen with a laser emitting at longer wavelengths. Therefore, multifrequency laser emission with a flatter intensity distribution could be obtained by changing the dye laser wavelength from 450 nm to 500 nm.

Figure 2 shows the emission spectrum of a high-power multifrequency laser (200 mJ x 10 Hz = 2 W), which is generated by pumping with a single-beam elliptically-polarized second harmonic emission (400 mJ) of a Nd:YAG laser: the two-color laser system can be avoided by using an elliptically-polarized laser system. Such a high average power laser might be useful for laser lightening, e.g. projection of the laser beam onto the building wall. Modelocked titanium:sapphire laser emitting at 800 nm has a higher peak power (200 MW) due to a shorter pulse width, resulting in a 6 times higher power. As shown in Fig. 3, equally-spaced laser emissions can be obtained very efficiently and dominantly due to a longer pump wavelength and a nonlinear behavior of this phenomenon, though vibrationally-shifted rotational lines are also weakly observed.

Figure 4 shows the Commission Internationale de l'Eclairage x-y chromaticity diagram of the multifrequency laser emission. The area surrounded by the connected lines, which indicates the color reproduction region, is 95 % of the whole (curved) area, which is much larger than the value given by the use of an argon ion laser (53 %), a He-Cd laser (64 %), or the dye-mixture laser (66 %).[2] This indicates that the spectrum and then the color of light can be more faithfully reproduced by the multifrequency laser emissions generated in this study.

References

1. T. Imasaka, S. Kawasaki, and N. Ishibashi, Appl. Phys. B 49, 389-392 (1989).
2. Y. Saito, N. Nakai, A. Nomura, and T. Kano, Appl. Opt. 31, 4298-4304 (1992).

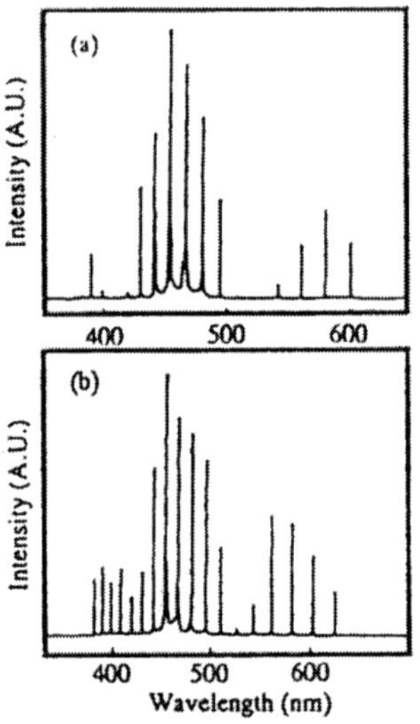

Fig. 1 Spectra of the multifrequency laser emission. (a) a single pass (b) 5 passes

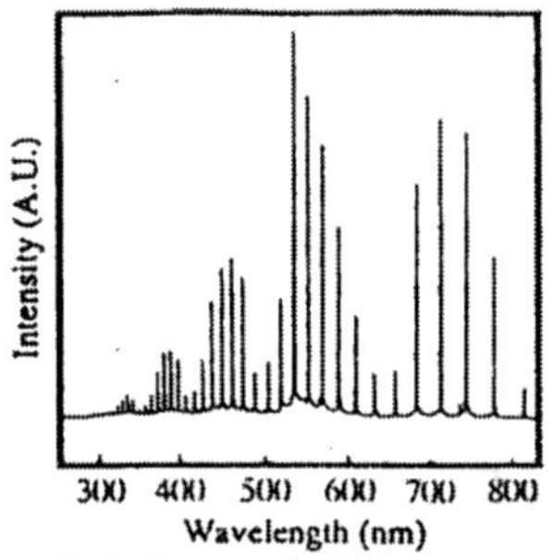

Fig. 2 Spectrum of multifrequency laser emission generated by SHG of Nd:YAG laser.

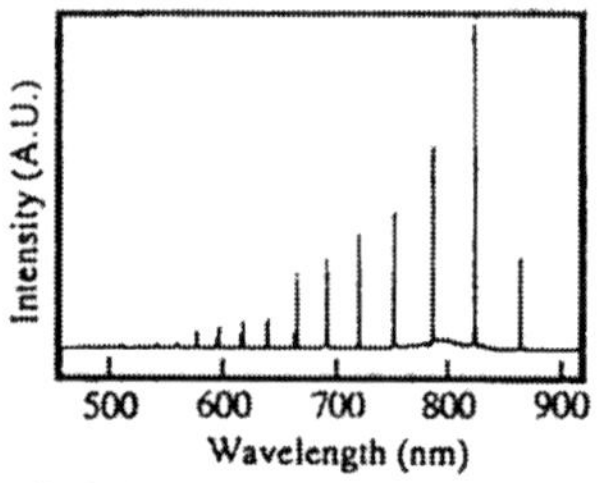

Fig. 3 Spectrum of the multifrequency laser pumped by Ti:Sapphire laser.

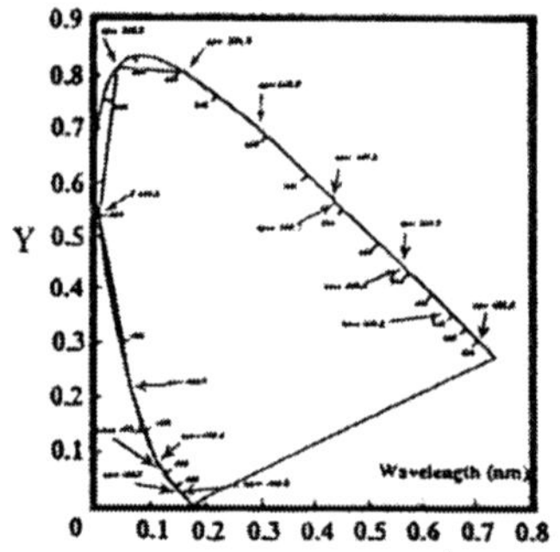

Fig. 4 Commission Internationale de l'Éclairage x-y chromaticity diagram of multifrequency laser emission.

THE RELATIONSHIP BETWEEN RADIANT EFFICIENCY AND E/p IN XeCl* ,KrCl* DIELECTRIC BARRIER DISCHARGE LAMPS

Yoshinori Aiura , Hiromitsu Matsuno , Tatsushi Igarashi and Tatsumi Hiramoto
USHIO inc. Research and Development, 1194 Sazuchi, Bessho-Cho Himeji City, Japan

Introduction

A lot of papers on characteristics of XeCl* and KrCl* dielectric barrier discharge lamps have been published .[1)2)3)] Although the lamps have an advantage of a high radiant efficiency , a relationship between the efficiency and electrical characteristics has not been discussed .

In this study it has been investigated what factors are dominant in order to design high-efficiency lamps filled with Xe/Cl_2 or Kr/Cl_2 .

Experiments

The lamps consist of two concentric transparent quartz tubes . An external mesh-electrode and an inner foil-electrode are set on the external surface of the outer tube and the inner surface of the internal tube respectively . The length in the axial direction of the annular space is 150mm and that of the two electrodes is 100mm . The annular space between the two quartz tubes is filled with discharge gases . The inside diameter of the outer tube is fixed at 24.5mm . The outside diameter of the inner tube has been varied in order to change the annular gap ***d*** . Sinusoidal voltages are applied between the two electrodes with a frequency of about 20 kHz . Electrical characteristics and radiation intensity have been measured for the lamps kept being connected with an evacuating system during a series of experiments of a definite ***d*** . So a new gas can be filled into the lamps without cutting off the filling-tips after measuring electrical and radiative properties of a lamp of a ***p·d*** .

Ignoring the effect of space change in the annular space , electrical characteristics were calculated from measured Lissajous figures as follows :

$$V_{gap} = V_S \times \frac{C_{glass}^*}{C_{glass}^* + C_{gap}^*} \qquad (1)$$

$$E = \frac{V_{gap} \cdot C_{gap}^*}{2\pi\varepsilon_0 r_0} \qquad (2)$$

where Vs , Cglass* , Cgap* ,E, r_0 denote applied voltage between the two electrodes , the capacitance of the dielectric quartz walls per unit length in the axial direction , the capacitance of discharge space per unit length in the axial direction, the electric field strength and radius of the middle point of the annular space from the lamp axis .

The discharge gas of Xe/Cl_2 or Kr/Cl_2 emits excimer radiation at 308 or 222nm respectively . A photo-detector was used to measure the radiation intensity . This is composed of a band-pass filter transmitting only the excimer radiation , a fluorescent plate converting the excimer radiation into visible light and a Si-photodiode . The sensitivity of this detector has been referred to a halogen lamp calibrated in Electrotechnical Laboratory of the Ministry of International Trade and Industry and a deuterium lamp calibrated in Prof. Nakagawa's laboratory in Saitama University .

Results and discussion

The value ***E/p*** is plotted as a function of ***p·d*** in Fig.1 . It may be shown for ***E/p*** to depend upon *p·d* , not *d* . The difference in *E/p* between Xe/Cl_2 lamps and Kr/Cl_2 lamps is very small . If the breakdown voltage is used in place of *E/p* , this relation is similar to Paschen's curve .

The radiant efficiency is shown as a function of *E/p* in the lamps filled with Xe/Cl_2 and Kr/Cl_2 in Fig.2 . As shown in Fig.2 , the efficiency increases steeply at low *E/p* and decreases gradually at high *E/p* as *E/p* increases . The maximum of the efficiency exists .

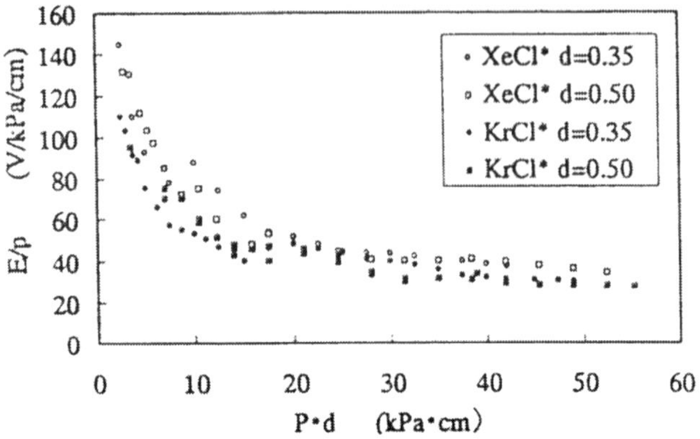

Fig.1 Relationship between E/p and p•d in XeCl* lamps and KrCl* lamps

A similar trend of the efficiency with the rise in *E/p* has been theoretically obtained for Xe_2* lamps .[2)3)] In genaral , it may be said that the electrons are accelerated in the electric field in the region of low *E/p* ,and gain an energy which is insufficient for the rare gas and Cl atom to form the excimers . However , at high *E/p* electrons can excite both the rare gas and Cl atoms , which form the excimers . At very high *E/p* the electrons gain sufficient energy for exciting both the atoms to too many high levels and ionizing the atoms . Many species are formed , which are independent of the excimers , or quenching the excimers . An optimum value of *E/p* for the radiant efficiency exists as a result . The maximum efficiency in KrCl* lamps occurs at a smaller *E/p* compared with that in XeCl* lamps . This may be explained by the smaller momentum transfer collision cross-section for Kr atoms to electrons of several electron volts which may be the lowest energy for inelastic collision proceeding to formation of KrCl* . However , although quantitative discussions on the relation between the radiant efficiency and *E/p* is different , the formation process of the KrCl* excimers is inefficient compared with that of the XeCl* excimers at high *E/p* .

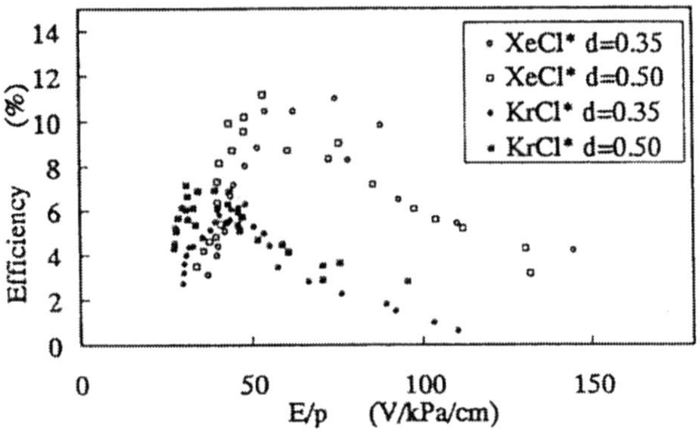

Fig.2 Efficiency as a function of E/p in XeCl* lamps and KrCl* lamps

Conclusion

An appropriate value of *E/p* is necessary for the radiant efficiency to be optimized .

References

1) C.von Arx , U. Kogelschatz ,LS:6 , 79:P (1992)
2) B.Gellert and U. Kogelschatz , Appl. Phys. , B52 , 14-21 (1991)
3) B. Eliasson and U. Kogelschatz , Appl. Phys. , B46 , 299-303 (1988)

INVESTIGATION OF THE ELECTRICAL AND OPTICAL PROPERTIES OF DIELECTRIC BARRIER DISCHARGES

S. Müller and R.-J. Zahn
Institute of Low-Temperature Plasma Physics
Robert-Blum-Straße 8-10, D-17489 Greifswald, Germany

Dielectric barrier discharges (DBD's) are initiated by ac-voltages in gaps where one or both of the electrodes are covered by a dielectric layer. These discharges are efficient UV/VUV-light sources capable of producing selective incoherent excimer radiation /1,2/.
Individual microdischarges (filaments) as well as large-area radiation sources with a large number of statistically distributed microdischarges were investigated. In this work we present some results of optical and electrical measurements. In particular the spatial and temporal discharge development was of interest. A 1-D and 2-D model was developed additionally to explain some of the observed phenomena. In the model the discharge cell was considered to be electrically equivalent to a nonlinear gaseous conductor in series with a capacitance formed by the cell.

Fig.1 shows a schematic of our cylindrical discharge device. Besides, plane panels and discharge gaps with glass covered electrodes were used. The discharge was driven by a sinusoidal power amplifier as well as square waves in a range between 10 and 375 kHz respectively 1 and 20 kHz.
The gases used in this investigation were rare gases, Penning-mixtures and Kr/He/Cl_2. The discharge gap had a width between 1 and 5 mm. The pressure was varied between 0.1 and 1 bar. Our experimental studies were performed using a streak camera from Hamamatsu (spectral sensitivity: 115-850 nm) with a slow speed unit as well as a shutter camera (CCD detector and MCP). The applied MCP's could be gated in less than 5 ns. The spectral distribution in the UV/VUV range was investigated by a 1m ARC spectrometer.

The intensity of the excimer emission is influenced by a lot of parameters. Fig.2 shows the influence of the pulse width of the square wave sustaining voltage. One reason for this behaviour are different stages of the discharge development. In the case of sinusoidal voltages, the discharge development is sometimes not completed in succeeding half periods.

Fig.3 shows the time resolved VUV spectral distribution of a filament in Xe. With increasing time, after the first discharge activity, a gradual shift of the intensity occurs from the first continuum to the second one. The same trend was found with increasing pressure.

Examples of the spatial and temporal discharge development can be seen in Fig.4 and 5. Fig.4 shows a streak image; tomographic images of a filament at different stages of development are given in Fig.5. Mean stages of the discharge development are the Townsend phase, the space charge wave, the streamer phase and the afterglow phase as well as the discharge spreading at the electrodes. Additionally, oscillations are observable in some phases.
In Fig.6 we show a result of our 2-D model. The set of images in Fig.6 shows the origin of excimer radiation at different times in the box during the decay of the excimer for the reaction scheme

$$\text{Kr-Cl}^* \Longrightarrow \text{Kr} + \text{Cl} + h\nu\ (220\ \text{nm}).$$

The spreading at the anode is caused by field perturbations due to wall charges. A spreading at the cathode is found only by taking into account a secondary electron emission from the whole cathode surface due to photons.

/1/ U. Kogelschatz: 9th Int. Symp. on Plasma Chem., Pugnochiuso, Italy, Sept. 4-8, 1989, p. 1-8
/2/ H. Müller, M. Neiger, et al.: LS 5, York, England, Sept. 110-14, 1989, p. 171-172

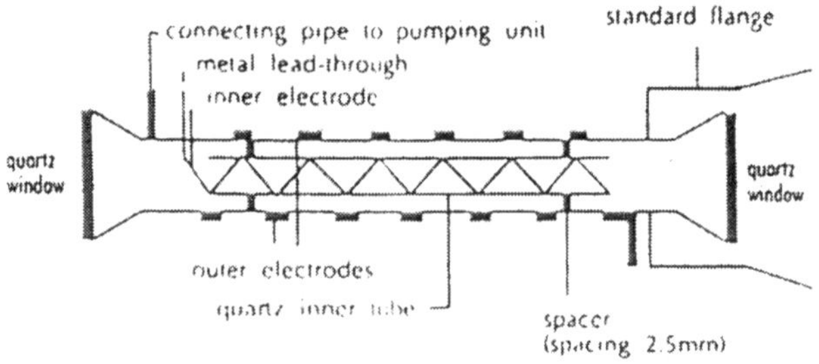

Fig. 1: Cylindrical discharge device

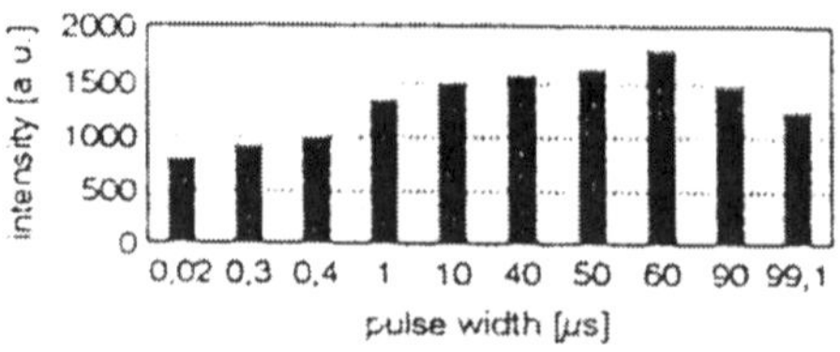

Fig. 2: Dependence of excimer emission intensity (170 nm) on pulse width

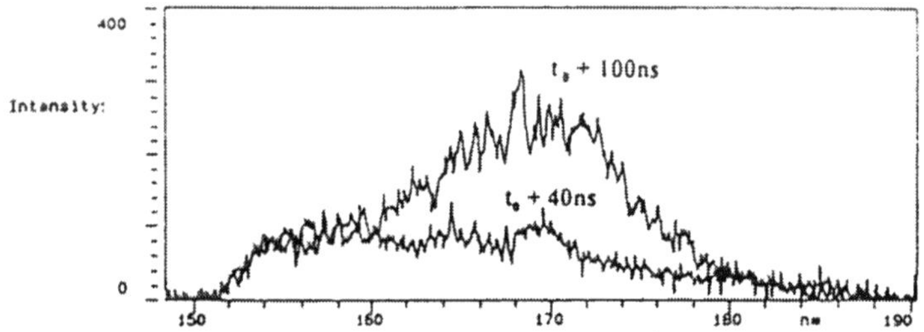

Fig. 3: Time resolved spectral distribution of a filament in Xe at different times

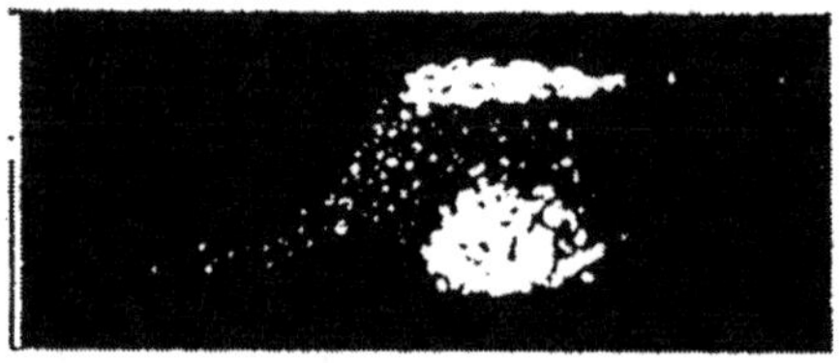

Fig. 4: Streak image of the discharge developement; $Kr/He/Cl_2$, 400 mbar, streak duration: 88 ns, anode: bottom

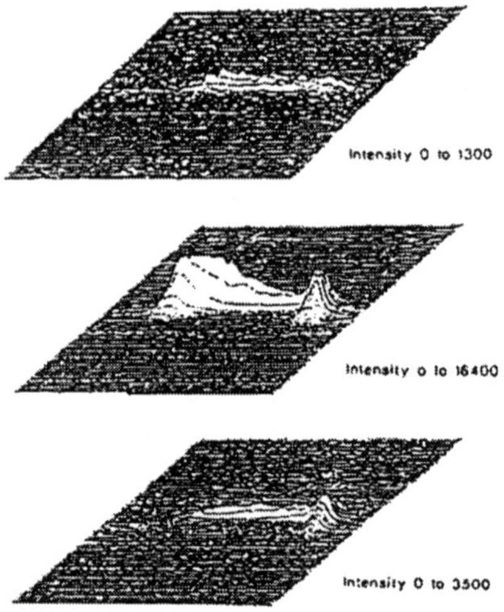

Fig. 5: Tomographic images of a Kr filament at different times; top to bottom: t_0 + 7 ns, t_0 + 46 ns, t_0 + 100 ns, anode: left side

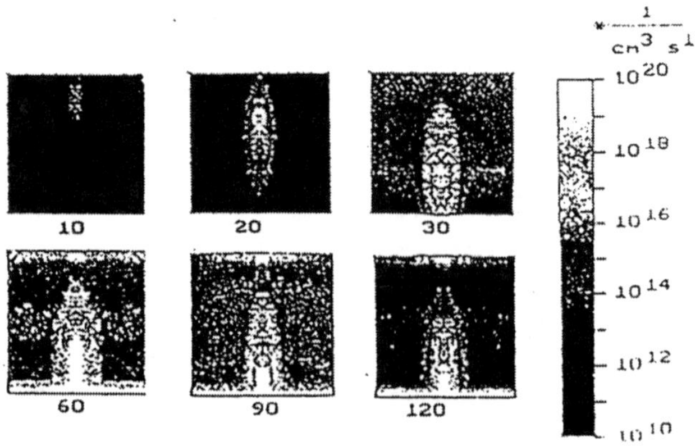

Fig. 6: Origin of excimer radiation; anode: top

Modeling of a Barrier Discharge Excimer Lamp

A.Oda, Y. Sakai, P.Ventzek, H. Tagashira, H. Matsuno* and T. Igarashi*
Department of Electrical Engineering, Hokkaido University, Sapporo 060 Japan
*Ushio Electric Corp., Himeji 671-02 Japan

Historically, dielectric barrier discharges have been used as ozonizers but are now enjoying renewed interest as UV light source plasmas such as excimer lamps.

In this paper, we present a model of the Xenon excimer lamp shown in Fig. 1. It is a cylindrical chamber with an outer radius of 1cm and an inner radius of 0.9 cm. The high voltage is applied at the center and is shielded from the plasma by a dielectric (quartz) that is 0.1 cm in radius. Typical rf voltages are 2-3 kV (peak to peak) at frequencies up to 100 kHz. The fill gas is pure Xe at pressures from 10 to 300 Torr.

Our model consists of 2 parts: a 1 dimensional microdischarge model and a 2 dimensional lamp model. The microdischarge model is an extension of the self-consistent fluid model for ArF excimer lasers developed by Akashi et al.[1] This model consists of a solution of the drift-diffusion equation, Poisson's equation and an energy equation with the transport parameters determined from the Boltzmann Equation. The two-dimensional model is a lamp scale model. Microdischarges are initiated quasi-randomly between the barrier electrodes according to the local E/P_0, where E is the electric field and P_0 is the gas pressure at 0 C, and proximity to seed electrons (ie. proximity to the nearest previous microdischarge). After a microdischarge occurs, depending on the applied electric field, the liberated charges are deposited on the appropriate electrode. The potential in the gap is the re-calculated in 2-dimensions to determine the probability, as a function of axial location, that a microdischarge will occur.

At present, we have considered the step response of the barrier discharge detailed above. The calculation is begun with n_e and n_p (the electron and ion densities having a pre-determined spatial distribution throughout the gap). For the following figures, a step voltage of 3000 V is applied to the center electrode. The pressure in the chamber is 100 Torr. Fig. 2 shows the gap I-V characteristics. As is typical for this pressure regime, the microdischarge current has a 50 ns FWHM and the gap voltage drops rapidly over that same period. After the voltage drop across the gap has effectively decayed to zero, the charged species are lost by diffusion only at a much smaller rate. The accumulated charge on the anode and cathode are shown in Fig. 3. The anode is rapidly charged negative in the first 200 ns of the discharge. The electric field at the cathode, the grounded dielectric is nearly zero. This is because of the low mobility of the ions. They cannot reach the cathode before the gap voltage drop decreases to a low value due to the charge-up of the anode. The spatio-temporal evolution of the gap electric field is shown in Fig. 4. After, 50 to 150 ns the gap electric field drops from 5000 V/cm to less than 1000 V/cm. This is the reason for the behavior shown in Fig. 3. In summary, when the step positive voltage is applied, the electrons rapidly charge the anode, lowering the gap voltage drop, and preventing the ions from charging the cathode.

1. H. Akashi, Y. Sakai and H. Tagashira, J. Phys. D, 27, 1097-1106 (1994)

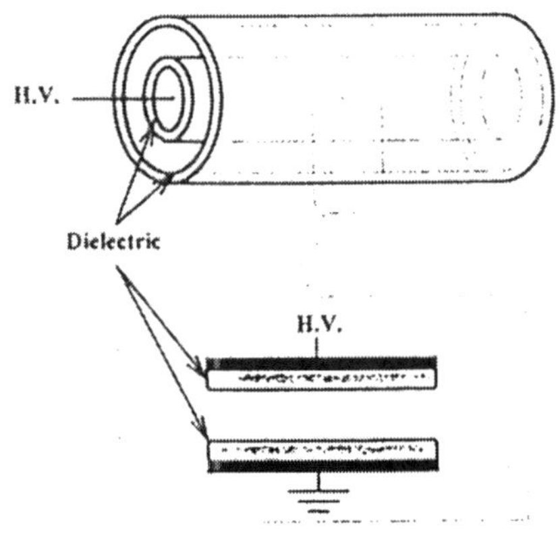

Fig.1 Xenon Excimer Lamp

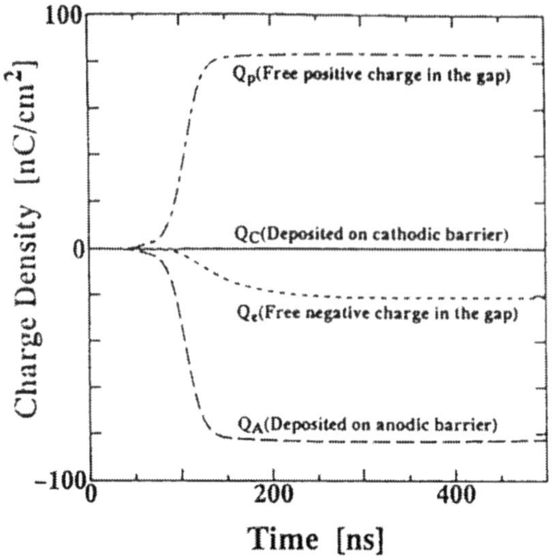

Fig.3 Inventory of Free and Surface Charge Density

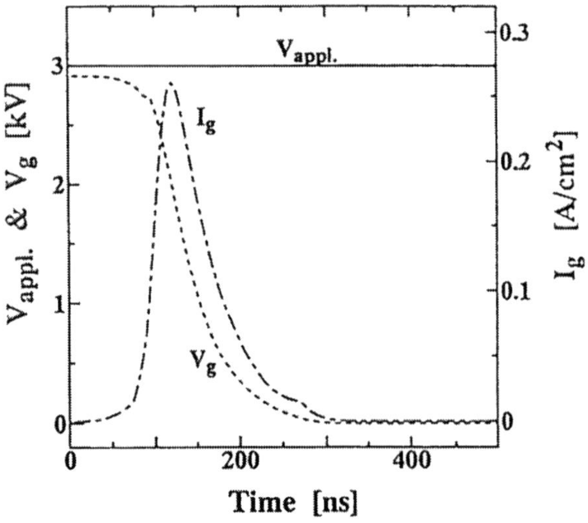

Fig.2 I-V Characteristic

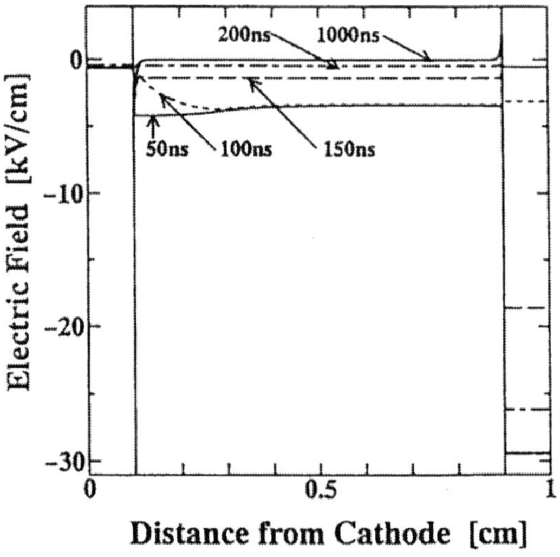

Fig.4 Spatio-temporal Evolution of Electric Field

The Pulsed Discharge Sources With VUV Continuous Spectrum Radiation

Shaolong Zhu, Xiaoyong Zhou and Wei Zhou
Institute of Electric Light Sources, Fudan University
Shanghai, China 200433

1. INTRODUCTION

Vacuum ultraviolet is refered to the electromagnetic wave with wavelength ranging from 0.2 to 200nm. Because of the strong absorption by atmosphere, it can only be transmitted in vacuum. It is one of the important research subjects in modern science and technology.

Synchrotron radiation is undoubtedly the most powerful VUV sources, but obviously it is not accessible to most laboratories and industries. Under certain conditions the low pressure gas discharge may produce resonance radiation in VUV region. Based on this principal some VUV line sources has been designed and investigated.[1] This paper is dedicated to a new VUV source with continuous spectrum.

2. DESIGN PRINCIPAL AND STRUCTURE

It is well known that in the gas discharge radiation the continuous spectrum is originated from molecular band, recombination and decelaration of charged particles. In most discharge conditions, the velocities of charged particles are not high enough, so that the last one is negligible in VUV region. The intensity of recombination radiation is proportional to the charged particle density, and is shifted towards shorter wavelength with increasing temperature. Since high pressure and high current density discharge produces high charged particle densities, it is desirable for the VUV source to work under very high pressure and high current, which is also in favour of the molecular band radiation. Pulse discharge with very short duration may increase the current density, and also the electron temperature. Therefore, we design the new VUV source to be under high pressure, short pulse, high voltage and high current.

The VUV source and the storage capacitor are integrated together to reduce the discharge inductivity and consequently to compress the pulse width to the order of 10^{-7} sec. The electrode separation is about 1~2 mm. The filling gas of VUV source is H_2, Ar, N_2 and so on, filling pressure is 0.2~0.9 MPa. The VUV output window is a piece of MgF_2 single crystal. The VUV source is working under self-breakdown. The repetition rate is about 10 kHz depended on the breakdown voltage and the power supply.

3. CHARACTERISTICS

The axial spectral radiance of the VUV source is measured by a prior-calibrated VUV spectral detecting system.[4] It shows the atomic resonance line L_α (121.6nm) is absent because of the self-absorption in high pressure discharge plasma. The continuous spectrum of the VUV source may be divided into two parts: in the region of 140~160nm, there is hydrogen molecular bands, related to the transition from $1s\sigma\ 2p\sigma\ ^1\Sigma_u^+$ to $(1s\sigma)^2\ ^1\Sigma_g^+$ and from $1s\sigma\ 2p\pi\ ^1\Pi_u$ to $(1s\sigma)^2\ ^1\Sigma_g^+$ groups. Because of the very small separations and the line broadening of the molecular lines, the spectrum is regarded as continuous. In the region of 170~200nm there is a hydrogen molecular continuum, originated from the transition from stable states $1s\sigma\ 2s\sigma\ ^3\Sigma_g^+$ to unstable states $1s\sigma\ 2p\sigma\ ^3\Sigma_g^+$.

Axial integrated radiance L_e may be obtained by integration of the $L_{e\lambda}$. It is found at 8 atm of hydrogen, the axial integrated radiance from 120 nm to 200 nm is about 11 mW/Sr · cm². Therefore it is a useful VUV continuous spectrum source.

It is worth to point out that different continuous spectrum may be obtained simply by changing of the filling gas kind.

REFERENCES

[1] Shaolong Zhu, Li Sun, Wei Zhou and Yiming Li: Journal of Fudan University (Natural Science), 32, 3 (1993) 247 (in Chinese)

[2] J.M. Meek and J.D. Graggs: Electrical Breakdown Of Gases, Oxford University Press, London, 1953

[3] I.S. Marshak: Pulsed Light Sources, Consultants Bureau, London and New York, 1984

[4] Shaolong Zhu, Wei Zhou and Yiming Li: Journal of Fudan University (Natural Science), 30, 3 (1991) 241 (in Chinese)

DOPED AND UNDOPED XENON SHORT ARC LAMPS WITH EXTREMLY HIGH SPECTRAL RADIANCE

E. Smolka, A. Schnabl, F. Schilling
Heraeus Noblelight, Heraeusstraße 12-14, 63450 Hanau, Germany

Many applications like confocal microscopy [1], photodynamic therapy, incoherent laser pumps, photolithography, atom absorption spectroscopy (AAS) and flat-panel display technology are searching for new economic and compact light sources with extremely high spectral radiance and good stability and lifetime. Xenon short arc lamps have the best potential to achieve these requirements. Therefore we chose this lamp type and improved its features concerning pressure and electrode design. Additional dopings increased the intensity with its line radiation in important spectral ranges extremely or should improve lifetime through a regenerative tungsten cycle.

To intensify the radiance of the lamps we started investigating its pressure dependance. Lamps with cold fill pressure from 10 up to 22 bar and operating pressures from 50 to 85 bar (Palm et al. [2]) were built, but showed rapidly the limiting factor of this parameter. Lamps with fill pressures higher than 13-19 bar ignite very difficult and tend to explode during warm-up or very early in lifetime (< 50 h). So we fixed our fill pressure between 13 to 19 bar, depending on the wattage (500 to 30 W) and the size of the quartz bulb. It was not possible to measure the increase in radiance exactly, due to high variation in the intensity of new lamps (> 50%), but showed the same tendency like the results of Zander et al. [3]. Varying the fill pressure in an open lamp they found out, that the radiance is proportional to the pressure $p^{0,8}$.

Recently Hoppstock [4] reported a strongly constricted plasma with very high radiance in xenon short arc lamps with TlJ dopant. We got this constriction or "super hot spot" also without any dopant, simply by choosing the right electrode design. Increases in radiance up to 10 times the maximum radiance of a standard commercial lamp were achieved, while stability of the hot spot decreased dramatically. The spherical plasma spot on the cathode with an diameter less than 300 μm jumped spontaneously up to 1000 μm or started rotating around the cathode tip for more than 5 minutes. Variation of the lamp current in different lamp types finally showed the reason: Decreasing the lamp current we observed in every lamp the existence of a hot spot mode as soon as cathode temperature was to small to emit electrons thermically [5]. The arc builts up a huge cathode fall that bursts the electrons out of a small cathode spot by field emission. Power densities up to 1 MW/m^3 are coupled into the plasma directly in front of the cathode. The spherical plasma hot spot is created by the expanding hot gases and plasma temperatures more than 17000 K are reached. While the diffuse, thermal arc starts at an rather big cathode area that is fixed through the temperature profil of the cathode and its slow temperature behaviour. The field emission spot jumps to every small scratch on the cathode surface and changes the place again as soon it is melted or vaporized away and has therefore no constant spot stabilisation.

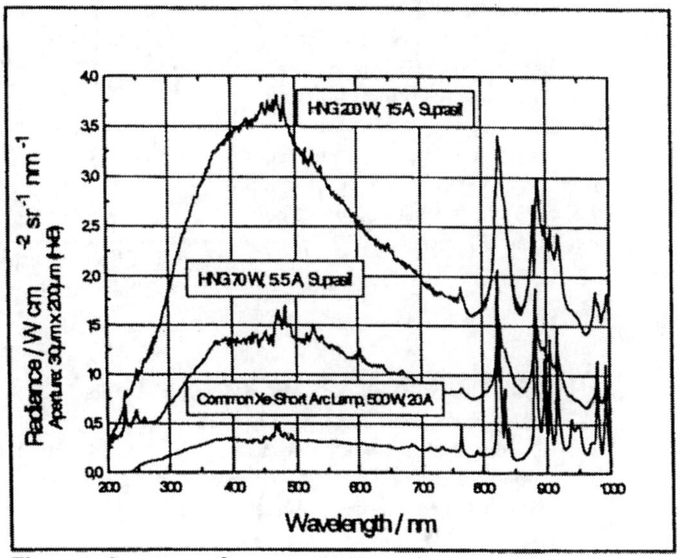

Figure 1: Spectrum of a common xenon short arc lamp compared to the developed lamps with extremly high radiance

The way to reach maximum radiance and satisfacory stability at given, fixed lamp current and design is now straight forward: The cathode tip must be formed in such a way, that the current of the lamp cannot be lessened substantially beyond the given current without a breakdown in stability and the transition into field emission. If this is accomplished the arc is in the highest current density that is stable and reaches luminaces up to 360.000 cd/cm² and spectral radiances larger than 1 W/(sr nm cm²)from 250-1000 nm. The spectrum gets a strong shift in the UV, where the continuum output now can be even higher than the strongest IR-line at 800-1000 nm (see Figure 1)!

To increase the lamps spectral radiance for example in the wavelength below 200 nm, we doped it with selenium and arsenic. Figure 2 shows the spectrum in the interesting region from 190 to 200 nm where the important lines especially for the AAS lie (As: 193,7 nm, 197,2 nm, 199,0 nm, Se: 196,03 nm). The spectral lines of As and Se appear strongly and their spectral radiances are 10^3-10^4 times a hollow cathode lamp!!

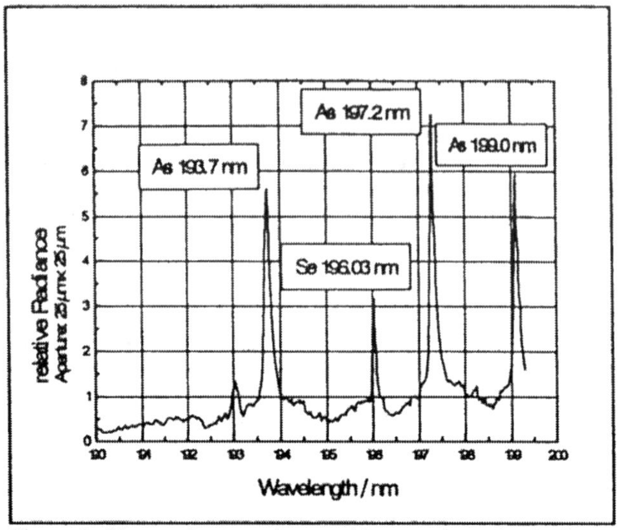

Figure 2: Xenon short arc lamp with arsenic and selenium

Figure 3 finally shows the lifetime behaviour of intensity of the described lamps measured along the whole arc between cathode and anode. The line for pure xenon and xenon/TlJ lamp is the mean value from two species while the graph for the xenon/As and xenon/Se lamp results out of three species. The individual lifetime results for all lamps were very close together. The xenon/TlJ lamp was built to introduce a regenerative tungsten cycle into the lamp to overcome a disadvantage coupled to the extremely high radiance. The current density and therefore the plasma temperature close to the cathode is extremely high and creates a very high thermal stress on the cathode material that is strongly evaporated. After 300-500 hours the cathode tip vanishes, the arc becomes unstable and the quartz bulb starts blackening. In contrast to experiments from Hoppstock [4], who reported the sucessful operation of the regenerative tungsten cycle in his xenon short arc lamps with TlJ dopant, we could not state the proper function of the tungsten cycle. On the contrary we found a worse lifetime (like in the case of the Xenon/Se lamp) that could not be explained till now.

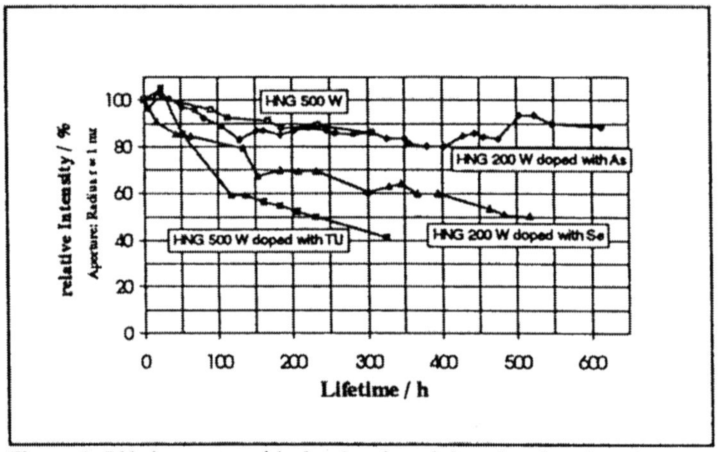

Figure 3: Lifetime tests with the developed doped and undoped xenon short arc lamps

[1] Hell, Smolka et al., J.of Microscopy, Vol. 163, Pt 2, August 1991, pp 179-187
[2] Palm et al., Ann. Phy. 18, 17, 1956
[3] Zander et al., Applied Optics, Vol. 28, No. 18, September 1989
[4] Hoppstock, Phd Thesis, University Karlsruhe, 1987
[5] Günther, Phd Thesis, University Bonn

HIGH-RADIANCE LAMPS

Tatsumi Hiramoto
Ushio Inc., 1194 Sazuchi, Bessho-cho Himeji city, Japan

1. Introduction

Discharge Lamps have recently been used as a light source in an optical instrument in the wavelength region of UV-B, -C and VUV. It is required that the half-width of a spectral line used is extremely narrow, because of difficulty in designing fully achromatic optical system in this range of wavelength. However, its intensity is required to keep a short exposure time or a large exposure area.

In this report, physical properties of such high-radiance lamps are deduced and some lamps are shown as an example.

2. Physical properties of high-radiance lamps

From the equation of radiative transfer in LTE, the spectral radiance L_λ and radiance L_T on the surface of a lamp in the normal direction are respectively expressed as Eqs. (1) and (2),

$$L_\lambda = \int_{-R}^{R} \kappa_\lambda B_\lambda exp\left(-\int_{r}^{R} \kappa_\lambda dr'\right) dr, \tag{1}$$

$$L_T = \int_0^\infty L_\lambda d\lambda, \tag{2}$$

wher λ, R, r, B_λ, κ_λ denote wavelength, the lamp radius, the distance from the lamp axis, Planck function and the absorption coefficient being expressed as,

$$\kappa(\lambda, T) = \frac{\pi e^2 f n_l \lambda_0^2}{4\pi\varepsilon_0 mc^2} P(\Delta\lambda)\{1 - exp(-hc/\lambda kT)\}, \tag{3}$$

where λ_0, f, n_l are the unshifted center wavelength of the line, the oscillator strength and the density of absorbing species respectively. The normalized line profile function is denoted as $P(\Delta\lambda)$.

The physical image of Eq. (1) can be describted as follows : radiation emitted at r reaches the lamp surface with absorption of the factor of $exp\left(-\int_r^R \kappa_\lambda dr'\right)$ on the way. Physical properties of high-radiance lamps are deduced as follows : in general, the spectral line used shoul have a large f, and n_l should be very high in the arc and steeply decreases with reduction in the temperature in the periphery. High-arc temperatures are required. Moreover, it is better that a buffer gas has a large polarizability and a low ionization energy.

In the case of a line with a limited half-width, n_l or k_{λ_0} should be kept to be an optimum value, and a steep-temperature distribution is required. For the buffer gas a small polarizability and a high ionization energy are required.

3. Examples of high-radiance lamps

(1)Lamps emitting a line with a large probability of the transition of which lower level is highly excited. E. g., high-pressure or super high-pressure mercury lamps emitting the 436, 365nm lines

(2)High-current Lamps, filled with species of high vapor pressure and low ionization energy, which emit resonance lines of ions[1]. E. g., Cd^{+} lamps operated with high currents, which emit the 214.4nm line.

(3)Lamps emitting strong atomic lines within the arc and almost all the atoms are in the form of an allotropic molecular gas in the periphery of the arc. E. g., lamps containing S emitting the 181nm line

In the above three types of lamps, self-controled distributions of the density of absorbing species are established inside the lamps.

4. Numerical example

Calculated spectral radiances of the resonance line of Cd^{+} (214.4nm) from a short-arc lamp as well as from long-arc lamps are shown in Fig. 1. Both of a parabolic distribution and a constant distribution were assumed for the temperature inside the lamps. The distribution measured by H. Burck et al.[2] was used for a short-arc lamp. This calculation has been done using the quasi-static approximation of line broadening. The ratio of L_T of the long-arc lamp with the parabolic distribution to that of the lamp with the constant temperature is 0.258, which is as large as twice compared with that of NaD lines of HPS lamps filled with Xe gas. The high ratio due to less self-absorption is shown. According to the reference 1) the resonance radiation at 228.8nm of Cd atom in the short-arc lamp is not observed except in the far wing. This is a piece of evidence of high radiance of ionic resonance lines from the lamps.

5. Conclusion

It is concluded that even if the spectral width is limited to be very narrow, the spectral line with a large oscillator strength can be of high radiance under the following condition : the lamp is operated at high electrical power densities and one chooses the absorbing species of which density varies with strongly depending upon the temperature.

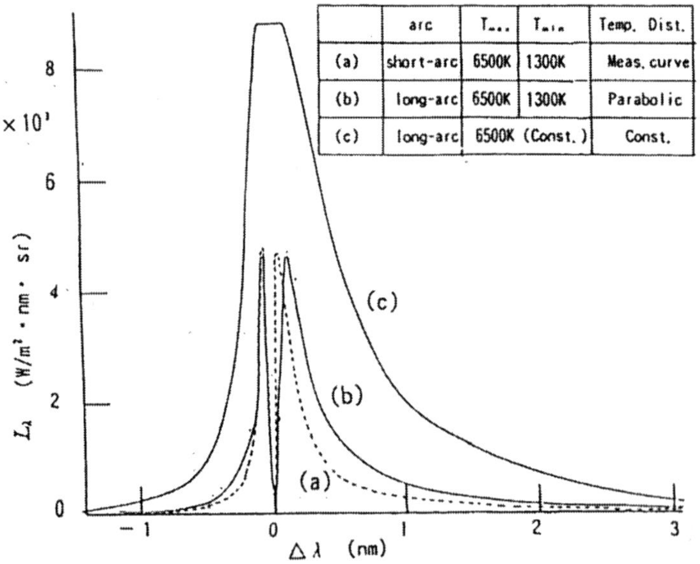

	arc	T_{max}	T_{min}	Temp. Dist.
(a)	short-arc	6500K	1300K	Meas. curve
(b)	long-arc	6500K	1300K	Parabolic
(c)	long-arc	6500K (Const.)		Const.

Fig. 1 Calculated spectral radiance of lamps filled with Cd/Kr. The lamps shown in the inserted table are shown by (a), (b) and (c).

References

1) S. Yasuda et al. ; EP 0599229 (1992)
2) H. Burck and K. Dshi-yü ; Exp. Techn. Phys., 7. 39 (1959)

DENSITY OF EXCITED ATOMS AND VUV RADIATION IN THE PULSED XENON MEDIUM PRESSURE DISCHARGE

E. Kindel , C. Schimke*
Institute of Low-Temperature Plasma Physics, Greifswald, FRG
* Ernst-Moritz-Arndt-University, Greifswald, FRG

The time resolved densities of the metastables Xe*($1s_5$) and the resonance atoms Xe*($1s_4$) in a xenon discharge tube (i. d. 6.5 mm) has been measured using the hook method in the pressure range of 1 - 40 Torr.
The spectral behaviour of the positive column was investigated with an 0.2 m VUV monochromator (ARC,VM502) and a solar-blind photomultiplier (Hamamatsu R1081).
The peak pulse current was usually between 100 mA and 300 mA. Repetition frequency and pulse duration varied between 1 and 30 kHz and 5 and 300 μs, respectively. The time resolved line intensities at λ = 828,01 nm ($2p_5 \rightarrow 1s_4$) and λ = 823,01 nm ($2p_6 \rightarrow 1s_5$) are shown in Fig. 1. It can be clearly seen, that in the afterglow the intensities at first increase and then exponentially decay . This behaviour can be explained by the dissociative recombination between molecular ions and electrons [1], according to the relation

$$Xe_2^+ + e^- \rightarrow (Xe_2^*)_{unstable} \rightarrow Xe + Xe^* + W_{kin} \rightarrow 2Xe + h\nu + W_{kin}$$

In this process an overpopulation of the $1s_4$ and $1s_5$ states is obtained (see Fig.2). These states play the role of precursors in the formation of excimer states by three-body collisions between an atom in an excited state and two atoms in the ground state [2].
The build-up of the first and second continuum (see Fig. 3) at 150 nm and 170nm respectively, results from a transition of a weakly bound molecular state ($^1\Sigma_u^+$or $^3\Sigma_u^+$) to a dissociative ground state $^1\Sigma_g^+$, as described by the reactions

$$Xe(1s_4) + 2Xe(^1S_0) \rightarrow Xe_2^*(^1\Sigma_u^+) + 2Xe(^1S_0),$$

$$Xe_2^*(^1\Sigma_u^+) \rightarrow 2Xe + h\nu(150nm),$$

$$Xe(1s_5) + 2Xe(^1S_0) \rightarrow Xe_2^*(^3\Sigma_u^+) + 2Xe(^1S_0),$$

$$Xe_2^*(^3\Sigma_u^+) \rightarrow 2Xe + h\nu(170nm).$$

In dependence on the duty cycle the process of generation of VUV radiation was investigated. In Fig.4, for example, a typical time resolved course of radiation in the wavelength range from 140 nm to 180 nm is given.

[1] Pfau, S., Rutscher, A., Beitr. Plasmaphysik 12 (1972), 53
[2] Millet, P. et al., J. Chem. Phys. 69 (1978)1, 92

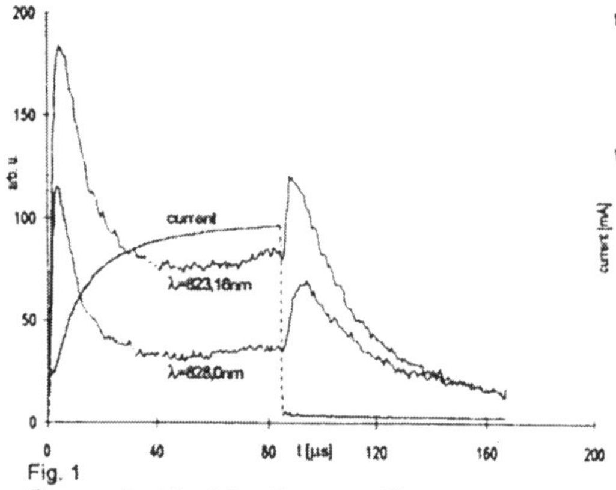

Fig. 1
Time resolved line intensities at p=40Torr, f=6kHz and at a peak current of 130mA

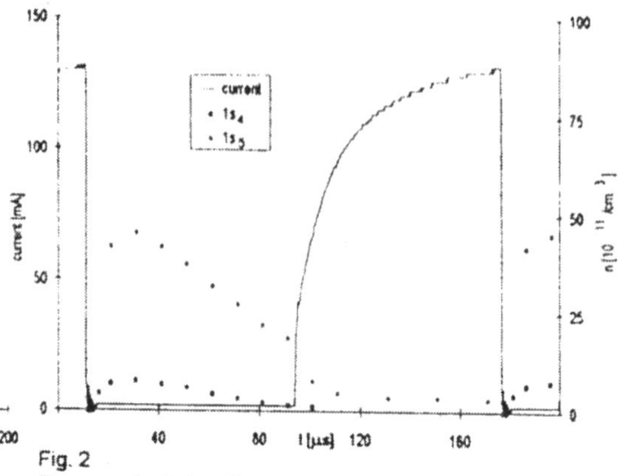

Fig. 2
Time resolved densities at p=40Torr, f=6kHz, and at a peak current of 130mA

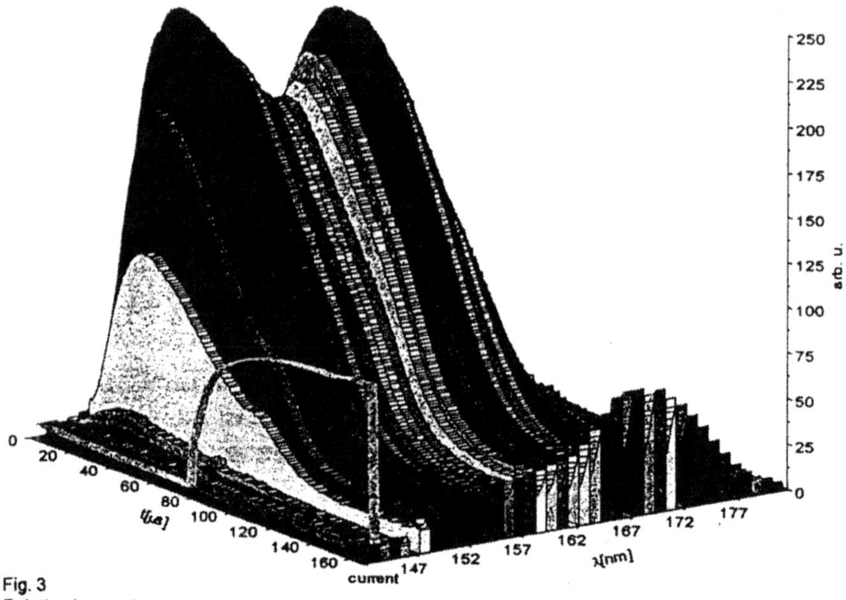

Fig. 3
Relative intensities, time and wavelength resolved at p=40Torr, f=6kHz and at a peak current of 130mA

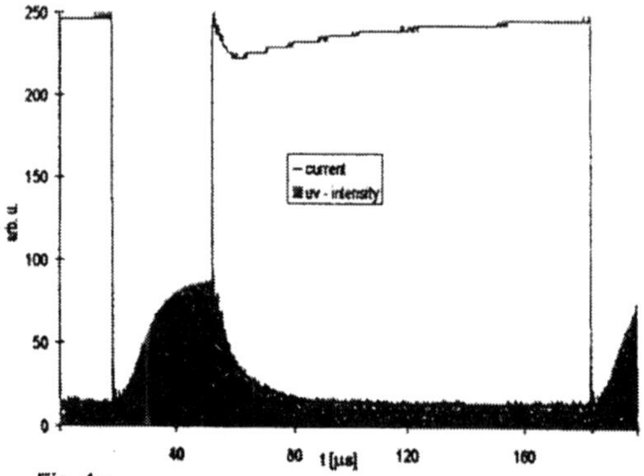

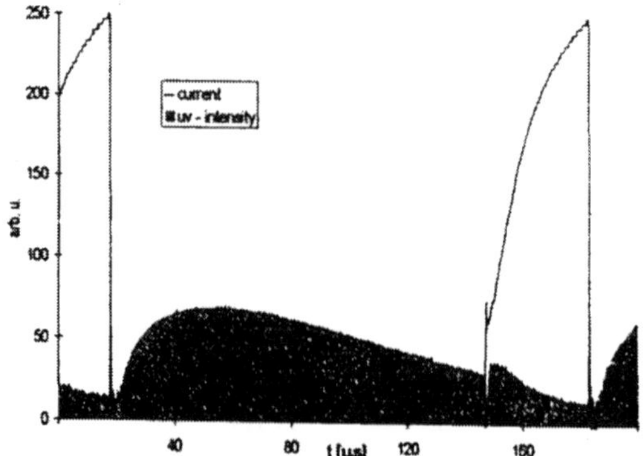

Fig. 4b

Fig. 4a
Relative continuum intensity between 140 and 180nm at p=20Torr, f=6kHz and at a peak current of 250mA
a) duty cycle 4:1
b) duty cycle 1:4

EMPIRICAL MODELLING AS A METHODOLOGY FOR LAMP AND PROCESS DESIGN

B. Preston
GE Lighting Ltd.
Melton Rd.
Leicester LE4 7PD
England

Introduction

In todays competitive environment the ability to generate profit from a product is greatly affected by its cost of manufacture, its quality and the time taken to develop it. It is generally accepted that a very high percentage of both the cost and the quality of a product are built in at the product and process design stages. These factors dictate that product and process design methodologies have great importance for manufacturers. Only in a limited number of cases in the lamp industry has it been possible to make use of predictive mechanistic models to facilitate product or process design. Within my own organisation, mechanistic models are used successfully in the development of incandescent lamps[1] and high pressure sodium lamps, using code developed by Dakin et al.[2] In the majority of cases where design cannot be implemented analytically, an empirical approach, based on statistically planned experiments is often used. Statistically planned experiments were first employed in the cotton and woollen industries in the 1930's.[3,4] The techniques appear to have been developed and applied, in the next 30 years by statisticians working in the chemical industry. During the last 10-15 years their use has grown, to become a tool of engineers in many industries. A major influence on this growth has been Professor Taguchi,[5] who has promoted the use of fractional factorial design arrays for the purpose of designing quality into products. This paper reveals nothing new, rather it is a distillation of ideas and techniques, which I have found useful while being actively engaged in the areas of product and process, design and improvement. It draws on a mixture of what might be termed classical 'design of experiment' techniques and some of the more recent ideas of Professor Taguchi.

So what precisely can 'design of experiment' methodology achieve? I have found, or observed, statistically designed experiments to be a cost effective and timely technique:

In product development to:

- identify key design parameters.
- optimise product designs.

In process development to:

- identify key process parameters.
- optimise processes.
- determine control adjustments.

In product and process improvement to:

- solve quality and reliability problems.
- improve process yields.
- characterise products and processes.

Given the usefulness of the technique, it is surprising how few lamp related studies have appeared in the literature. I am aware of only three.[6,7,8] Interestingly, two of these studies are concerned with performance optimisation in metal halide lamps as a function of dose. Metal halide lamps certainly appear to be a subject best tackled using empirical methods. Mechanistic models developed to date, have not been able to emulate existing metal halide products to the level of precision required for reliable product engineering.

In discussing the approach in more detail, I will address the following topics:

1. The choice of experimental design and how it is influenced by the required outcomes.
2. Execution of the experiment.
3. Analysis of the data.
4. Detecting outlier data points.
5. Judging the validity of the model.
6. Exploring and utilising the model.

The routine use of these techniques for the production of engineering models, as aids to the design and development process has been greatly enhanced with the development of good quality software, particularly statistical software packages, which run on personal computers. A brief review of a number of useful packages is given in Wadsworth.[9] I have found that a combination of Statgraphics[10] and Microsoft Excel[11] provide nearly all the facilities required to analyse data from experiments and explore the resulting models.

The Choice of Experimental Design

Many investigations start at what is a very explanatory stage, when not much is really known about the system under investigation. Predictor variables are identified, but it is often not known if they are significant or not. In this case my usual starting point is to employ 2 level orthogonal experimental designs. If the number of variables is only two or three, then the preferred design is a 'full factorial ' design. All possible combinations of the factor levels are tested. For 2 variables this amounts to 4 test cells (2^2), and for three variables 8 test cells (2^3). The anatomy of a 2^3 factorial design is shown in Figure 1. The first three columns Fig (1a) show all possible combinations of the three predictor variables A, B and C, the '+' signs representing the high levels of the factors and the '-' signs the low level of the factors.

Figure (1)

A	B	C	Response
-	-	-	R1
+	-	-	R2
-	+	-	R3
+	+	-	R4
-	-	+	R5
+	-	+	R6
-	+	+	R7
+	+	+	R8

Fig (1a) - Design Matrix

A	B	A*B	C	A*C	B*C	A*B*C	Response
-	-	+	-	+	+	-	R1
+	-	-	-	-	+	+	R2
-	+	-	-	+	-	+	R3
+	+	+	-	-	-	-	R4
-	-	+	+	-	-	+	R5
+	-	-	+	+	-	-	R6
-	+	-	+	-	+	-	R7
+	+	+	+	+	+	+	R8

Fig (1b) - Analysis Matrix

The balance in each column should be noted, each column contains 4 '+' signs and 4 '-' signs. This is the 'design matrix'. Fig (1b) shows the levels of the interaction terms. It is seen that these are 'generated' by multiplying the corresponding row signs in columns A, B and C. All 7 columns together with the response column, form the 'analysis matrix'. Full factorial designs at two levels will generally indicate only major trends in the data. This is a consequence of the fact that only a linear equation can be fitted to the data. This is outweighed by the economy of effort and the following valuable features:

- They allow the experimenter to look for and see patterns in the data. This is a great aid in further building a model.
- Model parameters can be calculated using simple arithmetic.
- They provide good estimates of model parameters, in fact providing the smallest possible variances that can be produced by any design covering the same space.
- They can be used to find and analyse interactions.
- They can, as we shall see later, be used as building blocks to develop more sophisticated models.

- They form the basis for two level fractional factorial designs, which provide further economy of experimentation by cutting down on the number of test cells, as the number of predictor variables increase. It is important to understand what is being traded when employing these designs.

For a full factorial, it is not only possible to estimate the effect of the predictor variables (main effects), but also the effect of every conceivable interaction between them. In a 3 factor experiment (main factor A, B and C) the highest interaction term will be a three factor interaction term, A×B×C. While in a 8 factor experiment, the highest interaction term will be an eight factor interaction. Clearly such interaction terms are unlikely to have real physical significance. If these higher order terms can be ignored, then suitable subsets of full factorial experiments can be chosen, in which it is always possible to estimate all the main effects clear of two factor interactions. Main effects are only confounded with high order interactions. Whether two factor interactions are confounded with one another, or only with higher order interactions, depends on the size of the fraction used.

If it is thought that interactions do not exist, or at the early stages of an investigation are considered to be best ignored, then advantage can be taken of the orthogonality of the design matrix to saturate every column with a main effect variable. While these designs allow the investigation of n variables for only (n+1) experimental runs, main effects are confounded with second order interactions, and such designs should be used with real caution. I will return to this point later. Details of generating fractional factorial designs can be in found Box and Draper.[12] Stratgraphics will also generate fractional designs for up to 16 variables.

These exploratory experiments then identify which variables are important (or unimportant) and often a direction for further experimentation. Even at this stage, if the results indicate say an improvement in a process, factor levels can be set to realise this improvement. Often parameters which do not figure in the model (have any effect), can at least be chosen for lowest cost.

Many physical processes, unless they involve oscillatory phenomena or changes of state, can be approximated over limited regions by second order equations. Second order equations can be fitted to experimental designs with factors at three or more levels, allowing for optimisation of response functions. I have found two classes of designs particularly useful for the optimisation phase of the experimental programme.

The first of these designs, known as composite designs, run factors at five levels. The designs consist of a factorial or fractional factorial 'hyper cube' part and an added star part.

The second, known as Box-Behnken designs, run factors at 3 levels. These two classes of design are from a group of designs known collectively as second order response surface designs.

Table 1 shows the required number of runs for each of the designs for 3 to 6 variables. The efficiency of the designs, when compared with full factorials is apparent. At this stage of the process it is important that the number of variables is limited to the minimum necessary, otherwise the complexity and cost of the experiment rapidly escalates. Again, details of these designs can be found in Box and Draper[13] and Box and Behnken.[14] Statgraphics offers Response Surface Designs for 2 to 8 variables. The choice between 5 level variables and 3 level variables for a design, often revolves around the practicality and cost of providing variables at 5 levels.

Table 1

Design Matrix Type	Number of Predictor variables			
	3	4	5	6
Central Composite	*16*	*26*	*27/44*	*46*
Box - Behnken	*15*	*27*	*46*	*54*
Full Factorial	*27*	*81*	*243*	*729*

The Number of Experimental Cells as a Function of the Number of Predictor Variables for Three 2nd Order Designs

Execution of the Experiment

In choosing factor levels, it is important that levels are chosen so that the response signal is well above 'noise' levels. These 'noise' levels are often known from engineering or manufacturing experience. This can mean being fairly bold in selecting a wide separation in levels, usually well outside of normal manufacturing tolerances. It is vital that the order of the experimental runs is randomised. Where possible, components should be withdrawn from their populations at random. These precautions help to eliminate systematic biases, which may occur in the experiment. The topic is discussed by Nelson[15] in some detail.

All the experimental protocols should be set out in writing. Everyone concerned with the experiment needs to be familiar with their role. Material required for the experiment should be prepared in advance, so that the experiment is executed as rapidly as possible.

During the experiment it is important to check that all has been done according to plan. That an eye has been kept open for the unexpected.

Analysis of the Data

For simple 2 level factorial experiments, calculation of effects is straight forward, and details can be found in Box, Hunter and Hunter.[16] To see how to calculate effects we need to refer again to Figure (1b). For any factor we just calculate the difference between the sums of the responses indicated by the + and - signs in the factor column. In the case of unreplicated factorials, Daniel[17] has provided a graphical method of analysis, easily used by engineers. Effects are plotted in rank order on a half normal grid. From the plot, judgements about the reality of observed effects can be easily made, together with an estimate of the residual standard deviation. In his book[18] *'Application of Statistics in Industrial Experimentation'*, Daniel provides useful insights into the use and analysis of factorial experiments.

With the software available for modern personal computers, the preferred method of analysis of data from designed experiments is linear multiple regression. Statgraphics, for instance, provides two multiple regression routines. MREG a routine that calculates a regression for all input factors, and STEPWISE a routine that allows one factor at a time to be inserted, or removed, from the regression.

The STEPWISE routine is the preferred approach. Run in forward mode, the routine begins with no variables in the model, adding one at a time, as long as the new variable adds significance to the model. The system also checks at each stage to see that previously selected variables are still significant. The routine eliminates variables from the model that become insignificant. Run in backward mode, the routine begins with a model that contains all the variables and eliminates them one at a time. It checks at each stage that previously removed variables remain insignificant, and re-enters variables into the model, if they become significant as other variables are removed. The routine also provides for manual insertion or removal of factors. This is useful for re-inserting into the regression, main effect factors which otherwise have only appeared in interaction terms, a recommended procedure, or inserting factors on scientific or engineering grounds. This latter step is unlikely to be important in a designed experiment. A stable system produces the same model whichever procedure is adopted. Stepwise regression and the partial 'F' tests used to determine significance is fully discussed in Draper and Smith.[19]

A vital step to undertake before a regression analysis is undertaken is to scale the variables. This is usually done by subtracting the average from each value and dividing by some scale factor. This could be the range or the standard deviation. This transformation is carried out for two reasons. Firstly the procedure destroys correlations, which will exist between main factors, squares of main factors and interactions. It also ensures that all the variables are of the same order of numerical size, so that any computer rounding errors will not effect the results. Once the significant variables have been identified running the regression with real variables, rather than scaled variables, will result in model equations being in the correct engineering units.

Statgraphics computes all regression coefficients, together with parameters indicating the statistical significance of the model coefficients. It provides an analysis of variance (ANOVA) for the model and will compute ANOVA statistics for all the regression factors. It also provides extensive graphical facilities for examining the residual errors.

Detection Of Outliers

In my own experience, and in that of others, the occurrence of maverick data points is not as uncommon as we would wish. Daniel[17] reports that bad data points appear with relative frequencies from 0.01 to 0.1, depending on the complexity of the experimental situation, and on the experience of the experimenter. In the last 15 years, some excellent techniques for the detection of outliers based on a least median of squares regression[20,21] have been developed.

A method of using these routines is to run both multiple least squares regression and multiple least median of squares regression. Generally, if the two reasonably agree, the data can be judged consistent. If the two disagree, the residual analysis of the least median of squares analysis will identify data points with large residuals. These can be removed from the data set and the analysis re-run. In the majority of cases once this has been done, the agreement between the two techniques becomes much closer. The identified data points must then be carefully investigated. Two courses of action are possible. After due consideration, the points may be included or excluded on engineering/investigative grounds. Alternatively, we could choose to build models with and without the identified points. When the empirical model is finally tested against new experimental 'confirmation trials', a reason for inclusion, or exclusion, will have been established.

The Validity of the Model

On completing the regression analysis we are in a position to write down empirical equations for each of the responses we have measured in the experiment. The question arises, can we make a judgement on the predictive power of the model? The assumptions of the techniques we have applied are that our residuals, differences between actual and fitted data points, are normally distributed with mean zero and a constant standard deviation. If our model is 'satisfactory,' our residuals should display the above properties, or something close. Residuals should be checked for normality. They should be plotted against both response and predictor variables and against the time sequence of the experimental points. Any systematic behaviour detected in the residuals is a cause for concern. All these plots can be made in Statgraphics quickly and simply. Box and Draper[22] show how systematic changes in residuals can suggest data transformations to improve the model and regularise the residuals. An excellent discussion on the examination of residuals can be found in Draper and Smith.[23]

The R^2 for the regression fit is an important parameter. $R^2 \times 100$ is the percentage of the total experimental variation explained by the model. A percent explained greater than 50% is a desirable characteristic of a useful model.

A guide to model adequacy that I have found useful has been given by Box and Wetz.[24] They suggest that the model 'F' ratio, the ratio between the regression mean square and the residual error mean square, should exceed the chosen tabulated significance level by a factor of about 10. F values at these levels produce satisfactory models.

Exploring and Utilising the Model

Once the validity of the model is established, we want to put it to use in understanding and optimising our product or process. Response functions are generally plotted against pairs of predictor variables, producing either contour or 3-dimensional surface plots. When a number of response variables have been measured, it is usually a requirement to 'optimise' the responses. We are occasionally required to maximise or minimise just one response variable. Frequently we are called upon to set a number of response variables at desirable levels of output and compromise a number of responses to obtain a 'best' overall performance. When models might be produced in a six dimensional space, say a response variable and 5 predictor variables, a simple easily available technique for producing desirable compromises is required. Microsoft Excel, versions 4.0 and 5.0, contain a solver routine which can be utilised for this purpose. This routine allows the user to set a target cell within a worksheet to a maximum, minimum or specified value. This cell will be set as one of the response variables. The target is reached by changing the values in arrange of 'changing' cells, that represent the predictor variables. The whole process is constrained by placing upper and lower bounds on the changing cells - the experimental boundaries. Up to a further 100 constraints, either constants or functions of predictor or response variables, can be included in the calculations. Repeated operation of the solver can lead to a 'feel' for the whole design space and an efficient optimisation of the model. Constraints can be used with imagination to produce a variety of useful plots.

Conclusion

I would like to close with a quote from the book by Draper and Smith.[25] "*No scientist should be persuaded to abandon his scientific insight and principles in favour of some computerised statistical screening procedure. The use of multiple regression techniques is a powerful tool only if applied with intelligence and caution*".

References

1. Howe S.H. and Mottram D. (1983). A new method for predicting the initial performance of minature incandescent lamps. J.I.E.S. Vol. 13 No 1. pp 122-127
2. Dakin J.T., et al (1989) Anatomy of a vertical metal halide discharge, J. Appl. Phys. Vol 66, No 9, pp 4074-4088
3. Tippet L.H.C. (1935). Some applications of statistical methods to the study of variation of quality in the production of cotton yarn. Supplement to the Journal of the Royal Statistical Society, II, pp 27-55
4. Daniels H.E. (1938). Some problems of statistical interest in wool research. Supplement to the Journal of the Royal Statistical Society, V, pp 89-112
5. Taguchi, G., (1986). Introduction to Quality Engineering. Asian Productivity Organisation, UNIPUB White Plains, New York.
6. Partlow W.D. & Kussmaul K.L. (1982). Experimental characterisation and statistical modeling of Na-Sc iodide lamps. J. I.E.S. Vol. 11 No. 3. pp 153-161.
7. Krasko Z.K. (1982). The "design-of-experiment" method and development of new metal halide lamps. J. I.E.S. Vol. 11 No. 3. pp 162-169.
8. Bigio L., Kubicki B. & Codella P.J. (1992). The 6th Int. Symp. on the Science and Technology of Light Sources. Budapest
9. Wadsworth H.M. (editor) (1990) Handbook of Statistical Methods for Engineers and Scientists. McGraw Hill. London. Chp. 13, pp 32-34.
10. Manugistics Inc., 2115 East Jefferson Street, Rockville, MD 20852, USA.
11. Microsoft Corporation.
12. Box G.E.P., and Draper N.R. (1987) Empirical Model-Building and Response Surfaces, Wiley. New York. Chp. 5, pp143-181
13. Box G.E.P., and Draper N.R. (1987) Empirical Model-Building and Response Surfaces, Wiley. New York Chp. 15, pp502-527
14. Box G.E.P., and Behnken D.W., (1960). Some new three level designs for the study of quantitative variables. Technometrics, Vol. 2, pp 455-457.
15. Nelson P.R. (1990) Design and Analysis of Experiments, pp 14.1-14.2 in Handbook of Statistical Methods for Engineers and Scientists. ed. Wadsworth H.M., McGraw Hill. London.
16. Box G.E.P., Hunter W.G., and Hunter S.J., (1978) Statistics for Experimenters. Wiley. New York. Chp. 10,. pp 306-342.
17. Daniel C. (1959). Use of half-normal plots in interpreting factorial two-level experiments. Technometrics Vol. 1, No 4., pp 311-341.
18. Daniel C. (1976). Application of Statistics to Industrial Experimentation. Wiley New York.
19. Draper N.R. and Smith H. (1981) Applied Regression Analysis, Second Edition. Wiley. New York. Chp. 6, pp 307-312

20.Rousseeuw P.J. (1984) Least Median of Squares Regression. J. of the American Statistical Association. Vol. 79 ,No 388, pp 871-881

21.Rousseeuw P.J. and Leyroy A.M. (1987). Robust Regression and Outlier Detection. Wiley. New York.

22.Box G.E.P., and Draper N.R. (1987) Empirical Model-Building and Response Surfaces, Wiley. New York Chp. 8, pp, 268-301

23.Draper N.R. and Smith H. (1981) Applied Regression Analysis, Second Edition. Wiley. New York. Chp. 3, pp 141-183.

24.Box G.E.P. and Wetz J. (1973). Criteria for judging adequacy of estimation by an approximating response function. University of Wisconsin Statistics Department. Technical Report No. 9.

25.Draper N.R. and Smith H. (1981) Applied Regression Analysis, Second Edition. Wiley. New York. Chp. 8, p 422.

REVIEW OF DISCHARGE MODELLING

Graeme G. Lister

OSRAM SYLVANIA INC., Danvers, MA 01923, USA

1. INTRODUCTION

Mathematical modelling of gas discharges has a long and distinguished history, dating from the end of last century. Many experimental observations could be explained using purely theoretical models, but it also became clear at quite an early stage that these models were inadequate to describe a number of more complex phenomena. As early as the 1940s, prior to the invention of the digital computer, simple numerical models were developed which added insight and understanding to some fundamental processes in gas discharges (e.g. Morton 1946). Advances in computer technology in recent years have been such that quite sophisticated numerical models can now be run on a desktop PC, while a number of problems are still sufficiently complex to require many hours of time on super-computers.

The principal successes of numerical modelling applied to industrial applications of gas discharges have occurred when multi-disciplinary groups, including physicists, chemists and engineers have combined theoretical and experimental programmes to address specific issues relevant to optimisation or design of these systems. This has been true in applications as diverse as lighting, semi-conductor processing and plasma deposition.

Many advances made in discharge research through modelling have been greatly assisted by the parallel development in plasma diagnostic techniques. The application of laser spectroscopy to gas discharges has made a major impact on the understanding of the cathode region of low pressure discharges (Lawler 1995), and advanced Langmuir probe techniques play an important role in understanding of dc and rf discharges (Godyak *et al* 1992).

A complete model of a gas discharge would include the following components:

a) *an electrical (or plasma) model* describes the behavior of the charged particles in the discharge. Since their motion forms part of an electric circuit, the natural boundary conditions are the electrodes and the components of the real external circuit.

b) *a chemical and thermodynamic model* describes the interaction between the different atomic and molecular species, together with the effects of gas flow and heat transfer.

c) *a surface model* describes the interaction of the various species in the discharge with the walls, electrodes and, where appropriate, surfaces to be processed.

This description is clearly an over simplification, since each of these components is dependent to some extent on the other two. To date, no numerical model has been able to incorporate all of these components; most models concentrate on one or at most two of them. The emphasis of the present paper is on the modelling of the plasma, with aspects of the chemistry and thermodynamics introduced where appropriate, to illustrate their influence on the plasma.

The task of reviewing the full range of discharge models for different applications would be impossible in the space available here. This paper will concentrate, therefore, on a number of particular examples which will illustrate the ways in which plasma modelling can be used in an industrial environment.

2. MODELLING OF PLASMAS IN GAS DISCHARGES

Plasmas used in industry are generally classified into three groups:

- *low pressure discharges* operate at gas pressures up to a few tens of torr and the mean electron energy is considerably higher than that of the ions
- *thermal plasmas* are high pressure discharges (1 to several tens of atmospheres) in which the ions and electrons are in local thermal equilibrium (i.e. the ion and electron energies are comparable)
- *non-thermal plasmas* are high pressure discharges which have a highly energetic electron component

Examples of these plasmas for the production of light are the standard fluorescent lamp (low pressure), HID lamps (thermal) and barrier discharges for UV production (non-thermal).

A number of numerical techniques are available for modelling plasma systems (Lister 1992), and the main characteristics are discussed in Section 3 below. The choice of technique for solving a particular problem will depend on the physical properties of the process to be analysed, such as: whether the process can be adequately described by a steady state representation or whether a time dependent solution is required; the number of physical dimensions required for the results to be meaningful; the frequency of the important particle interactions in the plasma; and the degree to which certain aspects of the problem can be represented by an analytic solution. The model should neither be so simple that we have "thrown the baby out with the bath water", nor so complex that the user loses sight of the objectives of the exercise in a welter of numerical "noise".

3. TECHNIQUES FOR PLASMA MODELLING

Particle Methods

In principle, the most accurate way of treating the plasma is to follow a representative sample of particles in their self-consistent electromagnetic fields (particle-in-cell (PIC) methods) and apply statistical techniques to describe the collision processes with other particles in the plasma (Monte Carlo collision (MCC) methods). These methods require the tracking of large numbers of particles and often require extensive computer time. PIC/MCC for gas discharges have been reviewed by Birdsall (1991).

PICC/MCC methods are in principle more flexible than other methods, since three dimensional particle motion and angular scattering due to elastic and inelastic collisions are easily included. They have been most succcessful when applied to regions where the only a few particle collisions occur over the dimension of interest, and where electromagnetic fields are varying rapidly, such as the cathode fall region of a low pressure discharge. However, the availability of data on collisional processes is a practical limitation, and an example of this is illustrated in Section 4.

The Boltzmann Equation

Particles may be classified into "boxes" of six dimensional "velocity space" (three coordinates in space and three in velocity). The size of boxes and the population in each defines the "distribution function" , whose evolution in time is described by the Boltzmann equation. Distribution functions may be defined for each species in the discharge and may, in many cases, be measured directly by experiment to test the accuracy of the models. Models which solve the Boltzmann equation have been particularly useful in describing the electron energy distribution function in both low pressure and non-thermal plasmas.

Fluid Models

The complete solution to the Boltzmann equation requires a seven dimensional representation of the plasma - time plus three dimensions in each of space and velocity. We can reduce the number of dimensions by averaging over velocity to obtain "fluid like" quantities such as particle density, particle flux (current density) and average energy. The resulting equations must often be solved in parallel with equations for the electromagnetic fields. One of the advantages of fluid models is that they use averaged coefficients for atomic and molecular processes, for which data is more readily available and reliable than individual particle cross sections. They provide a powerful tool for understanding many plasma processes.

There are a number of numerical techniques for solving fluid equations and the technique adopted depends on the relative importance of the physical processes represented by the terms in the equations - i.e. the distance and time scales over which the physical parameters vary.

Global and Analogue Circuit Models

Highly sophisticated models are not always necessary to obtain valuable information about plasma systems. Global ("zero dimensional") models use assumptions about the plasma properties to compute empirical values of the plasma parameters. Although information about spatial variation is lost, these codes can be extremely useful in designing the electric circuit or deciding on power requirements.

The design of efficient electrical circuits is one of the most important aspects of any gas discharge system which in industry, and an understanding of the way the plasma impedance responds as a component of the circuit can be a useful tool in reducing the costs and improving performance. Relatively simple models have been used to successfully describe the current-voltage characteristics of low pressure dc (Gottscho *et al* 1988) and rf (Bletzinger 1990) diodes and inductively coupled discharges (Wharmby 1994). Equivalent circuit models have also been used to describe the leader current in corona discharges in air (Fofana and Béroual 1995) for lightning applications.

'Hybrid' Models

In practice, models which describe plasma systems often use a combination of two or more of the above models. Particle methods or Boltzmann equation solutions are useful for modelling electron behavior in high electric fields, while fluid models are generally adequate for ion and neutral particle behavior. Examples of the use of hybrid codes will be discussed in Sections 4 and 5.

Time Dependence

Time dependence adds an extra complication to plasma modelling, which usually increases the computer resources required to obtain useful solutions. There are a number of numerical methods which can be applied to these models, which are the subject of a number of text books and will not be discussed further here.

In selecting a numerical method, it is important to analyse some of the basic physical processes in order to optimise the method for the problem at hand. Time dependent solutions require integration of a set of plasma equations in discrete "time steps", and the size of the time steps plays a major role in determining the amount of computer tine required. It is important to distinguish between the concepts of *stability* and *accuracy*. A *stable* method is one which is not subject to *numerical* instabilities - that is, the computed physical quantities do not grow exponentially as a result of discretising the problem in time and space. The *accuracy* of the solution depends typically on the fastest physical process represented by the system of equations to be solved. Even if a solution is stable, it may not be accurate, because the time steps are too large. Physical insight can

The first example comes from fluid models of fusion plasmas, which were initially developed in the late 1950s. The first of these described plasma "pinches" (Hain *et al* 1960), in which either an external field of an axial current was used to contract the plasma, with the aim of raising the ion temperature to promote fusion. These models described some aspects of the pinch extremely well. However, attempts to model fast pinch experiments were hindered by the large contrast in densities, with a near vacuum region outside the central core of plasma (Lister 1975). The fastest time scale in the system is that of Alfvén waves, whose velocity is infinite in vacuum. Although these waves are of little interest to the main physical processes described in the code, it was necessary to use an extremely small time step to prevent numerical errors in describing these waves in the near vacuum region from influencing the results for the main plasma, and producing large amplitudes numerical noise. A small artificial "viscosity" was therefore introduced to damp the unwanted oscillations, while not adversely affecting predictions of the model in the central plasma. Later models included more complex physics, such as the presence of neutral gas outside the pinch, which reduced the problem.

The second example is the solution of the time dependent Boltzmann equation for electrons in low pressure discharges. The time scale limiting the standard method of solution is that of the velocity of the fastest electrons in the high energy tail of the distribution. However, the fastest *physical* time scale of interest is the shortest mean collision time of the electrons, which has a minimum as a function of energy, and is much less than the maximum energies of interest. Sommerer *et al* (1989) exploited this property in developing a convective scheme for solving the Boltzmann equation, with considerable saving of computer time.

4. THE DC GLOW DISCHARGE

The physics of the low pressure dc glow discharge provides an illustration of the application of each of the methods described above. Electrons emerging from the cathode are accelerated through a positive space charge region of high electric field, the cathode fall, into the relatively weak electric field of the negative glow. There is an over production of ions in the negative glow, which is compensated by a dark region of low ionisation in the Faraday dark space, followed by a region of constant axial electric field, the positive column, which is separated from the anode by the anode dark space and sometimes a glow at the anode.

The positive column

In the positive column, the collisional ionisation of gas atoms and molecules is balanced by the loss of charged particles to the walls by diffusion in the radial electric field produced by the space charge at the walls. Most models of the positive column assume that plasma parameters vary only in the radial direction, and can therefore be described in 1D. Collisional-radiative models have been successful in characterising a number of parameters in glow discharges, supported by laser spectroscopy and Langmuir probe measurements. Such models are the subject of another paper in this symposium (Fujimoto 1995).

An important factor in modelling the positive column is the electron energy distribution function (eedf). The high energy tail of the distribution is depleted by inelastic collisions, and this effect is countered by electron-electron collisions. In many discharges, the electron density is insufficient ($<10^{12}$ cm^{-3}) for electron-electron collisions to play a dominant role, and the distribution diverges from Maxwellian. A number of different approximations have been used to avoid the solving the complete Boltzmann equation. including a "two temperature" approach (Vriens *et al* 1973), in which the electron temperature in the high energy tail is different than in the main part of the distribution, and an analytical approximation (Maya and Lagushenko 1990). Yousfi *et al* (1990) compared results of solving the Boltzmann equation with these two methods for mercury-

and an analytical approximation (Maya and Lagushenko 1990). Yousfi *et al* (1990) compared results of solving the Boltzmann equation with these two methods for mercury-rare gas discharges and found the analytic solution gives results which are quite close to the detailed solution.

The DC Diode

In low pressure discharges of interest in plasma processing, the distance between the electrodes is often much shorter than the electrode width; they usually operate in the *abnormal glow*, in which the discharge covers the cathode, and a one dimensional model is appropriate. These discharges generally have little or no positive column, and models have concentrated on the cathode fall and negative glow.

Early mathematical models of the cathode fall (cf. Townsend 1916, von Engel and Steenbeck 1934), assumed that all electron transport properties depend only on E/p (the local field approximation), where p is the gas pressure and E is the local value of the electric field. In particular, the current density of electrons in the cathode fall J_e was assumed to vary axially according to

$$dJ_e/dx = \alpha J_e$$

where x is the axial distance from the cathode and α is Townsend's first constant. Predictions of these models were, however, contradicted in experimental measurements by Druyvesteyn (1938). A numerical model developed by Morton (1946), was able to explain this discrepancy by solving a simplified version of the Boltzmann equation. In Figure 1, the experimental measurements of the ratio of the discharge current to the primary current are compared with numerical results using the Morton calculations and those obtained from the local field approximation. Morton's calculations agree remarkably well with experiment for gas pressures below about 0.2 torr, but break down at higher pressures due to the simplifications in the collision model adopted. This result was possibly the first to illustrate the importance of "fast" electrons which experience few collisions in the cathode fall after leaving the cathode.

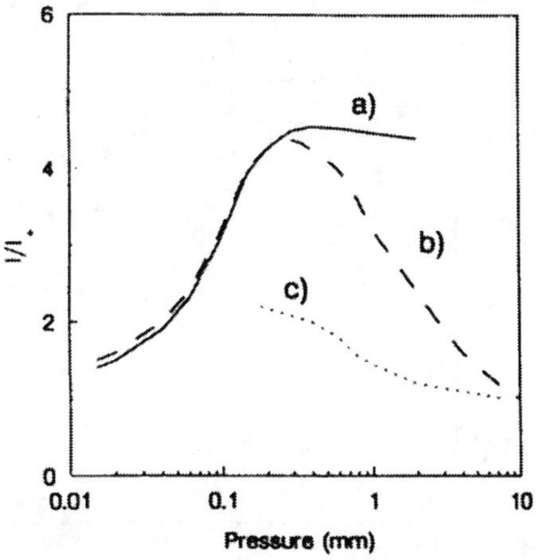

Figure 1. Ratio of discharge current to primary current as a function of pressure ($V_c = 100V$) a) Boltzmann equation (Morton (1946) b) experiment c) local field approximation

Since the early 1960s, a great deal of numerical modelling has concentrated on understanding of the cathode fall region, particularly with the aid of PIC/MCC methods. In many gases, such as helium, fast electrons are preferentially scattered forward (Labahn and Calloway 1964) and hence 1D models have been able to reproduce essential features of many experimental results. However, detailed comparison often depends on the scattering model adopted (Surendra *et al* 1990).

Electrons traversing the cathode fall may be broadly classified into fast or "beam" electrons, which reach high energies, and "slow" electrons which lose energy through a number of inelastic collisions with gas atoms and molecules, or are created during ionisation processes. Particularly in or near the negative glow, the density of slow electrons may be much higher than that of the beam, and models which follow individual particles become extremely computer intensive. A number of "hybrid" codes have therefore been developed, (two electron group model, Boeuf 1990), including a separate model for the slow electrons, which are assumed to be in equilibrium with the local electric field. Gottscho *et al* (1988) compared results obtained from local field and two electron group models, and found that, although the local field approximation cannot reproduce the spatial variation in the discharge, a model using this approximation can reproduce the V-I characteristics fairly well.

One of the fundamental processes yet to be fully understood is the behaviour of ions between the cathode fall and negative glow. Den Hartog *et al* (1988) used a combination of modelling and experiment to address this problem in helium discharges. The electric field and the gas density were measured in the cathode fall using laser optogalvanic spectroscopy (LOG), and measured values were used to compute the electron motion in the cathode using Monte Carlo methods. Laser induced fluorescence (LIF) and absorption spectroscopy were used to determine metastable atom densities. An empirical value for the ratio $J_e(0)/J_i(0)$ of ion to electron current at the cathode was obtained from a separate ion model.

The results of this comparison are illustrated in Figure 2, for a number of values of the discharge current J_D. The result demonstrates the importance of the scattering model in MCC calculations. Since fast electrons are mainly scattered in the forward direction, a model which assumes isotropic scattering clearly distorts the results and may in fact be inferior to a 1D model in which electrons are assumed to be undeviated by collisions. The authors conclude from the model that the ion current at the boundary of the cathode fall and negative glow is negligible in comparison to the discharge current.

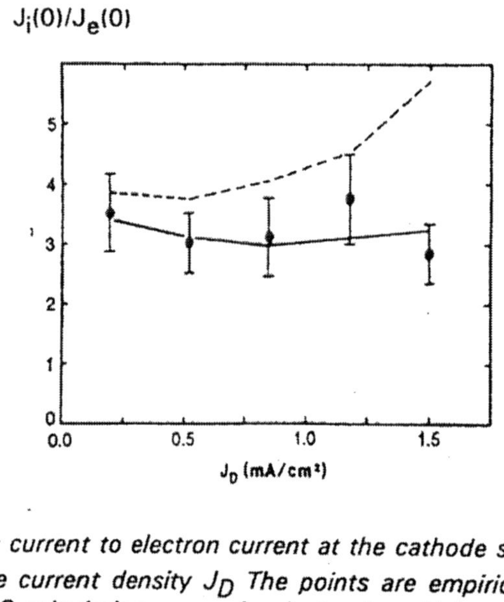

Figure 2. Ratio of ion current to electron current at the cathode surface $J_i(0)/J_e(0)$as a function of discharge current density J_D The points are empirically derived values, the dotted lines from MCC calculations assuming isotropic collisions and the solid lines from MCC calculations assuming anisotropic collisions. (From den Hartog et al (1988), Figure 12)

4. HIGH FREQUENCY DISCHARGES

Discharges maintained by the application of high-frequency electromagnetic fields are widely used in a number of industrial applications of plasmas, including surface processing and lighting. They have been applied to both high and low pressure discharges and generally allow higher degrees of ionisation and dissociation and achieve better control over ion energies than dc discharges. Further, in most cases they can be operated without electrodes, which is an advantage for many applications.

Examples of high frequency discharge are the rf diode or capacitively coupled discharge (CCD) and the inductively coupled discharge (ICD). In the CCD, an alternating potential is applied between the electrodes of a glow discharge. In the ICD, an alternating magnetic field from a coil produces a ring like discharge and the volume filled by the plasma can be compared to a single short-circuited secondary turn of a transformer. The emphasis on the present paper is on ICD. For a review of CCD modelling, the reader is referred to Surendra (1995), who has compared test results from a number of codes and discussed the domain of applicability of each.

Inductively Coupled Discharge (ICD)

ICD operate in several different geometries, characterised by the position of the rf coil relative to the plasma - the coil may suround the discharge, be contained inside the discharge or form a flat disk beneath or above the discharge. The electromagnetic fields consist of two components, corresponding to the fields induced by the oscillating current and the static electric fields along the coil which capacitively couple to the plasma. In norrmal operation, the inductive fields ('H' mode') provide the main source of electron density in the discharge During H mode operation, there remains a substantial capacitive component which produces a space charge sheath near the high voltage end of the coil, resulting in the presence of high energy electrons and ions.

The first model of an ICD was published by Thomson (1927), who computed the minimum magnetic field at the coil B_0, required to sustain a ring discharge. The relevant condition is an analogue to the Paschen Law for glow discharges, and shows that for low pressures, $B_0 \propto 1/p\omega$, where p is the gas pressure and ω is the applied frequency, while at high pressure $B_0 \propto p\omega$. This result illustrates the difference in starting procedures for low pressure discharges and induction arcs.

In the H mode, ICD has many similarities to the positive column of a glow discharge, and fluid models have been successful in reproducing many of the plasma parameters. These codes consist of an *electromagnetic* model, and a *plasma* model. The two models are solved independently - the electromagnetic model assumes an approximate value for the electron density (i.e. the electrical conductivity) to compute the electromagnetic fields and local power densities in the plasma. The plasma model uses this information to re-calculate the actual electron density and other plasma parameters, and the process is repeated until a convergent solution is obtained. Problems associated with fast particle production due to the capacitive coupling cannot be treated adequately using these models, which require similar techniques to those for the cathode fall region of the negative glow of a dc discharge.

If the coil is assumed to be sufficiently long relative to the radial dimensions of the discharge for end effects to be unimportant, the model can be reduced to one dimension and the physical values depend only on the radial position *r*, as in models of the positive column. The coil in this case is represented by a sheath current. Useful information regarding the profiles of plasma parameters can be obtained from these models (Lister and Cox 1992), as is illustrated in Figure 3, for a discharge with external coil. The electron density is peaked in the centre, but the current and power density form a "ring" in the discharge. 1D models have been particularly successful in predicting the circuit characteristics of ICD (Wharmby 1994).

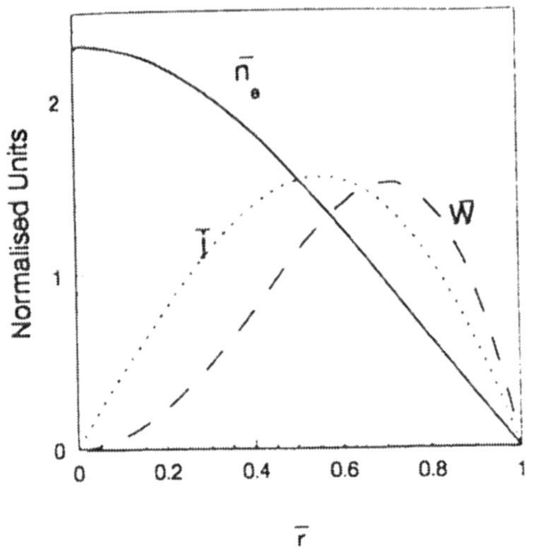

Figure 3. Radial variation of electron density n_e, current density j and power density W predicted by a 1D model of a typical cylindriucal ICD discharge with external, coil

However, 1D codes cannot model the closed magnetic field configuration in ICD and, for a more complete picture of the electromagnetic fields 2D models, are required. The 2D fields in a vacuum can be solved analytically (short solenoid approximation) and this provides an alternate limit to the 1D model. As the electron density in the ICD increases, the discharge tends to be shielded from the coil due to the "skin effect". If the plasma density were constant, corresponding to a constant electrical conductivity σ, the skin depth δ for exponential decay of the fields from the coil is

$$\delta = (2/\mu_0 \sigma \omega)^{1/2}$$

In a discharge of radius *a*, if a/δ is small, the vacuum field solution will be a good approximation.

There is an extensive literature of modelling thermal ICD plasmas (the "induction arc"). Mostaghimi and Boulos (1989) developed a 2D model and concluded that the limit for the vacuum field approximation was $a/\delta \approx 1.5$. In 1D models, only the magnetic field parallel to the code is included. The 2D code showed that even for configurations with very small radial magnetic field components, the axial field is modified substantially and the power deposition profile (cf. Figure 3) is displaced outwards. Induction arcs typically run at 7000-8000K, and 2D flow is also important, particularly in discharges where the skin depth is small. In this case, the stability of the plasma is strongly affected by the physics of the near coil region, and models have successfully predicted the effects of gas rotation perpendicular to the coils axis (or "swirl") as a cooling mechanism in high pressure ICD (Benoy 1993).

The Microwave Plasma "Ball" Reactor

Another high frequency discharge is the microwave resonant cavity, in which the wave length *c/ω is* small compared to the discharge dimensions. In the microwave plasma "ball" reactor, used for depositing high quality industrial diamond, hydrogen gas at a few tens of torr pressure and containing a small amount of hydrocarbon such as methane is fed into a variable length chamber. Microwave power is introduced from an antenna through a wave guide to the cavity. These reactors are limited by stability problems - as the power is increased, the discharge sometimes moves from an area just above the substrate to another part of the vessel, reducing the efficiency of diamond growth and even damaging the vessel walls and window.

This instability can be explained by the skin effect and its onset has been quantitatively predicted using a 2D electromagnetic-plasma model (Coe *et al* 1994) of the form described in the section on ICD modelling . A typical result showing the 2D electron density contours in an experimental microwave cavity is illustrated in Figure 4 for two different values of the microwave power. The shift, or "mode hopping" of the discharge at the higher power is clearly visible, and occurs at a value close to that observed experimentally.

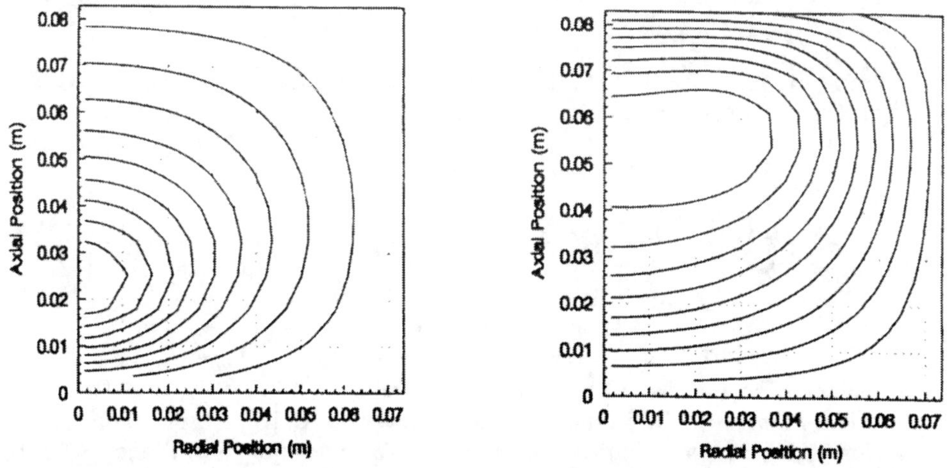

Figure 4. 2D electron density contours in a mw plasma "ball" reactor for different total mw power in a 30 torr hydrogen discharge. a) 0.5 kW b) 1 kW

5. BARRIER DISCHARGES

Non-thermal discharges may be achieved at high pressures (>100 torr) through the production of microdischarges or "streamers", which are plasma filaments produced by highly localised space charge waves. The short lifetime of the streamers (tens of nanoseconds) means that ion motion is undisturbed, but the source of high energy electrons can initiate chemical reactions efficiently. Discharges between metal electrodes (for example a wire and a cylinder) are called "corona"; the introduction of a dielectric barrier between one or both electrodes and the discharge is known as a "barrier discharge". The function of the barrier is to spread the microdischarges over the volume of the discharge and to limit the current in each streamer. Barrier discharges can be efficient producers of UV radiation (Kogelschatz 1993).

The local field approximation is valid for electrons in streamers and hence their time development can be followed using fluid models, including the solution of Poisson's equation (Pietsch *et al* 1993). To accurately model the chemical processes in the discharge, it is essential to have a good approximation to the electron energy distribution function (eedf), which is far from Maxwellian. Pietsch *et al* used Monte Carlo simulations to compute the eedf - since the electric fields are computed in the fluid simulation, there is no need for a full PIC calculation, and the number of particles followed was around 15000. These calculations showed that the eedf relaxes to equilibrium in a few *picoseconds*, and hence the local field approximation is easily justified.

The front of a streamer is similar to a shock in fluid dynamics, and particular care must be taken to ensure that the physical diffusion is not obscured by numerical diffusion. To overcome this difficulty, Morrow and Cram (1985) adapted the flux corrected transport (FCT) method to modelling corona streamers. 2D models using this method, such as that of Pietsch *et al* (1993), have been successful in describing all stages of the barrier discharge, from the dark discharge, through the avalanche, onset of the streamer front, cathode fall development and final decay.

CONCLUSIONS

Numerical modelling of gas discharges has made important advances in increasing our understanding of physical processes. Plasmas are complex systems and the task of modelling all the possible processes, even with the ever improving state of computer technology, is an impossible one. Therefore, it is necessary to identify the principal factors affecting each particular plasma process and to develop appropriate models. This paper has endeavoured to show ways in which this has been achieved in a number of areas of gas discharge research.

ACKNOWLEDGEMENTS

I am grateful to Professor J E Lawler of the University of Wisconsin, USA and Dr S E Coe of Central Research laboratories, London, UK for permission to use Figures 2 and 4 respectively.

REFERENCES

Benoy D A,1993,*Modelling of Thermal Argon Plasmas,* PhD Thesis, Eindhoven Technical University

Birdsall C K,1991,*IEEE Trans. Plasma Sci.* 19 65

Bletzinger P,1990, *J. Appl.Phys.* **67** 130

Boeuf J-P,1990, in *Physics and Applications of Pseudosparks,* (eds M A Gunderson and G Schaefer,Plenum, New York) p255

Coe S E, Bailey D S, Savage J A and Rodway D C,1994,*Proc. SPIE Conf. on Window and Dome Technologies and Materials IV, 26-28 July 1994* San Diego, SPIE Vol. 2286, p186.

Den Hartog E A, Doughty D K and Lawler J E,1988,*Phys. Rev. A* 38 2471

Druyvesteyn M,1938,*Physica*,1938 5 875

Fofana I and Bécoual A,1995,*J.Phys.D:Appl.Phys.* **28** 305

Fujimoto T,1995, pesented at this symposium

Godyak V A, Piejak R B, Alexandrovich B M,1993,*J.Appl.Phys.* 73 3657

Gottscho R A, Mitchell A, Scheller G R, Schryer N L, Graves D B and Boeuf J-P,1988, *Proc. 7th Symp. on Plasma Processes,* (eds. G S Mathad, G G Schwartz and D W Hess; Electrochem. Soc., Pennington, NJ) Vol 11 88-22 p1

Hain K, Hain G, Roberts K V, Roberts S J and Köppendörfer W,1960,*Zeit. für Naturforschung,* **15a**, 1040

Kogelschatz U,1993,in *Non-Thermal Techniques for Pollustion Controlo, Part B* (eds. B.M. Penetrante and S.E.Schultheis; Springer, Berlin) p339

Labahn R W and Callaway J, 1964,*Phys. Rev.* **135A** 1539

Lawler,1995,presented at this symposium

Lister G G,1975, in *Third Eur. Conf. High Beta Plasmas* (ed. D.E. Evans;Pergamon) p341

Lister G G, 1992,*J. Phys. D:Appl. Phys.* **25** 1649

Lister G G and Cox M, 1992,*Plasma Sources Sci. Technol.* **1** 67

Morrow R and Cram L E,1985,*J.Comp.Phys.* 57 129

Morton P L,1946,*Phys.Rev.* **70** 358.

Mostaghimi J and Boulos M I,1989,*Plasma Chem. Plasma Proc.* 9 25

Pietsch G J, Braun D and Gibalov V I,1993,in *Non-Thermal Techniques for Pollustion Controlo, Part A* (eds. B.M. Penetrante and S.E.Schultheis; Springer, Berlin) p274

Sommerer T J, Hitchon W N G and Lawler J E,1989,*Phys. Rev.* A **39**, 6356

Surendra M, Graves D B and Jellum G M,1990,*Phys. Rev A* **41** 1112 (1990)

Surendra M,1995,*Plasma Sources Sci. Technol.* 4 56

Thompson J J and Thompson G P,1927,*Conduction of Electricity Through Gases* Vol 2 (Cambridge) p431

Townsend J S,1915, *Electricity in Gases,* (University Press, Oxford)

Vriens L,1973,*J.Appl.Phys.* **44** 3950

von Engel A and Steenbeck M, 1934,*Elektrische Gasentladungen* Vol. 2 (Springer, Berlin)

Wharmby D O,1994, *48th Gaseous Electroniucs Conference*, Gaithersburg

HID-MODELING AND EXPERIMENTAL ASSESSMENT

Helmar G. Adler, Richard W. Liebermann, and Richard Speer
OSRAM SYLVANIA INC., 100 Endicott Str., Danvers, MA 01923, USA

The development of HID-models for accurate prediction of discharge behavior and spectral power distribution (SPD) is one of the major goals in this field of research. Whereas in the past most models were designed to compute either the energy transport or the spectral output of the lamps we have combined both approaches in simulation models.

Lowke's 1-D and 2-D models for vertically operating lamps have been adapted to an algorithm that iteratively determines self-consistent radiation fields.[1,2,3] All b-b, f-f, and f-b radiative mechanisms in the 50 - 2000 nm spectral range for each atomic and ionic gas species in mercury and/or MH arc lamps have been taken into account. The present algorithm is able to handle more than 1000 grey spectral lines and up to 25000 frequency points within a computer time of several minutes. In this contribution spectral, thermal and electrical measurements of high pressure mercury plasmas in vertically operated arc discharge lamps are reported and compared to both 1-D and 2-D simulation models. The lamp models will be described and areas that need future work will be outlined.

In support of previous work,[4] the present results indicate that precise descriptions of the plasma's spectral absorptivity are required in order to obtain the correct arc energy balance and predict the correct color and luminosity of the light output. More especially, we find that absorption line shapes of the mercury ground state transitions play an important role. Side-on SPD-calculations show that conventional convoluted dispersion/quasistatic line shapes are adequate,[5] provided the proper broadening parameters are known. Radiation transfer formulae as developed by Yos for infinite-length, axially-symmetric positive arc columns, seem to be sufficient for determining net emission coefficients as a function of temperature.[6] However, a more sophisticated 3-D radiation model will be required to determine the proper lamp color and efficacy.

To be able to make quantitative comparisons between measured and simulated spectra, detection systems have to be intensity and wavelength calibrated. A scanning 1m-spectrometer (spectral resolution FWHM=012nm) with pmt-detection and digitizing equipment has been used for data aquisition. Instrument profiles at different wavelengths have been measured to be able to convolute these with the calculated spectra from the simulation models. The spectral radiances were absolutely calibrated by means of a tungsten ribbon lamp. Low pressure discharges with narrow lines at well-known wavelengths have been used for wavelength calibrations.

Inverted radial intensity distributions of optically thin lines were used for determination of radial temperature profiles. In addition, intensity and line shape features of several optically thick lines have been analyzed for determining core temperatures and radial temperature profiles close to the discharge axis with the advantage that no Abel-inversion was necessary. Results of several methods.[7,8] have been compared. The half-widths of particular lines (Hg491.6nm and Hg577nm) were used to provide Hg-pressure measurements.[9] Wall temperature measurements were taken with an infrared camera system.

As a starting point for experiment-model comparisons we chose a 175W Hg-discharge in order to keep the spectral features as "simple" as possible. Besides the above mentioned parameters (core temperature, radial temperature profile, pressure) measured line shapes of particular lines (e.g. Hg365nm, Hg546nm etc.) and measurements reported by others have been compared with 1-D and 2-D simulation results.[10] These comparisons showed where the modeling efforts have to be improved. They allowed judgments about the accuracy of line broadening parameters and gA-values used in the simulations.

As an example of a lamp with a more complicated chemistry we chose a metal halide lamp with Na-Sc-I-Hg-Ar fill; we will show comparison spectra similar to those described above.

We find that only the approach of comparing simulations with quantitative spectroscopic measurements leads to advancements in modeling efforts and diagnostic capabilities.

---

[1] J. J. Lowke, R. J. Zollweg, and R. W. Liebermann, J. Appl. Phys., **46**, 650-66-, 1975

[2] J. J. Lowke, J. Appl. Phys., **50**, 147, 1979

[3] R. J. Zollweg, R. W. Liebermann, and D. K. McLain, J. Appl. Phys., **52**, 3293, 1981

[4] D. A. J. Mottran, L. Cifuentes, and S. A. Mucklejohn, 5th International Symposium on The Science & Technology of Light Sources, York, England, 1989

[5] H. P. Stormberg, J. Appl. Phys., 51, 1980

[6] J. Yos, et al., Technical Report AVCO Corporation Space Systems Division, Wilmington, MA: AVSSD-0063-66-RR, Contract NAS w-1188

[7] H. Zwicker in "Plasma Diagnostics", ed. Lochte-Holtgreven, North Holland Publishing Company, Amsterdam, 214, 1968

[8] D. Karabourniotis in "Radiative Processes in Discharge Plasmas", ed. J. M. Proud, L. H. Luessen, Plenum Press, New York, NATO ASI Series B, Physics Vol. **149**, 171, 1986

[9] A. Asselmann, M. Aubes, J. J. Damelincourt, J. Salon, J. Appl. Phys. **71**, 4739, 1992

[10] C. Allgeier, OSRAM GmbH, Munich, private communication, 1995

COLLISIONAL-RADIATIVE MODEL AND LOW PRESSURE DISCHARGE: POPULATION OF EXCITED ATOMS AND IONIZATION BALANCE

Takashi Fujimoto
Department of Engineering Science, Kyoto University
Kyoto 606-01, Japan

§1. Introduction

Excited level populations of atoms in a lamp discharge plasma have dual significance; The first is that they serve as the origin of the light emitted from the lamp and therefore they are the object of optimization for efficient lamps. The second is that they reflect the nature of the plasma and provide us with clues for understanding the characteristics of the plasma and the processes taking place in the plasma.

In the following we will examine several examples of level populations from both of the above aspects.

§2. Collisional-radiative model

We assume that the electron temperature T_e, density n_e and the ion density n_z are given. Let $n(p)$ denote the population density of a level which is specified by a group of quantum numbers as represented by p. We understand that p = g denotes the ground state. According to the collisional-radiative (CR) model [1], an excited level population is given in the form

$$n(p) = R_0(p; T_e, n_e)\, n_z n_e + R_1(p; T_e, n_e)\, n(\mathrm{g})\, n_e, \qquad (p \neq \mathrm{g}) \tag{1}$$

where $R_0(p; T_e, n_e)$ and $R_1(p; T_e, n_e)$ are called the population coefficients. These quantities are determined by the collisional and radiative processes taking place in the plasma and are functions of T_e and n_e. The second term is called the ionizing plasma component and the first one the recombining plasma component [2]. The corona equilibrium is a special case for the ionizing plasma component [3]. The LTE (Saha-Boltzmann) population is a special case of the recombining plasma component or of the combination of both the components [4].

Ionization and recombination of the group of atoms in the plasma, or of the plasma itself, is described in terms of the effective rate coefficients [1]; i.e.,

$$dn(\mathrm{g})/dt = -S_{CR}(T_e, n_e)\, n_e n(\mathrm{g}) + \alpha_{CR}(T_e, n_e)\, n_e n_z, \tag{2}$$

where $S_{CR}(T_e, n_e)$ and $\alpha_{CR}(T_e, n_e)$ are called, respectively, the collisional-radiative ionization and recombination rate coefficients. They include, bisides the contributions from the direct ionization or recombination processes, those from the indirect processes proceeding through excited levels.

§3. Ionization imbalance

Figure 1 shows examples of the population density distributions of excited helium levels of $p = n^3D$ calculated for typical plasma conditions of low pressure discharges and their breakdown into the two components [5]. It is seen that in the case of Fig. 1(a) the second term is predominant and the first term is smaller by about 3-5 orders for low-lying levels. In the

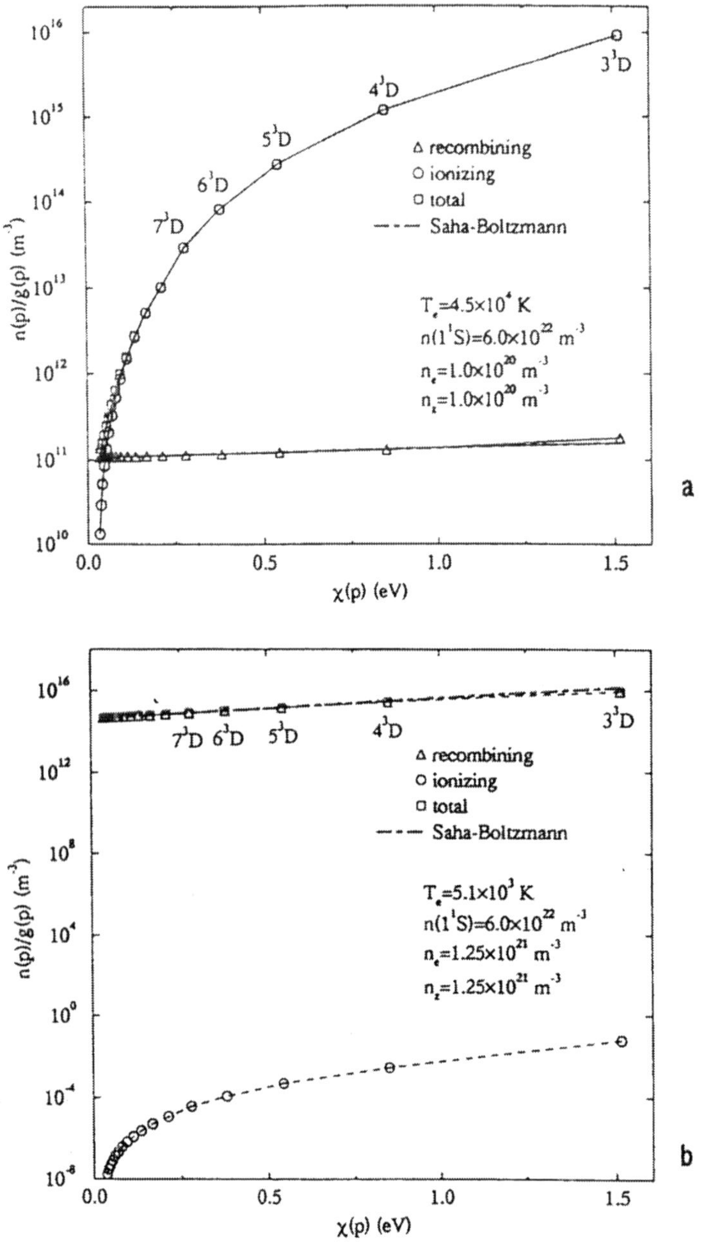

Fig. 1. Typical population distributions over helium excited levels. $n(p)/g(p)$ is the population density divided by the statistical weight.

latter case (Fig. 1(b)) the situation is reversed; the second term is smaller than the first by more than 10 orders. It may be interesting to see that the population densities of the low-lying levels ($p = 3^3D$ and 4^3D) are similar for both the cases, so that these two plasmas emit similar spectra in terms of quantity (intensity) as well as of quality (color).

Figure 2 shows the temporal development of the discharge current in a pulsed helium discharge and that of the emission intensiy of the 587.6 nm line ($2^3P - 3^3D$) [6]. The intensity has two peaks; the first is during the current rise and the second occurs just after the current ceases. By scan-

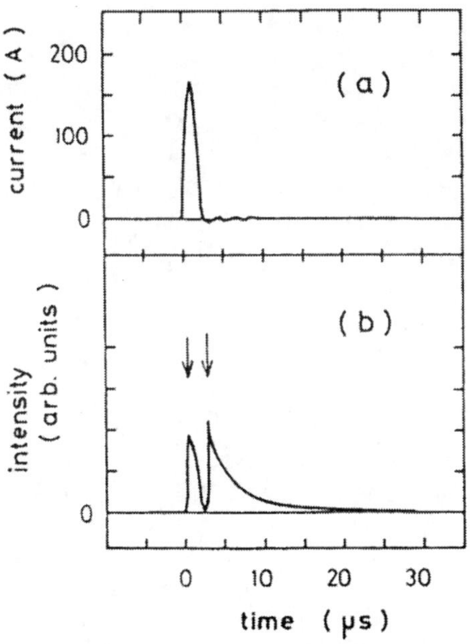

Fig. 2. The temporal development of (a) current of a helium discharge, and (b) emission intensity of a neutral helium line (2^3P – 3^3D). From ref. [6].

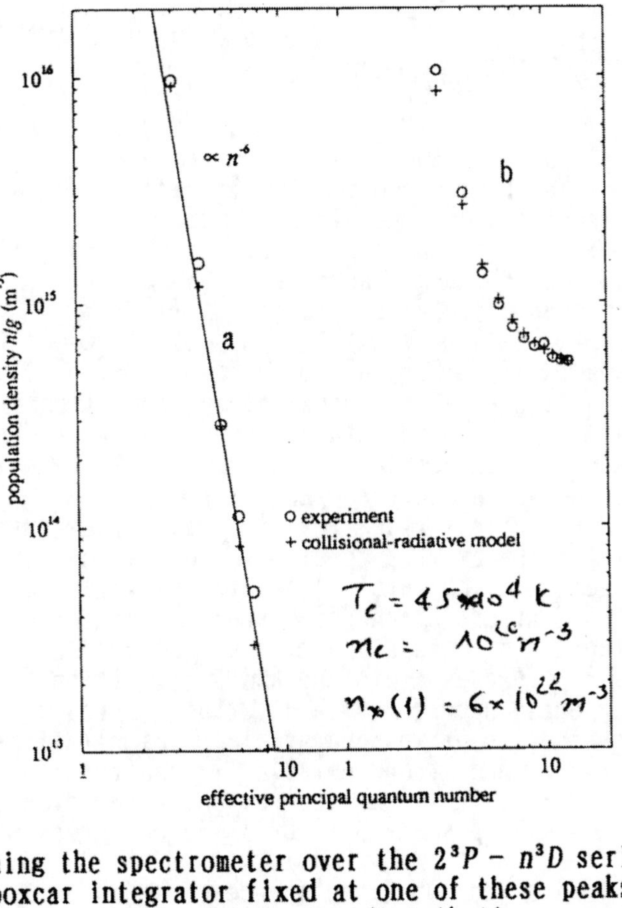

Fig. 3. The population distribution over the excited levels of helium. (a): for the first peak of Fig. 2(b). (b): for the second peak. From ref. [6].

ning the spectrometer over the 2^3P – n^3D series lines with the gate of the boxcar integrator fixed at one of these peaks, we observed the spectrum and determined the population distribution over the upper levels at these peaks. Figure 3 shows the distributions (○). The abscissa is the effective principal quantum number and the ordinate is the population divided by the sta-

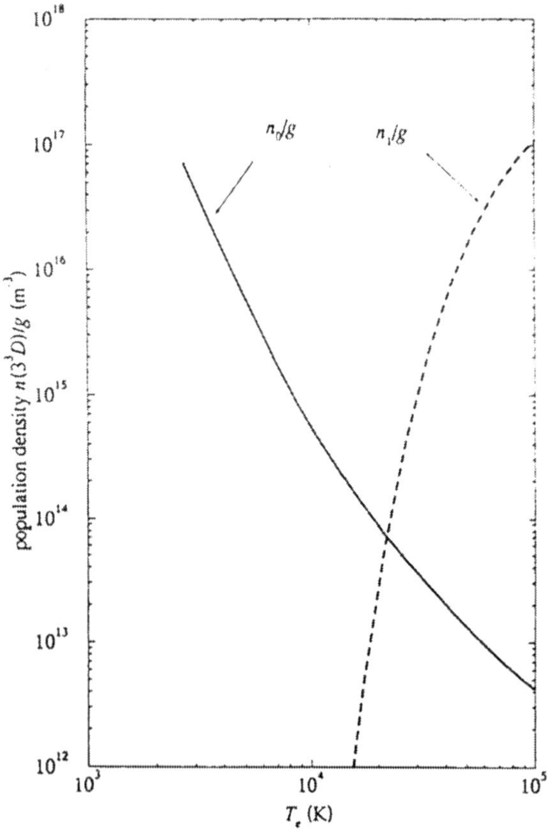

Fig. 4. The component populations as functions of T_e. $n(1^1S) = 6 \times 10^{22}\ m^{-3}$, $n_z = n_e = 10^{21}\ m^{-3}$. From ref. [6].

tistical weight, $g(n^3D) = 30$, both in the logarithmic scale. The calculated population, eq. (1), which has been shown in Fig. 1, is reproduced in this figure (+). It is thus clear that the plasma producing the former peak is a typical example of the ionizing plasma (Fig. 1(a)) and that for the latter peak is a typical recombining plasma (Fig. 1(b)).

Figure 4 shows the components of the population $n_0(3^3D) = R_0(3^3D; T_e, n_e)\, n_z n_e$ and $n_1(3^3D) = R_1(3^3D; T_e, n_e)\, n(1^1S)\, n_e$ calculated as functions of T_e, where we assume $n(1^1S) = 6 \times 10^{22}\ m^{-3}$ and $n_z = n_e = 10^{21}\ m^{-3}$. Under the conditions of this figure n_0 is approximately proportional to $n_z n_e$ and n_1 is proportional to $n(1^1S)$ and rather insensitive to a change in n_e. By using Fig. 4, we interpret the temporal development of the emission intensity or the peaks and the valley between them as shown in Fig. 2(b): During the period of discharge current rise, the electron temperature is high and $n_0 \ll n_1$ (Fig. 1(a)); the excited levels are populated mainly by electron collisions from the low-lying levels (ladderlike excitation) starting from the ground-state atoms. When the discharge current ceases, the electron temperature rapidly decreases and n_1 also decreases. When $n_0 = n_1$, the population $n(3^3D)$ reaches a minimum and a transition from the ionizing to the recombining phase occurs. With a further decrease in electron temperature, the recombination of low-energy electrons with ions dominates ($n_0 \gg n_1$; Fig. 1(b)), and strong emission is observed. In Fig. 4 the population for $T_e = 2.2 \times 10^4$ K is smaller, by about two orders of magnitude, than those at $T_e = 4 \times 10^4$ K or 5×10^3 K, which agrees with Fig. 2(b).

We now introduce the ionization imbalance parameter ξ which is defined as

$$\xi = \frac{[n_z/n(g)]}{[S_{CR}(T_e, n_e)/\alpha_{CR}(T_e, n_e)]} \tag{3}$$

This quantity gives a measure of the departure of this plasma from the ionization balance for which $\xi = 1$. The plasma corresponding to the first peak has $\xi = 5 \times 10^{-6}$, indicating that this plasma is a strongly ionizing plasma. The ionizing plasma component predominates over the recombining plasma component by a similar factor. The plasma for the second peak has $\xi = 1 \times 10^{17}$, which also is consistent with the predominance of the recombining plasma component. It is interesting to note that, for the condition of Fig. 4, the ionization balance $\xi = 1$ is established at $T_e = 1.8 \times 10^4$ K, which is rather close to $T_e = 2.2 \times 10^4$ K, giving the minimum of the population.

It is now proposed that the ionization imbalance parameter ξ be introduced in plasma research as one of the plasma parameters just like T_e and n_e. This parameter indicates the gross characteristics of the excited level populations.

§4. Ionizing plasma: Ladderlike excitation-ionization

As has been shown in Fig. 3, the excited-level populations in the ionizing plasma have the simple distribution

$$n(p)/g(p) \propto p^{*-6}, \tag{4}$$

where p^* is the effective principal quantum number of level p. In the case of Fig. 3, p^* is almost exactly equal to n, the principal quantum number of level n^3D. This minus-sixth power distribution has been originally derived for hydrogen [3]. We now take a hydrogen plasma for the purpose of illustration and denote a level by its principal quantum number n. This distribution is the result of establishement of the ladderlike excitation,

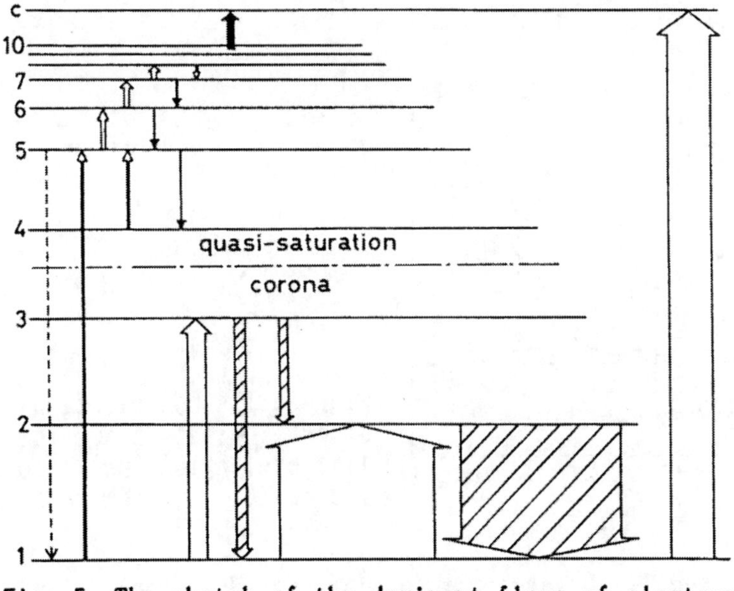

Fig. 5. The sketch of the dominant flows of electrons for the ionizing plasma of $T_e = 1.28 \times 10^5$ K and $n_e = 1 \times 10^{18}$ m^{-3}. The blank arrows denote collisional transitions and the hatched ones radiative transitions. From ref. [3].

$(n-1) \rightarrow n \rightarrow (n+1) \rightarrow (n+2) \rightarrow \cdots$, through the excited levels together with the approximate relationship for the excitation rate coefficient $C(n, n+1) \propto n^4$ and $g(n) = 2n^2$ for the statistical weight. See ref. [3] for more detail. The sketch of the dominant flows of electrons among the levels are shown in Fig. 5 for the ionizing plasma [3]; this is for the condition of $T_e = 1.28 \times 10^5$ K and $n_e = 1 \times 10^{18}$ m^{-3}. The ladderlike excitation chain is seen among the levels with $n > 4$. The lower end of the excitation ladder is given from a comparison between the radiative decay rate and the collisional decay (mainly excitation to the adjacent higher-lying level) rate from a level. The boundary level is defined as the level for which these two rates are equal, and this level is calld Griem's boundary level; This is because this level is essentially the same as the critical level for establishment of LTE as proposed by Griem [7]. This boundary level is approximately given as [8]

$$n_G \sim 480/n_e^{2/17}. \tag{5}$$

With an increase in the electron density from the low density limit, Griem's boundary level comes down, (In Fig. 5 this boundary level lies between $n = 3$ and 4.) and a larger number of levels contributes to the ladderlike excitation flow.

It is natural to expect that the ladderlike upward flow through the excited levels eventually results in ionization. This is actually the case. Figure 5 shows the magnitude of the ladderlike excitation-ionization flow with the solid arrow; this represents the flow through level $n = 10$. With the increase in the electron density, the contribution to the total or effective ionization from the ladderlike excitation-ionization increases. Figure 6 shows an example of the effective ionization rate coefficients as

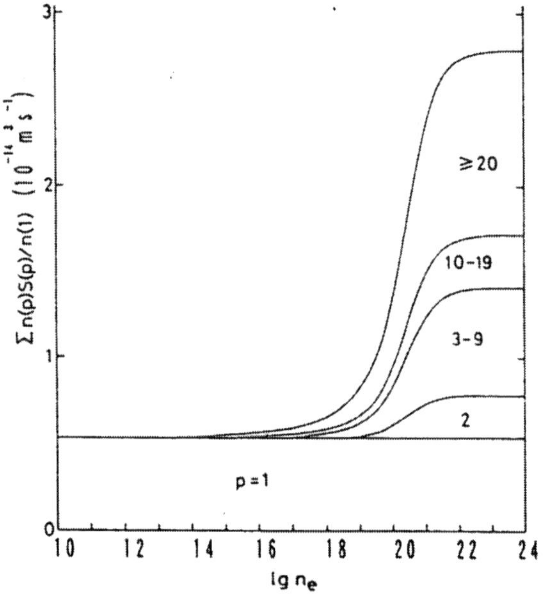

Fig. 6. The contributions from the direct ionization and the ladder-like excitation-ionization. $T_e = 1.28 \times 10^5$ K. From ref. [3].

a function of n_e and its breakdown into the individual ionization processes; the number in the figure is the principal quantum number of the last level as the final step of ionization.

The contribution from the ladderlike excitation-ionization may be approximately given as

$$R_L \sim \Sigma_{n>n_G} C(1, n), \tag{6}$$

where $C(1, n)$ is the rate coefficient for excitation from the ground state to level n, and $n > n_G$ means level n lying above n_G.

We now look at the relationship between the ionization cross section and the excitation cross section. We may assume that, near the ionization threshold, the ionization cross section has a linear dependence on the excess energy, *i. e.*, $\sigma_{ion}(E) = \alpha(E - R)$, where $R = 13.6$ eV is one rydberg. (Actually in the case of neutral atoms the depencence is the power of 1.17 of the excess energy, and for highly ionized ions the power

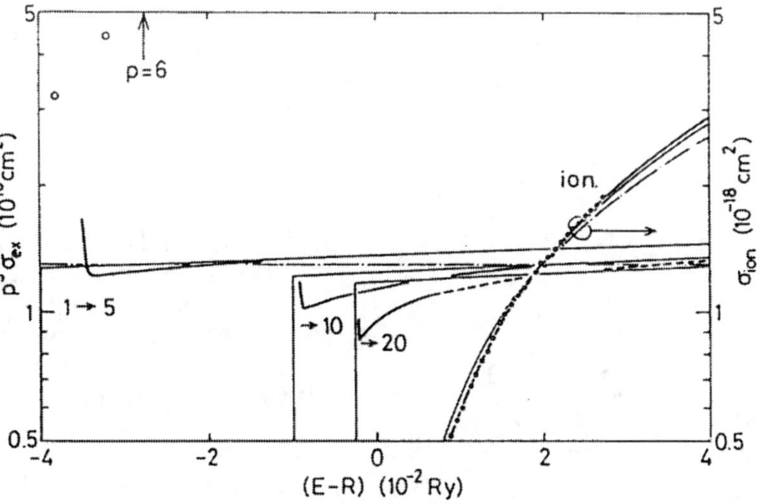

Fig. 7. Approximate relationship for ionization cross section and excitation cross section. From ref. [10].

tends to 1 [9].) Figure 7 shows the experimental ionization cross section (open circles) and its linear approximation (dash-dotted curve) [10]. From the quantum defect theory, we expect that the properties of the high-lying rydberg states continue smoothly across the ionization limit to those of the low-energy coninuum states. We may then expect that the cross section for excitation to level n is given by $\sigma_n(E) = \alpha 2R/n^3$. (See ref. [10] for more detail.) Figure 7 shows this approximation (dash-dotted line) together with the result of the Monte Carlo calculation on the basis of the extended Wannier classical theory [9].

A straightforward consequence of the above arguments is that, in order to take into account the contribution from the ladderlike excitation-ionization in evaluating the effective ionization rate coefficient, at least for temperatures for which the threshold regions of the cross sections are more important than the high energy tails, we may take the cross section for direct ionization and reduce the threshold energy down to the energy of Griem's boundary level.

The above arguments are valid also for elements other than hydrogen, at least qualitatively.

§5. Low-pressure mercury discharge

Wani [11] measured line intensities emitted from excited mercury atoms in a mercury-argon low-pressure discharge, and determined the population density distribution over the upper levels. Figure 8 reproduces the result. Two prominent features are seen; 1) Except for the lowest-lying level 8^3S, the distribution obeys eq. (4), where p^* has been replaced by x here. This is the clear evidence that the ladderlike excitation-ionization mechanism has been established among the higher-lying levels. 2) The levels in different configuration series, *i. e*, n^3S and n^3D, have approximately equal populations for equal effective principal quantum numbers, or

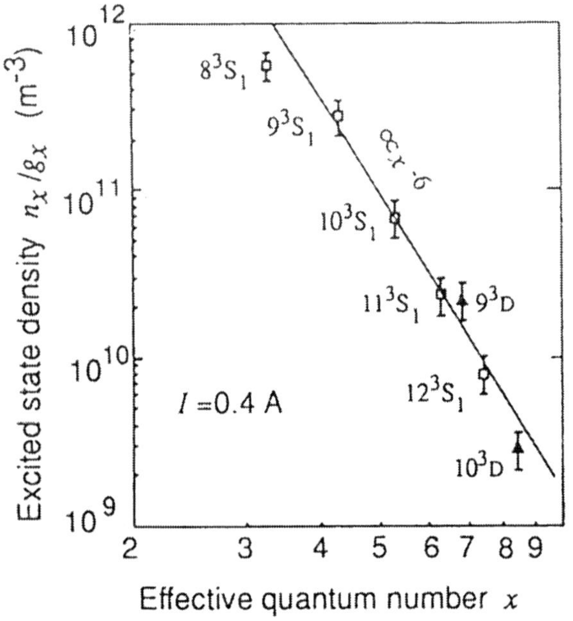

Fig. 8. The population distribution over the excited mercury levels in a low-pressure mercury-argon discharge. x is the effective principal quantum number. The x^{-6} distribution is the indication that the ladderlike excitation-ionization mechanism is established. From ref. [11].

the statistical population distribution is approximately established among these levels for equal energies. This situation is rather close to that for the case of hydrogen, where different l levels have the statistical populations in the same principal quantum number.

On the basis of the above findings, the author estimates the magnitude of the ladderlike excitation-ionization rate from the measured populations

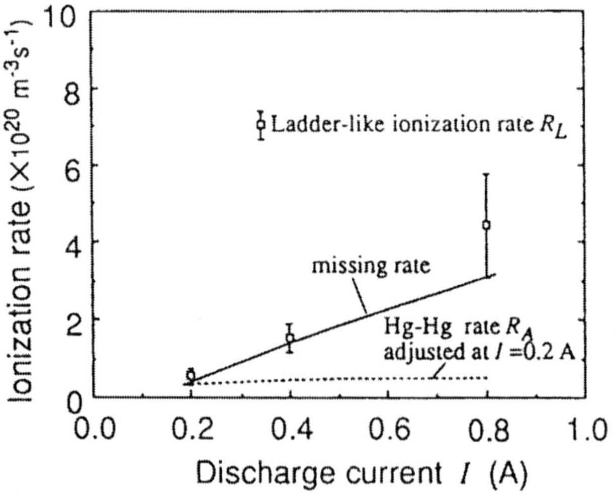

Fig. 9. The "experimental" rates for the ladderlike excitation-ionization estimated from Fig. 7 and eq. (7). The "missing rate" is also given. From ref. [11].

according to

$$R_L = 4p^{*2}\,[n(p)/g(p)]\,C(n,\,n+1)\,n_e \qquad (7)$$

instead of the approximation proposed in the preceding section. Here $C(n, n+1)$ is the rate coefficient for neutral hydrogen [12]. Figure 9 shows for several cases the experimental rate for the ladderlike excitation-ionization thus determined. Also shown is the "missing rate" for ionization which has been necessary to remove the difference, as shown in Fig. 10,

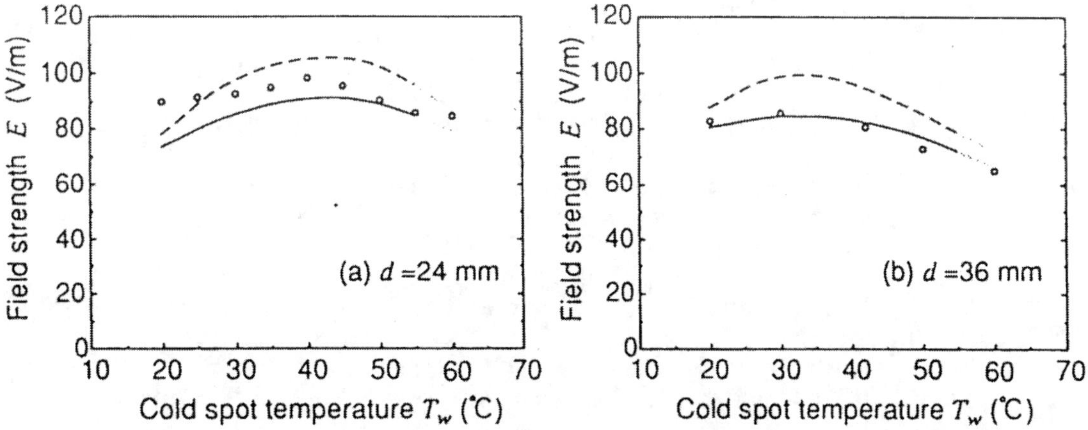

Fig. 10. The field strengths calculated with the ladderlike excitation-ionization taken into account (——), and without (- - -). From ref. [11].

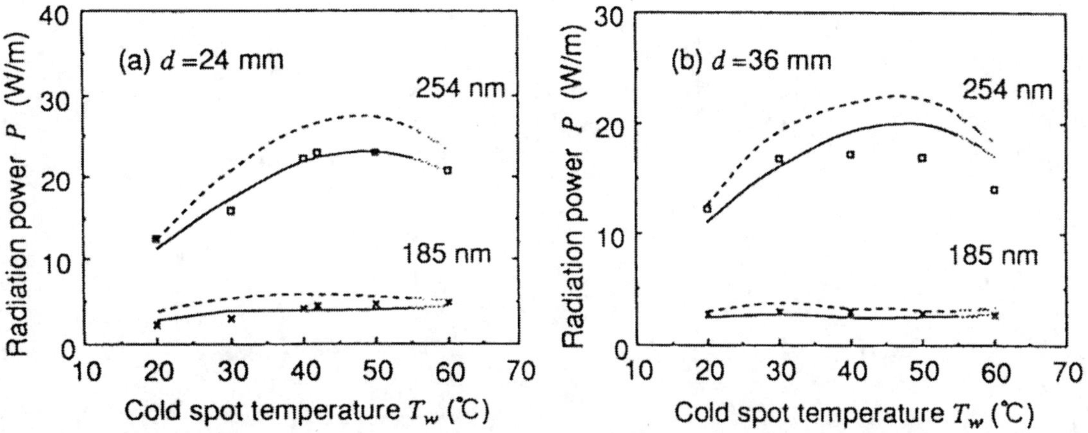

Fig. 11. Experimental and calculated radiation intensities. From ref. [11].

between the field strength predicted from the accepted ionization cross section and the measured one. Figure 11 shows the result of the inclusion of the ladderlike excitation-ionization rate for the resonance radiation powers for several discharge conditions. Agreement between the simulation and the experiment has improved.

Acknowledgments

The author is grateful to Dr. K. Wani for permission of reproduction of the figures 8, 9, 10 and 11. Americal Institute of Physics are acknowledged for permission of reproduction of Figs. 2 – 11. He also expresses his thanks to M. Goto for helping calculations.

References

1. D.R. Bates, A.E. Kingston and R.W.P. McWhirter, Proc. R. Soc. London **267**, 297 (1962); D.R. Bates and A.E. Kingston, Planet. Space. Sci. 11, 1 (1963); R.W.P. McWhirter and A.G. Hearn, Proc. Phys. Soc. London **82**, 641 (1963).
2. T. Fujimoto, J. Phys. Soc. Japan **47**, 265 (1979).
3. T. Fujimoto, J. Phys. Soc. Japan **47**, 273 (1979).
4. T. Fujimoto, J. Phys. Soc. Japan **49**, 1561 (1980); **49**, 1569 (1980).
5. T. Fujimoto, J. Quant. Spectrosc. Raditative Transfer **21**, 439 (1979); M. Goto and T. Fujimoto, to be published.
6. A. Hirabayashi, Y. Nambu, M. Hasuo and T. Fujimoto, Phys. Rev. A **37**, 77 (1988).
7. H. Griem, *Plasma Spectroscopy* (McGraw-Hill, New York, 1964) p.129.
8. T. Fujimoto, J. Phys. Soc. Japan **54**, 2905 (1985).
9. F.H. Read, J. Phys. B **17**, 3965 (1984); P. Hammond, F.H. Read, S. Cvejanovic and C.C. King, J. Phys. B **18**, L141 (1985).
10. T. Fujimoto and R.W.P. McWhirter, Phys. Rev. A **42**, 6588 (1990) (Appendix B).
11. K. Wani, J. Appl. Phys. **75**, 4917 (1994).
12. L. Vriens and A.H.M. Smeets, Phys. Rev. A **22**, 940 (1980).

SELF-CONSISTENT FINITE ELEMENT MODELLING OF AXISYMMETRIC HIGH-PRESSURE DISCHARGE LAMPS

H. Wiesmann, M. Neiger
Lighting Technology Institute (LTI), University of Karlsruhe, Germany

Introduction: Point light sources for projection television or for incoherent light transmission through quartz fibres have become of great interest. Such light sources can be approximated by high pressure discharge lamps with small electrode gaps (0.5 - 3 mm) and high operating pressure (4 MPa). The electrical input power varies from 50 W to some kilowatts and the lamps are typically filled with xenon or mercury. These operation conditions result in high electrical fields and high temperature gradients. For dc discharge lamps a hot spot is exhibited in front of the cathode.
Such high-pressure discharges are analysed using an axisymmetric model that not only describes the plasma itself but also includes the simulation of the physics of the electrodes. No explicit distinctions are made between special plasma regions like "space charge zone", "arc zone", etc. The finite element method is used to solve the relevant differential equations.

Theory: The time-independent high-pressure discharge in local thermodynamic equilibrium (LTE) is mainly described by the equation of charge balance and the energy balance equation of the arc.

$$\nabla \bullet i = \nabla \bullet (\sigma E) = -\nabla \bullet (\sigma \nabla V) = 0$$

$$\nabla \bullet (\lambda \nabla T) + iE - u = 0$$

(i = electrical current density, σ = electrical conductivity, E = electrical field, V = electrical potential, λ = thermal conductivity of the plasma, T = plasma temperature, u = emissivity)

The necessary material functions depend only on the local temperature and are derived using the Chapman Enskog method [1]. The physical behaviour of the electrodes is analysed by adding the heat flux equation of the electrodes. The emissivity and the cooling/heating due to the electron/ion current are included as boundary conditions.

$$\nabla \bullet (\lambda \nabla T) = 0$$

(λ = thermal conductivity of the electrodes, T = temperature of the electrodes)

To simulate non local thermodynamic equilibrium behaviour a diffusion differential equation can be included [2].

$$\nabla^2\sigma = -k_1\sigma\,(k_2 - \sigma/\sigma_0)$$

(σ = electrical conductivity (non LTE), σ_0 = electrical conductivity (LTE), $k_{1/2}$ = appropriate coefficients)

Solution method: The finite element method is used to solve the mentioned differential equations. The finite element method has been chosen because of the possibility to implement high order approximation functions (form functions) and to generate and handle meshes with strongly varying discretisations. These two points are very important to spatially resolve the electrical field and high temperature and power gradients. Another key advantage is the fact that various electrode shapes can easily be constructed.
A self-adapting mesh refinement has been developed to automatically reduce numerical errors. The location of the refinement is chosen on the base of local error indicators.

Results: Using appropriate boundary conditions for the electrical potential and the temperature at the electrode base point and at the quartz vessel a self-consistent model is obtained. Due to the self-adapting mesh generator and the use of high order form functions the hot spot and the existing high temperature gradients can be resolved self-consistently (see figure).

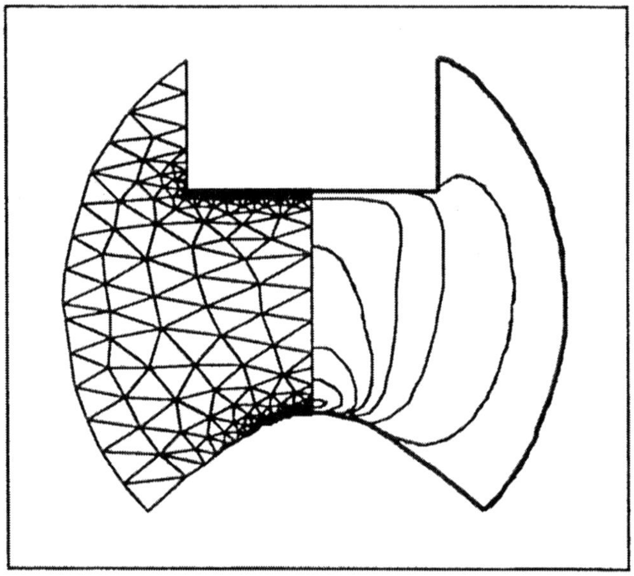

Fig.: mesh and temperature distribution of a discharge lamp

References:

[1] Palacin, A., Numerische Simulation von Edelgas-Hochdruck-Kurzbogenlampen, LTI, University of Karlsruhe, 1983, PhD thesis

[2] Fischer, E., Modelling of low-power high-pressure discharge lamps, Philips J. Res. 42, pp. 58-85 (1987)

The Development of the Simulation System for Analyzing HID Lamps

Hirokazu TAKAHASHI* Hisashi AZEGAMI*
Mototaka SONE* Yoshitaka KENMOTSU**

*Department of Electric and Electronic Engineering
Musashi Institute of Technology
1-28-1 Tamazutumi Setagaya-ku Tokyo 158, JAPAN
**Eye Lighting Systems Corporation
212-10 Akagidai Kwasato Kitasaitamagun Saitama Pref. 365, JAPAN

ABSTRACT

Generally speaking, it is very difficult to understand the mechanism which HID lamps light. The mechanism of Metal Halide lamps is especially difficult, because the structure of Metal Halide lamps have more complicated than one of other lamps (High Pressure Sodium Lamps, High Pressure Mercury Lamps). So, the condition to turn on Metal Halide lamps is not clearly understood. And, it is very hard to develop a ballast which is able to turn on Metal Halide lamps.

At present, the development of the ballasts is depends on the experience of expert engineers. So, it takes a long time to develop completely a ballast for the repetition of trial and error. And such trials and errors are repeated to turn on few lamps. Because each lamp has some differences from others on the producing process, it is not sure that a ballast can turn on all lamps produced. In fact, there are some cases which the lamps couldn't light for this difference.

In order to solve this problems, the condition to turn on the lamps has to be quickly analyzed and clearly understood. And the ballast which can match to the condition have to be developed. One of parameters to turn on the lamps is the electric power which the ballasts supplies. So, in order to analyze the electric power to turn on all lamp, this paper paid attention to the electric power supply, and developed the simulation system which could control the output power. Fig.1 showed the structure of the simulation system.

In Fig.1, the system which we developed consisted of a personal computer, an electronic ballast and a few circuits. The electrical ballast as the main circuit of the system consisted of the

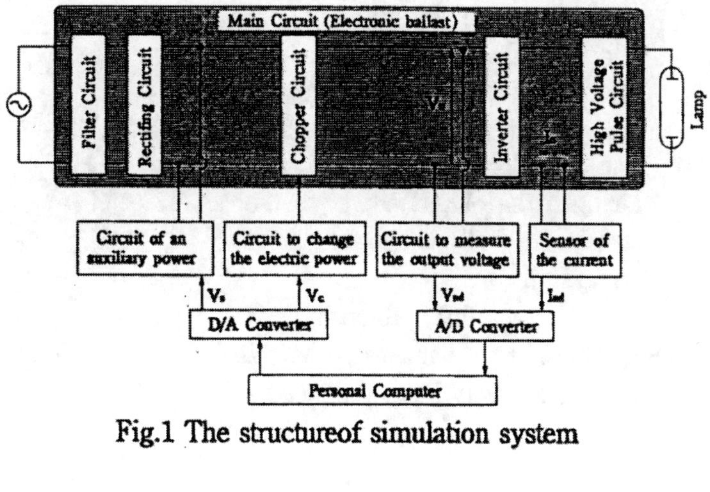

Fig.1 The structureof simulation system

filter circuit, the rectifing circuit, the chopper circuit to control the output power, the inverter circuit to change into rectangler wave, and the high voltage pulse circuit to generate the pulse of 2200 volts.

Because the chopper circuit fixed the output power of the electronic ballast, it joined the circuit which could change the electric power by the control signal, and the output power of the system could be controled by the personal computer in this system. Fig.2 showed the characteristics of the electric power which the system could supply to the lamp. It was possible that the maximum value of the characteristic was changed from 80 watts to 270 watts by this system in Fig.2. And the maximum voltage of the characteristics was 260 volts.

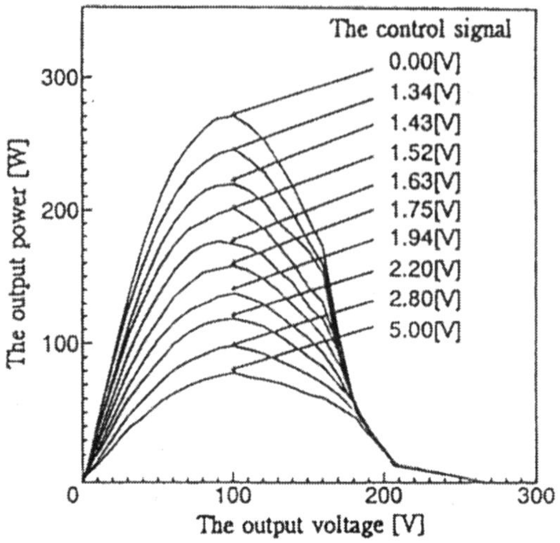

Fig.2 The output power of the simulation system

We expected to turn on every lamp by using the system developed, because it could change the output power into the good level. So, we tried the simulation to increase gradually the output power from minimum to maximum level untill one hundred Metal Halide lamps could light by using this simulation system, in order to analyze the enough power level to turn on those lamps. And we tried the similar simulations of different conditions showed in Table 1. Fig.3 showed the relation between the maximum value of the output power and the total number of lamps which could light in each condition. The standerd power of one hundred lamps was 150 watts.

In Fig.3, though all lamps were same kind of Metal Halide lamp, the enough power levels to turn on eachlamp were very different. And some lamps needed higher power level than the standard level of lamp. All conditions had the same tendency. Therefore, we thought that the output power had to be changed in order to turn on all lamps. So, this paper suggested a new algorithm to turn on all lamps. This algorithm has the rule which the system increased the output power by degrees until the lamp could light, and changed the output power to the standard level when the lamp could light.

If this algorithm is practical use, a ballast will be able to turn on every lamp produced.

Table 1. The simulation condition

Condition	Maximum voltage	Pulse voltage
Normal	260 V	2200 V
(B)	260 V	2700 V
(C)	320 V	2200 V

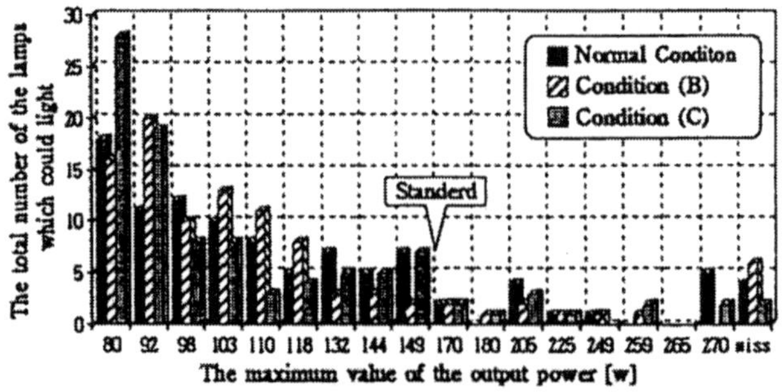

Fig.3 The distribution of the power to turn on the many lamps

COMPUTER MODEL OF THE GAS DISCHARGE IN A DEUTERIUM LAMP

G.R.Harris, A.Hartley and R.Devonshire

High Temperature Science Laboratories, Chemistry Department,
University of Sheffield, Sheffield S3 7HF, UK

A computer model of a low pressure (10 mbar, 300 mA), hot-cathode, DC deuterium gas discharge has been developed to simulate the important features of a commercially available deuterium lamp. These lamps emit continuum UV radiation in the region 180-370nm and find wide application in photometric and other equipment.

Using a fluid approach, and assuming quasineutrality and ambipolarity, a set of diffusion / reaction equations has been obtained that describes the weakly ionized plasma ($T_e \cong$ few eV) observed in the positive column region of such discharges. Included in the model are:

i) continuity equations for all important ground and excited state neutral and ionized species assumed present;

ii) energy equations for the electron and gas temperatures;

and iii) a diffusion equation for the current density.

Cross-sections for all electron-molecule reactions, including population of the 1st electronic excited state of D_2 (whose subsequent de-excitation-dissociation is responsible for the UV emission) by electron excitation of vibrationally excited D_2 ground state molecules:

$$D_2(X\ ^1\Sigma_g^+,\ \upsilon) + e \rightarrow D_2(a\ ^3\Sigma_g^+) + e$$

$$D_2(a\ ^3\Sigma_g^+) \rightarrow D_2(b\ ^3\Sigma_u^+) + h\nu$$

$$D_2(b\ ^3\Sigma_u^+) \rightarrow D + D$$

have been obtained from Celiberto *et al* [1]. The rate coefficients, and other physics inputs such as the diffusion and mobility coefficients, involve integrals over the electron energy distribution function f_e, initially assumed Maxwellian here. However, the analogy between D_2 discharges and Hg-noble gas fluorescent lamp discharges, where inelastic collisions deplete the high energy tail of the distribution, leading to a significant effect on calculation of the ionization term, implies a more sophisticated approach for f_e will be necessary. An adaptation of the Lagushenko and Maya [2] WKB analytic approximation for electron-Hg collisions to deuterium will be outlined.

Initial results will be presented for a limited choice of model species in axisymmetric, 1D (radial) geometry for steady-state solutions of the electron, ion and neutral density profiles, the electron and gas temperature profiles, and power density profiles for all important processes, including joule heating, ionization, radiation losses, recombination, heat conduction, neutral gas heating, etc. Comparisons of the simple model's predictions shall be made with measurements obtained in a parallel experimental program being undertaken to optimise deuterium lamp design within a Manufacturing Test Cell [3], before extension of the model to 2D geometry.

The radial and axial dimensions of the discharge in the compact D_2 lamp are similar. The plasma is channelled through and stabilised by an aperture in the hollow of a metallic "dimple" structure which normally is held at floating potential. The cathode is for the most part shielded by a metal box structure and is located off the optical axis defined by the anode and the aperture in the dimple. However, assuming a cathode structure with simple geometric form lying on the optical axis, and making the following simplifying assumptions for the cathode sheath, then a realistic 2D model for the lamp can be envisaged.

Briefly, it is assumed that mono-energetic beam electrons of known energy are emitted by the cathode [4], and that there is a given ratio of ion to electron current at the cathode and a given positive column electron density; no assumption of ambipolarity is necessary. A method for self-consistent calculation of the electron temperature $T_e(r, z)$ is described, where coupled diffusion equations for the electron energy, the plasma density, and the beam density are solved numerically using multigrid acceleration techniques. When numerical work is complete, a 2D description of the combined negative glow, Faraday Dark Space and positive column plasma regions of the discharge will result.

The authors gratefully acknowledge EPSRC and Cathodeon Ltd. for supporting this work.

References:

[1] R.Celiberto, P.Cives, M.Cacciatore, M.Capitelli and U.T.Lamanna, *Chem.Phys.Lett.*, **1990**,*169*, 69.

[2] J.Maya and R.Lagushenko, *Advances in Atomic, Molecular and Optical Physics* **1990**, *26*, 321.

[3] W.Yu, G.Gregory, P.Ingram and R.Devonshire (this volume).

[4] S.E.Coe and G.G.Lister, *J.Appl.Phys.*, **1992**, *71*, 4781.

AN INVESTIGATION OF THE HIGH PRESSURE SODIUM ARC SIMULATION MODEL WITH ENERGY BALANCE EQUATIONS

Toshihiko Ishigami
Toshiba Lighting & Technology Corporation
1-201-1, Funakoshi-cho, Yokosuka-shi, KANAGAWA 237 JAPAN

1. Introduction

It is desirable for the designer of high pressure sodium lamps to have a reliable model for the gas discharge. Such a model aids in obtaining a better understanding of the relationships between lamp parameters and lamp properties This study concerns with simulation model suitable for wider lamp parameters.

2. Model

2. 1 Calculation of arc temperature profile

The energy balance equations for the steady-state discharge is

$$\sigma E^2 = \mathrm{div}(-K\,\mathrm{grad}\,T) + U(T) \tag{1}$$

The relation between discharge current I and field strength E is

$$I = 2\pi E \int \sigma r\, dr \tag{2}$$

T : plasma temperature   σ : electrical conductivity   r : radius

K : thermal conductivity   U : net radiant power per unit volume

The temperature profile is calculated by a relaxation method [1] taking account of above energy balance equations, which involves calculating the radiation flux density as a function of radius for a given temperature profile using an integral solution of the radiation transport equation.

2. 2 Spectrum of the high pressure sodium arc

The spectrum is calculated from the solution of the radiative transfer equation for following four radiations by adopting the calculated arc temperature profile.

A. Na atomic radiation

Atomic spectra are calculated by the mechanism of resonance broadening and Stark broadening for Na resonance lines(D-lines) and nonresonance lines. [2]

B. Molecular radiation [3]

Four molecular spectra are calculated from determining absorption coefficients by comparison with measured data.

C. Radiation in the red wing of sodium D-lines

The red wing of the sodium D-lines shows an extra broadening when Na pressure and Hg pressure increase (high Ra type lamps). This radiation is calculated by the assumption of van der Waals broadening for the sodium D_1 and D_2 lines.[4,5]

D. The continuum radiation near 450nm

The continuum radiation near 450nm is increased also for high Ra type lamps. The origin of this radiation has not yet been identified.

This radiation is calculated by the assumption of NaHg molecular spectra (453, 470nm) suggested by Woerdman.[6]

3. Comparison of calculated results with measured data

Calculation results with this model have been compared with two kinds of measured data. One is the data of lamps rated near 400W with various lamp parameters. The other is the data of the high Ra type lamp with various input power(150 ~500W). Fig. 1 shows comparison of correlated colour temperatures for this lamp. The agreement of correlated colour temperature values in the high Na and Hg pressure region is attained by taking accout of radiations affected by Hg.

References

[1] J. J. Lowke, J. Quant. Spectrosc. Radiat. Transfer. Vol. 9 pp. 839-845(1969)

[2] J. J. de Groot , J. Phys. D: Appl. Phys., 8 pp. 651-662(1975)

[3] J. J. de Groot, Ph. D. thesis, Technical University of Eindhoven(1975)

[4] Chien Yu-Min, J. Appl. Phys., Vol. 51, No. 6 pp. 2965-2968(1980)

[5] H. P. Stromberg, J. Appl. Phys., Vol. 54, No. 8 p. 4340(1983)

[6] J. P. Woerdman, J. Phys. B: At. Mol. Phys., Vol. 18 pp. 4205-4221(1985)

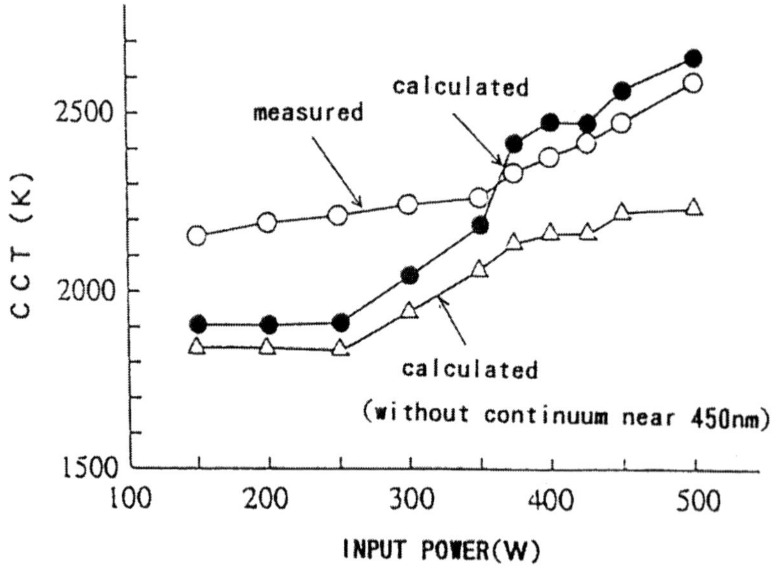

Fig. 1 Comparison of correlated colour temperature

EXPERIMENTAL MODELING OF THE FLUORESCENT LAMP

Guan-Chyun Hsieh
National Taiwan Institute of Technology
Taipei, Taiwan, R.O.C.

Ping-Shen Sung
Taiwan Electric Research & Testing Center, Inc.
Taipei, Taiwan, R.O.C.

Eel-Peeng Tsai
China Electric Mfg. Corp.
Taipei, Taiwan, R.O.C.

The characteristic of the fluorescent lamp has been widely described as a pure resistance in inverse proportion to the lamp current [1-4]. Due to the nonlinear feature of the fluorescent lamp, the fluorescent lamp characterizes its behavior as an impedance, which reveals more when the energized frequency is above kilo Hertz. In this paper, the empirical result shows that the lamp's characteristic inherently consists of a resistance and a capacitance in series. The resistance is the same as the aforementioned negative dependency to the lamp current. Interestingly, the lamp capacitance is in proportion to the lamp current, but is inversely proportionate to the operating frequency. A half-bridge series-resonant inverter in parallel-loaded with the fluorescent lamp is employed for lamp characteristic's measurement. The dynamic behavior of the fluorescent lamp is examined under the frequency ranged from 20 kHz to 50 kHz.

Figures 1(a)-(d) show the experimental behavior of the fluorescent lamp in steady-state operation. The measured responses reveal the adequate operation ranges for the fluorescent lamp, including the lamp current and the operating frequency. In Figs. 1(a)-(d), the dynamics of the fluorescent lamp are explored with the testing frequencies of 20, 30, 40, and 50 kHz. Fig. 1(a) is the measure of the lamp impedance versus the lamp current. The lamp impedance in high frequency is higher than in the low one. The lamp resistance shown in Fig. 1(b) varies inversely to the lamp current. The variation range of the lamp resistance is larger when the lamp operates in high frequency of 50 kHz. Figs. 1(c) and (d) show the lamp capacitance profiles, whose values are seriously dependent on the lamp current and switching frequency. The lamp capacitance inherently varies in proportion to the lamp current, but inversely to the switching frequency. Remarkably, the lamp capacitance is higher when the operating frequency is low. According to the experimental study, the actual model of the fluorescent lamp can represented as

$$Z(j\omega,i) = R(\omega,i) - j\frac{1}{\omega C(\omega,i)} \tag{1}$$

where $Z(j\omega,i)$, $R(\omega,i)$, and $C(\omega,i)$ are the lamp impedance, resistance, and capacitance of the fluorescent lamp respectively. They can be given from the figures 1(a)-(d) under the specified operating frequency. According to the experimental modeling, the dynamic characteristic of the fluorescent lamp operating in high frequency can be correctly estimated. It advantages the design of electronic ballast adequate for further study and application of dimmer system.

References

[1] R. R. Verderber, O. Morse, and F. M. Rubinatein, "Life of fluorescent lamps operated at high frequencies with solid-state ballasts," *Proc. IEEE IAS'85*, pp. 1724-1728, 1985.

[2] W. R. Alling, "Important design parameters for solid-state ballasts," *IEEE Trans. Ind. Appl.* vol. 25, no. 2, pp. , 1989.

[3] R. Abesamis, P. Black and J. Kessel, "Field experience with high-frequency ballasts," 1989.

[4] M. C. Cosby and R. M. Nelms, "Designing a parallel-loaded resonant inverter for an electronic ballast using the fundamental approximation," *Proc. IEEE APEC'93*, pp. 418-423, 1993.

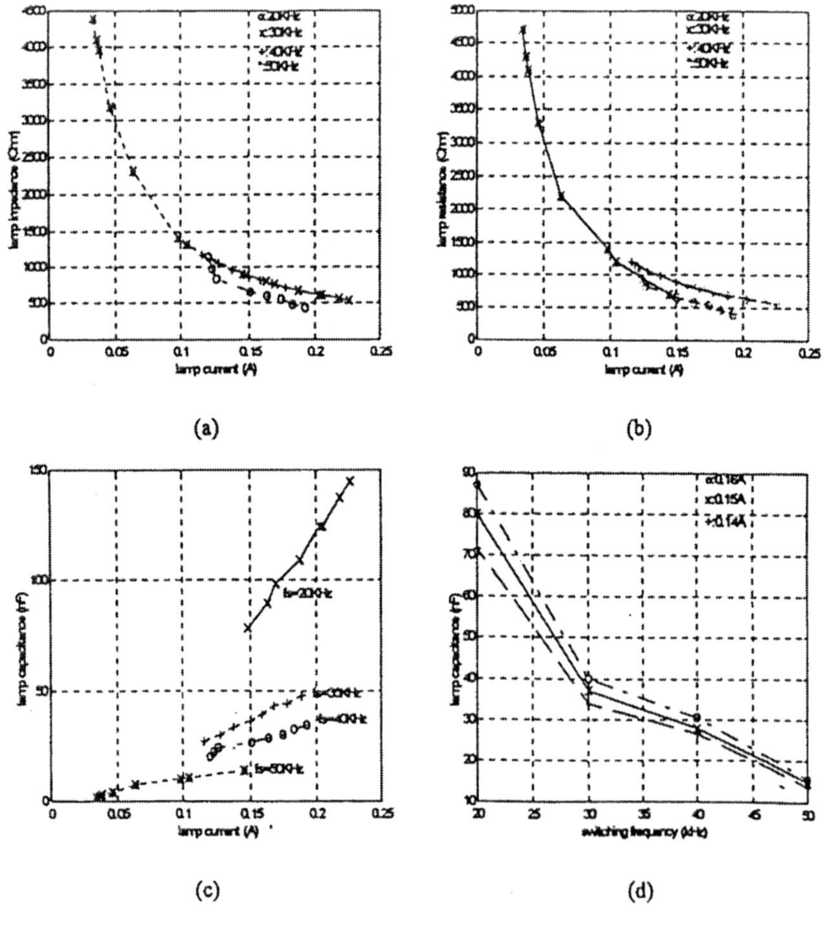

Fig.1 Dynamic behavior of the fluorescent lamp measured in steady-state operation
(a) The lamp impedance versus the lamp current
(b) The lamp resistance versus the lamp current
(c) The lamp capacitance versus the lamp current
(d) The lamp capacitance versus the switching frequency

A NALYSIS OF THE INDUCTIVELY COUPLED ELECTRODELESS DISCHARGE BY THE EQUIVALENT CIRCUIT

Yoshio WATANABE, Hirotaka MIYAZAKI
Kanagawa University, Yokohama, Japan

1. Introduction

The equivalent circuit for the inductively coupled electrodeless discharge including the excitation coil is introduced. Then, the comparison of the experiment with the calculation are presented.

2. Equivalent circuit

The system diagram is shown in Fig.1. The excitation coil wound uniformly over the tube surface with the range of l is connected with a high frequency power source through a matching circuit.

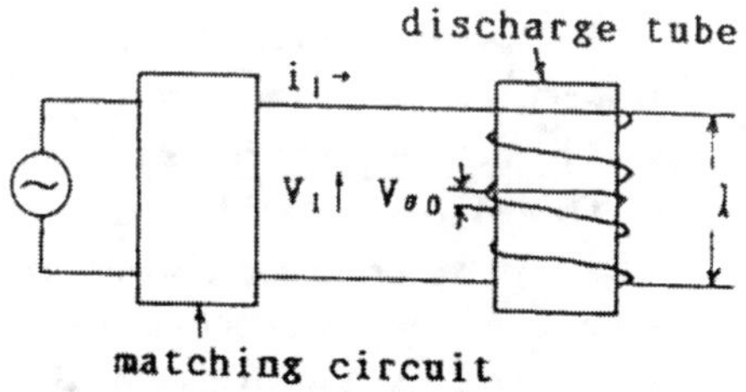

Fig.1 The system diagram.

The discharge tube is assumed as an infinite tube and the cylindrical coordinates are employed. Following equations are derived by Maxwell's equations with $\partial/\partial t = j\omega$.

$$-\frac{\partial}{\partial r}H_z = \frac{\sigma}{2\pi r}V_\theta + j\omega\frac{\varepsilon}{2\pi r}V_\theta \equiv g_\theta V_\theta + j\omega C_\theta V_\theta, \quad (1)$$

$$-\frac{\partial}{\partial r}V_\theta = j\omega\mu 2\pi r H_z \equiv j\omega L_\theta H_z. \quad (2)$$

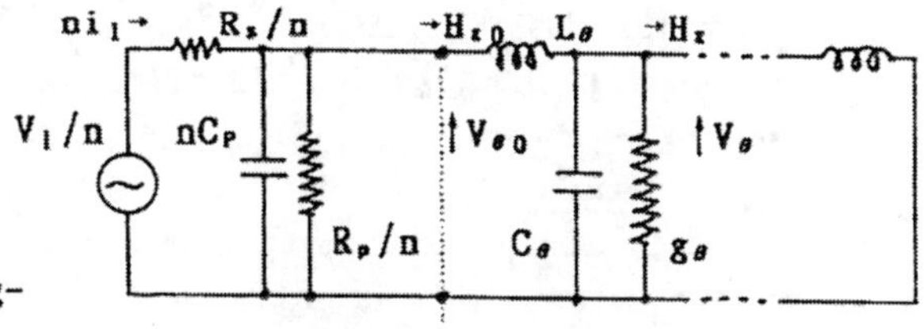

Fig.2 The equivalent distributed constant circuit

where H_z: axial component of magnetic field, V_θ: azimuthal component of voltage, σ: conductivity, ε: permittivity, μ: permeability. From Eqs.(1) and (2), the equivalent distributed constant circuit for H_z-V_θ is obtained and shown in Fig.2. The equivalent circuit for the winding region (per one turn) is also shown in Fig.2, where n: the number density of windings, R_s: the resistance including skin effect in the winding coil, R_p: the equivalent resistance due to the eddy current loss in the winding coil, C_p: the capacitance existed between the windings. The relationships between V_1, i_1 and $V_{\theta 0}$, H_{z0} on the outer surface of the discharge tube are given by,

$$V_{\theta 0} = V_1/n - R_s i_1, \quad H_{z0} = n(i_1 - j\omega C_p V_{\theta 0} - V_{\theta 0}/R_p). \quad (3)$$

3. Experiment by the simulation tube

The tube was made of glass with outer diameter of 55mm, inner diameter of 50mm and length of 200mm. The number of windings was 20 turns. To give the resistivity inside the tube, a graphite column (21.3×10^3, 100×10^3 Sm^{-1}) was inserted into the tube. One

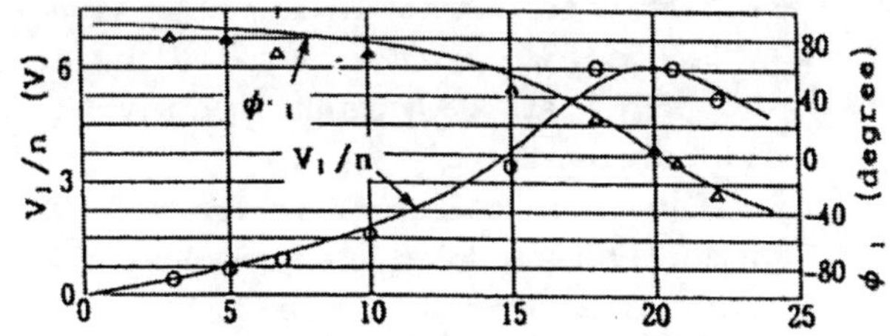

Fig.3 V_1, ϕ_1 vs. f (i_1=50mA, σ=0)

turn coil on the outer surface and a search coil were used to detect $V_{\theta 0}$ and the voltage induced inside the tube, respectively. The voltage and current are expressed in amplitude.

V_1 and phase difference, ϕ_1 (between V_1 and i_1), vs. operating frequency at i_1=50mA without the graphite column are measured and shown in Fig.3. The curves show resonance characteristics. With C_p=19.9nF, V_1 and ϕ_1 vs. operating frequency at i_1=50mA are calculated and shown in Fig.3. V_θ distributions at i_1=50mA and 3MHz were measured without/with the graphite column and shown in Figs.4 and 5 together with the calculated results, respectively. The lower measurement value than the calculation in Fig.5 is due to the lower effective resistivity of the graphite column in the experiment configuration.

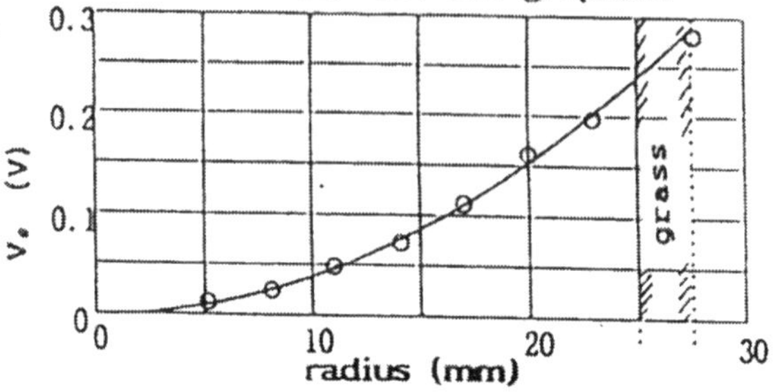

Fig.4 V_θ distribution at σ=0

4. Discharge experiment

The size of the discharge tube and the winding coil were the same. Ar, Ne and Kr were employed as operation gases. Under the discharge condition, $V_{\theta 0}$ and $\phi_{\theta 0}$ vs. i_1 were measured. The typical result obtained at 3 MHz with Ne of 133 Pa is shown in Fig.6.

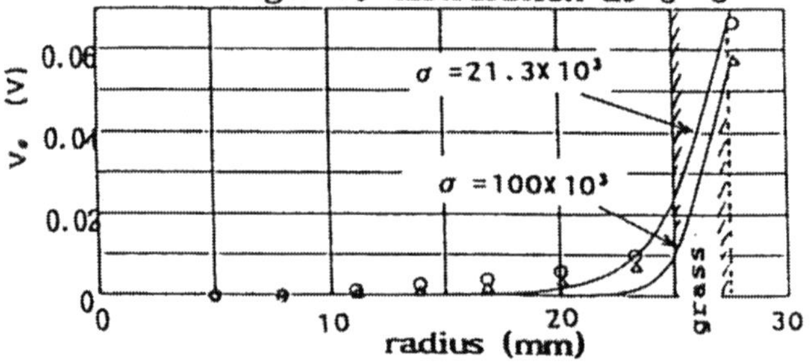

Fig.5 V_θ distribution with the graphite column

5. Discussion

The plasma conductivity under the discharge condition is estimated by comparing the experiment data of $V_{\theta 0}$ vs. i_1 with the calculation of those. For the calculation, the plasma conductivity distribution is assumed to be expressed as follows:

$$\sigma = \sigma_{max} x^2 \exp(1-x^2), \quad (4)$$

where $x=(r_0-r)/(r_0-r_m)$, σ_{max}: the maximum of the plasma conductivity, r_m: the radius where σ takes the maximum.

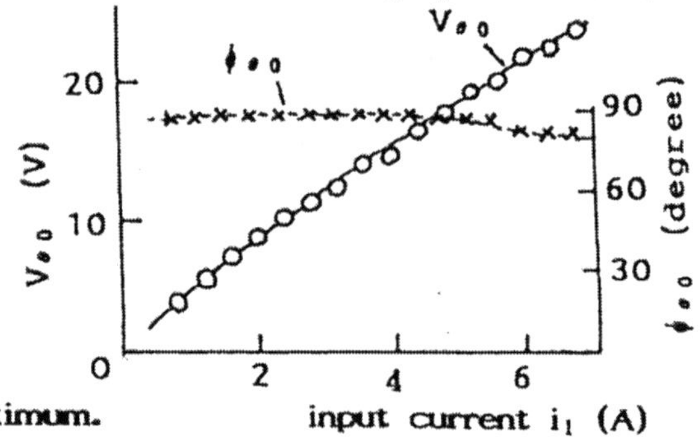

Fig.6 $V_{\theta 0}$ and $\phi_{\theta 0}$ vs. i_1

The results corresponding to the data shown in Fig.6 is shown in Fig.7 as a function of the induced current, i_2. The maximum of the plasma conductivity and the induced current in the tube under this condition are estimated to be around 500 Sm^{-1} at 30A/0.2m, respectively.

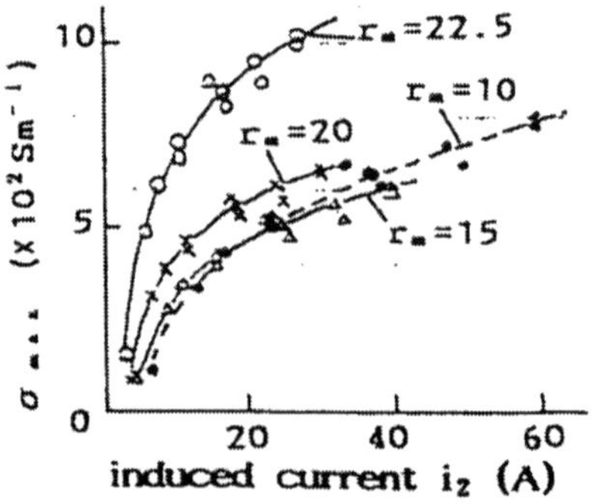

Fig.7 σ_{max} vs. i_2

6. Conclusion

The electrical characteristics of the inductively coupled electrodeless discharge will be estimated by the derived equivalent circuit.

MODELLING OF THE PHILIPS QL-LAMP

D.A. Benoy, K.T.A. Burm, J. Jonkers, J.A.M. van der Mullen, and D.C. Schram,
Department of Physics, Eindhoven University of Technology,
P.O.Box 513, 5600 MB Eindhoven, The Netherlands.

Abstract

The Philips QL lightsource (Fig.1) is a low pressure inductively coupled RF discharge. The main advantage of this system is the fact that it contains no electrodes so that the lifetime of the QL system is mainly determined by the electronic circuit of the powersupply-unit. A characteristic feature is that the induction coil is located at the centre of the lamp in order to avoid the shielding of the light. With this coil configuration the electromagnetic fields are also partially shielded outside the lamp. Furthermore the induction coil is wounded around a ferrite core to intensify the magnetic induction (Fig.2). The powersupply-unit generates a 2.65MHz RF current and is transmitted to the coil by means of a coaxial cable.

Figure 1: Photograph of the QL-lamp

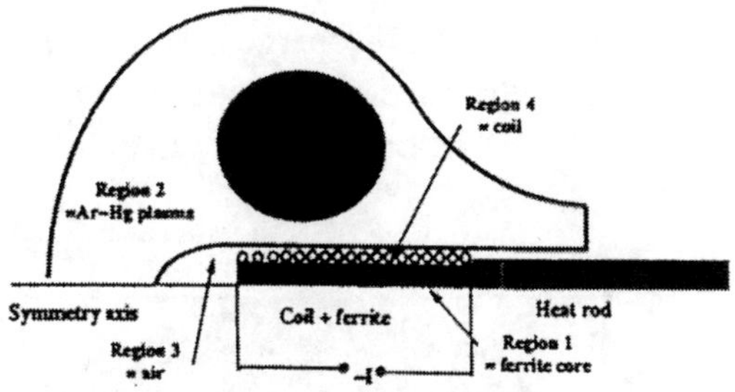

Figure 2: Schematic view of the QL-system

In this study the plasma simulation model developed at the Eindhoven University of Technology is used to investigate the mercury-argon discharge of the QL-system. The basic assumptions of this model are: 1) steady state, 2) axial symmetric geometry, 3) non-local thermal equilibrium and 4) the plasma can be described by a hydrodynamical approach. It is obvious that when we want to study the QL-system as a whole also the interaction between the coil and the plasma needs to be described. Distributions for the plasma density ρ, electron density n_e, electron temperature T_e, heavy particle temperature T_h and the electromagnetic field can then be computed by solving the following equations numerically,

$$\vec{\nabla} \cdot (n_e \vec{v}) - \vec{\nabla} \cdot (D_A \vec{\nabla} n_e) = n_e n_a S_{CR} - n_e n_+ \alpha_{CR} \tag{1}$$

$$\vec{\nabla} \cdot (\rho \vec{v} \vec{v}) = -\vec{\nabla} p - \vec{\nabla} \cdot \Pi + \vec{J} \times \vec{B} \tag{2}$$

$$\vec{\nabla} \cdot [(\frac{5}{2} n_e k T_e + E^+)\vec{v}] - \vec{\nabla} \cdot (\kappa_e \vec{\nabla} T_e) = \vec{J} \cdot \vec{E} - Q_{eh} - Q_{rad} \tag{3}$$

$$\vec{\nabla} \cdot [(\frac{5}{2}(n_e + n_a) k T_h)\vec{v}] - \vec{\nabla} \cdot (\kappa_h \vec{\nabla} T_h) = \vec{v} \cdot \vec{\nabla} p + Q_{eh} \tag{4}$$

where S_{CR}, α_{CR} are the ionization and recombination coefficients, Q_{eh} the energy exchange between electrons and heavy particles and Q_{rad} the radiative losses. The mathematical equations are expressed in generalized orthogonal curvi-linear coordinates so that a boundary fitted (QL-lamp geometry) grid system can be generated.

The QL-lamp system is considered as a ferrite-core transformer in which the discharge is the secondary circuit. The electromagnetic field must be calculated in the plasma (cf. Fig.2: Region 2) as well as in the ferrite-core (Region 1) in order to describe correctly the coupling between the plasma and the core. A two-dimensional self-consistent vectorpotential formulation

$$\vec{\nabla} \times \vec{\nabla} \times \vec{A} = -i\mu\sigma\omega\vec{A} \tag{5}$$

is used to calculate the electromagnetic field. With the electron density and electric conductivity distributions the integral plasma resistance and plasma self-induction coefficients can be determined. By calculating the geometric coupling coefficient also the resistance and self-induction coefficient of the primary circuit, which can be measured, are easyly obtained.

Since we deal with a low pressure discharge deviations of the electron energy distribution function (EEDF) from the Maxwell distribution must be taken into account in the calculation of S_{CR}. It is a good approximation to represent the EEDF by a two-temperature (tail and bulk temperature) Maxwellian distribution. The influence of the addition of mercury to the discharge on the n_e-distributions is investigated. In Fig.3 the calculated n_e profile in a pure argon (40Pa) discharge is shown. In this case the location of the active area of the discharge does not correspond with the actual location (cf. Fig.1 and the length of the ferrite core in the calculations is smaller than the ferrite core shown in Fig.1). In case of an argon-mercury

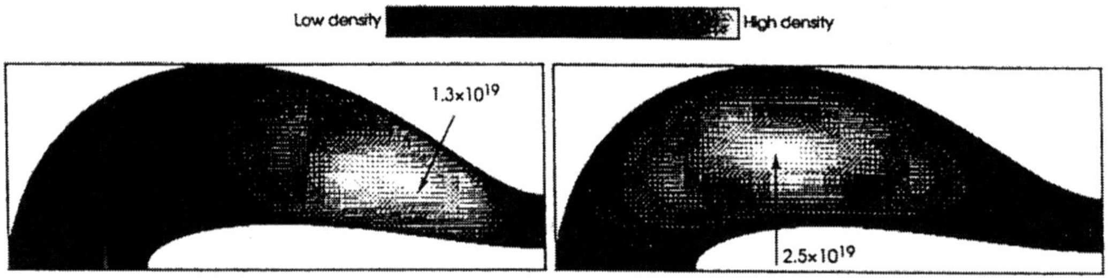

Figure 3: Pure argon, P_{calc} =55W, $n_e(\max) = 1.3 \times 10^{19}\mathrm{m}^{-3}$.

Figure 4: Ar-Hg mixture, P_{calc} =62W, $n_e(\max) = 2.5 \times 10^{19}\mathrm{m}^{-3}$.

mixture it is assumed that the kinetics (cf. S_{CR}) is only determined bu the inelastic collisions between the electrons and mercury and that argon is a purely buffergas, i.e. it determines the transport coeffcicients. Figure 4 shows a typical result of such a modelcalculation. As can be seen the active area is now more in agreement with the experimental findings. Still the description of diffusion deserves special attention. The results of the calculations will be compared with the experiment.

HALOGEN LAMP COMPUTER MODEL INCORPORATING LTCE AND SPECIES DIFFUSION

P. Heeley, D.D.Bruguier and R. Devonshire

High Temperature Science Laboratories, Chemistry Department,
University of Sheffield, Sheffield, S3 7HF, UK.

The ability to predict *a priori* the operating characteristics of incandescent lamps, particularly halogen lamps, is a long established goal in lamp science (see for example, T'jampens and van de Weijer (1966), Schnedler (1983), Correa (1986)). The ultimate aim is to develop a fully realised mathematical model of working lamps which has the ability to predict local concentrations and of all the species present in the lamp envelope at a given time, the mass transport of lamp materials, operating pressures, filament and envelope temperature profiles, etc. Such a computer based model would be an invaluable and cost effective tool for lamp design.

If one considers the chemical dimension of such a model then it is apparent that there are a large number of species present in an operating lamp. The filament, filament support, and envelope materials, and the halogen dose, buffer gas, and several common gaseous impurities interact in, effectively, a high temperature closed reactor stirred by strong convective gas flows. The total number of species present is measured in hundreds; the number of reactions occuring is correspondingly greater: the chemical complexity compares with that encountered in combustion science. To put this in context it should be noted that the current state of the art in computational fluid dynamics of reactive gas flows is to run an unconstrained optimisation of the four governing equations of reactive flows (mass, momentum, energy and species conservation) with less than 10 reactions.

To make progress with the problem it is clear that certain approximations must be made. To retain the kinetic features of the chemistry then the number of reactions must be reduced to a limited set of primary importance to the feature of lamp operation under study, e.g., the study of a 10 step reaction mechanism in an HBr dosed halogen lamp to be incorporated in a model based on the CFD program FLUENT, a study where, for example, tungsten transport was not considered (Abbott, Trindell and Devonshire (1993)).

To simplify the problem even further constraints are necessary. A valuable approximation is achieved if it is assumed that the chemically active species are only a small proportion (<1%) of the fluid, thereby contributing only a small (negligable) enthalpy contribution to the energy balances. In this case an accurate velocity and temperature field can be obtained for the whole computational domain using programs such as FLUENT. The problem then is to map the chemistry onto the fixed temperature field. If the relaxation times of chemical kinetic processes are assumed everywhere to be small compared to the characteristic times of diffusive and convective motion then the composition of the system at any given point may be considered to be a function of the temperature at that point only. A primitive approximation can then be made by fixing the local dose concentration (correcting for density only) and calculating the various local species concentrations by minimising the Gibbs free energy at that point using a thermochemical database software package such as MTDATA. Thus the local temperature is assumed to be an equilibrium temperature and the system is said to be in a state of Local Thermodynamic Chemical Equilibrium (LTCE). Such an approach is flawed fundamentally because it ignores the effects of differential diffusion of the species both before and after the Gibbs free energy minimisation.

The FLUENT code has an in-built model which is capable of incorporating the spatial distribution effects of diffusion for each of the species present. The model is particularly useful when dosage is small and ideality can be assumed. It requires the input of all the binary diffusion coefficients for the components of the mixture and these are used to calculate the necessary multicomponent diffusion coefficients. The binary diffusion coefficients are calculated internally to FLUENT using the Lennard-Jones parameters for the respective species. The effects of Soret diffusion can also be included easily in this model, again the required parameters are estimated from kinetic theory. This latter effect is particularly important when dealing with large temperature gradients such as those found in an operating lamp. A study of tungsten transport (to first order only) from a lamp filament to the envelope walls using this method (neglecting any other chemistry) is described in Correa (1986) and Heeley and Devonshire (1995).

The effects of diffusion cannot be ignored when dealing with species of widely differing molecular mass such as HBr, Br, H_2, H and N_2 the major components in the HBr system mentioned above. We are investigating a model which will achieve an LTCE prediction of the spatial distribution of species in multicomponent mixtures in the operating lamp which has been corrected appropriately for the effects of diffusion. In practise we are developing a model that combines FLUENT with MTDATA in an efficient manner.

References:

M.J.Abbott, D.L.Trindell, and R.Devonshire, "Computer Modelling of Halogen Incandescent Lamp Chemistry", in "High Temperature Lamp Chemistry III", The Electrochemical Society, **1993**, pp. 1-12.

S.M.Correa, "Fluid Flow and Heat Transfer in Incandescent Lamps", *Int. J. Heat Mass Transfer*, **1986**, *30*, pp. 663-672.

P.Heeley and R.Devonshire, "Computer Modelling of Tungsten Transport in a Halogen Incandescent Lamp", (this volume).

E.Schnedler, "Description of Tungsten Transport Processes in Halogen Incandescent Lamps", *Philips J. Res*, **1983**, *38*, pp. 236-247.

G.R.T'jampens, and M.H.A.van de Weijer, "Gas-filled Incandescent Lamps Containing Bromine and Chlorine", *Philips Tech. Rev.*, **1966**, *27*, pp. 173-179.

COMPUTER MODELLING OF TUNGSTEN TRANSPORT IN A HALOGEN INCANDESCENT LAMP

P. Heeley and R. Devonshire

High Temperature Science Laboratories, Chemistry Department,
University of Sheffield, Sheffield, S3 7HF, UK.

Tungsten transport is a central topic in any model of the operation and life of incandescent lamps and has been considered by numerous authors in the past. (T'jampens and van de Weijer (1966), Horster et al. (1971), Dettingmeijer et al. (1975), Dittmer and Niemann (1981), Schnedler (1983) and Correa (1986)).

Prompted by the work of Correa (1986), the present paper describes the results of an investigation of the effect of lamp geometry, and the physical properties of the gas fill, on tungsten transport from the filament to the envelope walls in lamps with halogen lamp size and geometry. The work is based on the commercial computational fluid dynamics (CFD) package FLUENT.

Two models were investigated. The first was an idealised lamp configuration equivalent to an infinitely long horizontal cylindrical envelope of uniform cross section with an axial filament of small, but finite, thickness. The second model was identical to that generated by Abbott *et al.* (1993) in their study of a DYS halogen lamp. This high fidelity model has a 3D geometry which incorporates all of the important envelope, filament, and filament support, geometric features and the temperature distributions at all the boundaries.

Halogen lamp chemistry is not considered at this stage of the work in either model. Such a study awaits the completion of a parallel activity to develop a model with the ability to predict Local Thermodynamic Chemical Equilibrium species distributions (Heeley and Devonshire (1995)).

The assumed boundary conditions are based on the work of Langmuir (1912) and Elenbaas (1948) who suggested that a thin stagnant layer of buffer gas exists around the filament which can be considered as being saturated with tungsten vapour. The mass fraction of tungsten at the filament can be calculated from a knowledge of the filament temperature and the corresponding tungsten vapour pressure. The mass fraction of tungsten at the bulb casing can be calculated in the same way. The no slip condition forces the tangential component of all velocities to be zero at all boundaries. The transport of the tungsten is driven by both concentration and thermally induced diffusion as well as the naturally occurring convection. The diffusion model available in FLUENT produces best results at low dose levels when ideal behaviour can be assumed. The model uses the binary diffusion coefficient of tungsten through the buffer gas and the Soret diffusion coefficient of tungsten to calculate the species diffusion vectors for both tungsten and the buffer gas. The necessary diffusion coefficients are estimated from kinetic theory expressions within FLUENT which use supplied Lennard-Jones parameters.

Figure 1 shows the distribution of tungsten vapour in the buffer gas of the DYS lamp model for a maximum filament temperature of 2,600 K. The distribution relates to the vertical plane, normal to the filament axis, and passing through its centre. The predicted distributions will be compared with the results of laser induced fluorescence studies of the tungsten atom distribution currently being carried out in the HTSL.

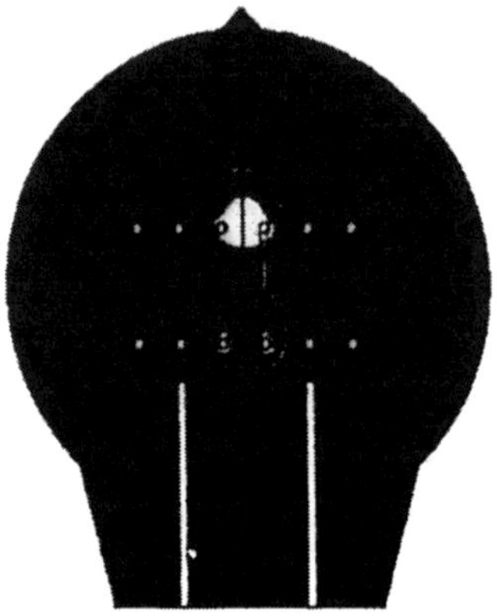

Figure 2 shows the distribution of tungsten deposited on the inner envelope wall of the DYS lamp along the line defined by the intersection of the vertical plane of Figure 1 with the envelope.

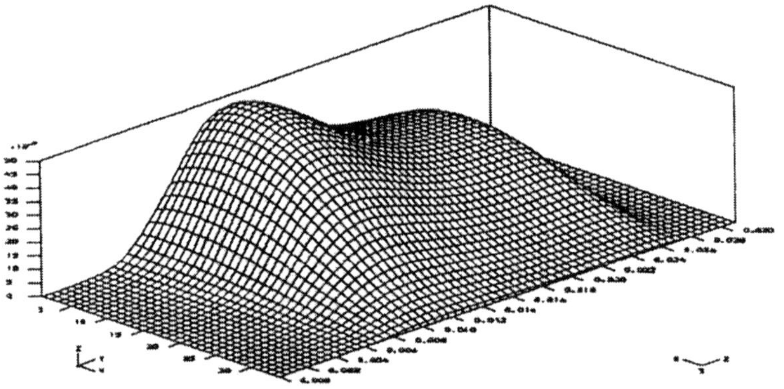

The differences in the tungsten deposition patterns predicted in the studies of the two models will be discussed in terms of lamp fill pressures, operating temperatures and features of the lamp geometries.

References:

M.J.Abbott *et al*, in "High Temp. Lamp Chem. III", **1993**, pp. 1-12.
S.M.Correa, *Int. J. Heat Mass Transfer*, **1986**, *30*, pp. 663-672.
J.H.Dettingmeijer *et al*, *Philips tech. Rev.*, **1975**, *35*, pp. 302-306.
G.Dittmer, and U.Niemann, *Philips J. Res.*, **1981**, *36*, pp. 89-113.
W.Elenbaas, *J. Appl. Phys.*, **1948**, *19*, p. 1148.
P.Heeley and R.Devonshire, (this volume).
H.Horster, E.Kauer and W.Lechner, *Philips tech. Rev.*, **1971**, *32*, pp. 155-164.
I.Langmuir, *Phys. Rev.*, **1912**, *34*, p. 401.
E.Schnedler, *Philips J. Res*, **1983**, *38*, pp. 236-247.
G.R.T'jampens and M.H.A.van de Weijer, *Philips Tech. Rev.*, **1966**, *27*, pp.173-179.

An elementary three-zones channel type arc-model of a high pressure Rare Gas -Rubidium Halide Discharge.

N. Sewraj, J. J. Damelincourt
Centre de physique des Plasmas et de leur Applications de Toulouse; 118, Rte de Narbonne; F-31062 Toulouse Cedex - France

Introduction

Metal Halide discharges using mercury buffer are widespread as infrared and ultraviolet sources in industry. However, such plasmas are prohibited in applications where hot re-ignition and rapid warm-up phase is of utmost importance. Moreover, because of its noxiousness mercury is most certainly meant to be substituted by some other element, notably a rare gas. Indeed, high presure Rare Gas Metal Halide (RGMH) discharges offer a rapid start-up phase and allow instantaneous re-ignition [1].

Up to now only a little attention has been paid to these RGMH discharges bearing a rather high chemical complexity. We performed some experimental works, [1], [2], [3], & [4], on high pressure Rare Gas/Rubidium Iodide discharges using either krypton or xenon. buffer Under our experimental conditions, the rubidium amount in the vessel is quite low (less than 10 μg).

For a better understanding of the high pressure RGMH discharge it seems interesting to elaborate a numerical code allowing coupling of a discharge's model with the circuitry one. In order to minimize the computing time we opted for simple "channel-type" model [6]. However simple it may be, the discharge model should still describe the main physical properties, and in our particular case, the radiation term. Details on the model can be found in reference 1. Here we only sort out the main points.

Classical channel type model (High pressure Mercury or Sodium discharges)

This model which expresses the energy balance (Eq. I) in the positive column, of a cylindrical plasma at Local Thermodynamic Equilibrium (L.T.E.), has been successfully applied to classical high pressure mercury [7] and sodium discharges [8].

$$P_{elect.} = P_{pos.\ col.} + \Sigma P_{rad.} + P_{therm.} \qquad (I)$$

P : power ; elect. : electrical ; pos. col. : positive column ; rad. : radiation ; therm. : thermal loss

The plasma is assumed to bear two distinct homogeneous zones. The active central hot cylindrical region is responsible for radiation emission as well as electron and heat production. The annular cylindrical colder mantle can only absorb the radiation coming from the core whose limit (reduced radius) is characterized by the β parameter. In order to account for the thermal loss, thermal flux (supposed constant in the colder region) is determined by using a specific temperature whose profile in the cold region is nearly linear [1].

As far as our discharge is concerned, the complex chemical equilibrium prevailing imposes some important modifications [1] & [5]. Nevertheless, we keep in mind the necessity of a simple interactive model.

Modified channel-type model (High pressure RGMH discharge)

Following the assumptions of the classical model, neutral population densities are much higher in the colder region. However, the neutral rubidium density is quite low near the wall [1] & [2]. Indeed, this is closely related to the role played by RbI which is stable at wall-operating temperature and particularly unstable at temperatures higher than 2500 K. The classical model can't account for that. Besides, other considerations such as the rather flat temperature profile as well as the radial shift of the maximum emission of rubidium lines, lead us to four major modifications of the classical model.

(i) We use a 3-zones channel : a hot active core playing the same role as its predecessor, a cold external mantle protecting the wall against neutral rubidium attack, and an intermediate cold region which can only absorb radiation. This leads to an effective radius given by : $R_{eff} = \beta_w R_w$, where $\beta_w = 0{,}95$ (determined experimentally [1]).

(ii) Since rubidium is present in small quantity, ionisation of the metal leads to a high inhomogeneity in the central part of the discharge. Thus, the most important part of the rubidium lines emission is shifted towards the periphery. So, the neutral rubidium density is corrected by the following formula (determiend on experimental basis) :

$$(n'_{Rb})_{core} = (n_{Rb})_{core}\left[1 + \left\{\frac{e^{-\rho_1 C}}{\beta C}\left(e^{\beta C} - e^{\rho_1 C}\right)\right\}\right] \qquad (II)$$

C and ρ_l are numerical dimensionless constants obtained experimentally (c = 26,3 & ρ_l = 0,65).

We remark that ρ_l corresponds to a reduced radius which hardly differs to the limit of emission of krypton lines. An analog correction is also neccessary for the colder region. Experimental results show that this corrective factor is about 0,5.
(iii) Following the first remark we take 2000 K for T_W. This value, relatively high compared to mercury discharges, corresponds to the actually measured temperature for a reduced radius $\rho = \beta_W R_W$.
(iv) With respect to the effective radius β is about 0,95. This leads to a relatively high temperature gradient in the intermediate zone. Following experimental results, thermal loss is corrected by a multiplicative factor neighbouring 0,3.

The circuitry model is based upon energy balance in the electrical network. Numerical resolution is based on a constant step Runge Kutta method [1].

Results

Validation of the model is realized by comparing experimental results [1] obtained with a Kr/RbI discharge operating at with a sinusoidal current at 50 Hz via a supply voltage supply coupled with a high impedance inductive ballast and working at 250 W. We consider emission of the most important emission lines namely Rb 780,0 nm & Rb 794,8 nm resonance ones. Experimentally it was found that the kryton to rubidium atomic ratio (assumed to be radially constant) is however time-dependent. Actually our model's structure can't account for such time variations. So we use is an overall value determined over one cycle.

Results obtained for an operating power of 350 W is in good agreement with experiments. For both power, rubidium is mainly responsible for ionisation in the hot region.

For higher power, about 1200 W and which can't be acheived experimentally, results are quite coherent and show an important increase in axial temperature. This indicates an enhanced ionisation of krypton, resulting to an important depopulation of rubidium in the core. Emission of the 811,3 nm krypton line becomes thus important.

It seems possible, provided that the cold temperature can be modified, to combine excitation and ionisation of both rubidium and krypton for specific applications where infrared radiation is desired.

Conclusion

Elementary and simple channel type arc-model works well for high pressure Mercury and Sodium lamps. In this work we show that such model can be extended to more complex discharges namely the high pressure rare gas-metal halide discharges, provided that some necessary and adequate modifications are achieved.

Acknowledgements

Authors are indebted to "Quantel" and "Verre & Quartz Flashlamps" for respectively their financial and material support for this work. We also thank K. Charrada and M. Aubès for their useful remarks.

References

[1] - N. Sewraj : Propriétés électriques et radiatives des décharges haute pression gaz rare/halogénures métalliques; Thèse de Doctorat d'Université, Université Paul Sabatier, Toulouse - France, (1994).
[2] - N. Sewraj, J. P. Ritter, J. J. Damelincourt, K. Charrada : Modulation of the chemical composition and the conductivity of a high pressure rare-gas/RbI discharge operating at industrial frequency (50 Hz); ICPIG XXI, Germany, (1993).
[3] - N. Sewraj, H. Elloumi : Comaparive study of electrical and radiative properties of high pressure Kr/RbI & Xe/RbI discharges; Gas Discharge XIth International Meeting, Japan, (1995).
[4] - N. Sewraj : Plasma composition of a rare-gas/RbI test discharge destined to the optical pumping of the Nd YAG laser rod; Eureka Eu226 7th International meeting, Munich - Germany, (1993).
[5] - N. Sewraj, L. Sevier, M. Aubès : Noble gas-metal halide high pressure lamps - Basic ideas, properties & possible developments; Eureka Eu226 8th International meeting, Budapest - Hungary, (1994).
[6] - W. Elenbaas : The high pressure mercury vapour discharge; Pg 55ff, Amsterdam - North Holland, (1951).
[7] - M. Stambouli : Modélisation d'une décharge de mercure haute pression - Application à l'analyse de circuits électriques comportant des lampes à décharge; Thèse de troisième cycle, N° 10, ENSET, Université de Tunis, Tunisie, (1984).
[8] - J. F. Waymouth : An elementary arc model of the high pressure sodium lamp; Journal of IES, 131-140, (1994)

UNSTABLE AND TURBULENT GAS FLOWS IN INCANDESCENT LAMPS: MEASUREMENT AND MODELLING

K.Joyce, R.Cairns, R.Devonshire and P.Heeley

High Temperature Science Laboratories, Chemistry Department,
University of Sheffield, Sheffield, S3 7HF, UK

The numerical modelling of the fluid flow behaviour of the filling gas in incandescent lamp systems (halogen and non-halogen) that has been published to date, including that from our own group, has assumed that the background fluid flow environment is laminar. The equations governing fluid continuity and conservation of momentum are assumed to possess stable solutions, i.e., solutions which suppress small random perturbations within a particular low range of Reynolds number, **Re**.

The dimensionless quantity **Re** is defined by **Re** = UL/ν, where U and L are velocity and length scales respectively, and ν is the kinematic viscosity. For simple pipe flow the onset of turbulent flow occurs for **Re** values above 2000. A useful brief discussion of the transition from laminar, through unstable regimes, to turbulent flow is given by Swinney and Gollub (1978). It is not obvious how to estimate **Re** for the recirculating gas flows found in lamps to judge *a priori* whether or not turbulence exists. The temperature gradients are much higher than in conventional fluid flow laboratory experiments and the values of U and L are ill defined. However, it is reasonable to imagine that as U increases the Reynolds number will pass out eventually of the range required by the solutions of the fluid flow equations to remain stable. Small random fluctuations, possible injected by lamp structures such as filaments and supports will no longer be suppressed and the flow patterns will begin to exhibit extremely complicated and unpredictable behaviour seemingly unrelated to the boundary conditions of the problem. This is the onset of turbulence.

There is experimental evidence from a number of sources for the existence of unstable / turbulent flows in incandescent. For example, in the study of Gillham *et al* (1975) such flows manifested themselves as fluctuations in the radiant output of a tungsten ribbon lamp used as a temperature standard.

This particular aspect of fluid flow must be considered seriously as a component of any realistic mathematical model of a working incandescent lamp. The laminar fluid flow inside the envelope is always convective and as such there will be turbulence left in the wake of the convective plumes. The transport of species is radically altered by the onset of turbulence as tracer visualisation experiments have shown (Blackwelder and Kaplan (1976) and Johansson and Alfredsson (1982)). Also, in models involving even basic chemistry, the way in which the reaction rates are modelled is drastically altered.

The major intrinsic disadvantage of any mathematical model of gas flows is their inability to describe the transition to turbulence, i.e. it is not yet possible to model the spontaneous degradation of laminar flow into turbulence. This is because the sets of equations used to describe laminar and turbulent flow are not limiting forms of each other as Re → 0, ∞. At present fluid flows can be modelled as laminar or turbulent, not both.

In the HTSL we have a research programme to investigate experimentally the presence of turbulent flows in operating lamps based on measurements of the probability density function (pdf) of local gas temperatures and refractive techniques with FFT analysis. Strong unstable / turbulent flows with well defined spatial distributions have been found in both conventional incandescent lamps and in halogen lamps.

The experimental work is supported by a modelling program based on the commercial CFD package FLUENT. We are able to compare the results (converged temperature and flow velocity fields) of laminar and turbulent cases for similar temperature boundary conditions and to compare, for example, the distribution of turbulent energy with the experimental observations. The turbulence model is based on the standard, semi-empirical k—e model.

The paper will address generally the issue of the differences between model predictions of lamp behaviour with or without the inclusion of turbulent flow regimes.

References:

H.L.Swinney and J.P.Gollub, "The transition to turbulence", *Physics Today*, **1978**, August, p41.

E.J.Gillham, D.Harmer and J.R.Moore, "Fluctuations in the light output from gas-filled incandescent lamps", *Nature*, **1975**, *258*, 416.

R.F.Blackwelder and R.E.Kaplan, "On the Wall Structure of the Turbulent Boundary Layer", *J. Fluid Mech.*, **1976**, *76*, pp. 89-112.

A.V.Johansson and P.H.Alfredsson, "On the Structure of Turbulent Channel Flow", *J. Fluid Mech.*, **1982**, *122*, pp. 295-314.

SPECTROSCOPIC DIAGNOSTICS OF GLOW DISCHARGES

J. E. Lawler and N. D. Gibson
Department of Physics, University of Wisconsin, Madison, WI 53706 USA

1. Introduction

Spectroscopic diagnostics which are most useful in low pressure lighting discharges are reviewed. Primary emphasis in this review is given to work in Hg-Ar glow discharges, although the techniques are much more broadly applicable. Representative examples of work using emission, laser induced fluorescence, optogalvanic, anomalous dispersion (hooks), and absorption spectroscopy are discussed.

2. Emission Spectroscopy

The importance of radiometrically calibrated emission spectroscopy as a glow discharge diagnostic should not be underestimated. Works of this type are often focussed on determining an overall power balance of the discharge (Koedam et al., 1963; Dorleijn and Jack, 1985). Typically a calibrated lamp is used as secondary radiometric standard for determining the absolute response of a spectrometer-detector combination. Improvements in these calibration lamps have made this type of work somewhat easier than in the past, but accurate absolute measurements of emission spectra continue to require great care.

The most widely used secondary standards of spectral radiance and irradiance in the near infrared and visible are tungsten filament lamps. Typically a grating spectrometer, possibly equipped with a detector array, is used in this type of work. The traditional tungsten strip or ribbon lamp is usually calibrated as a standard of spectral radiance (power / (solid angle x area x bandwidth)). A spectral radiance standard is imaged directly onto the entrance slit of the spectrometer with a coupling lens which illuminates most or all of the grating. The relatively low temperature tolerated by strip lamps causes these lamps to be deficient in ultraviolet radiation. The use of a tungsten strip lamp, even in the near ultraviolet, is fraught with difficulty due to error in the calibration from scattered light or leakage through the spectrometer of the massive amount of visible light from the tungsten lamp. A double or triple Raman spectrometer has the advantage of extremely low scattered light. The tungsten-quartz-halogen lamp operates a much higher temperature and thus produces much more ultraviolet. The complex and nonuniform structure of the filament makes this lamp best suited to serve as a secondary standard of spectral irradiance (power/(area x bandwidth)) at a fixed distance). Spectral irradiance standards are used to calibrate the response of spectrometer-detector system by simply placing the lamps at the recommended distance from the entrance slit without a coupling lens. Only a limited area of the grating is illuminated by the lamp, but any possible nonuniformity across the grating or mirror surfaces can be mapped by translating the lamp perpendicular to the optical axis of the spectrometer. Tungsten-quartz-halogen lamps are typically calibrated for wavelengths down to 250 nm, but often users find it more convenient to switch to a deuterium discharge lamp for calibrating at wavelengths below 300 or 350 nm. Error due to scattered light in the spectrometer is again the concern. Deuterium lamps produce most of their radiation in the 200 nm to 400 nm region. Deuterium discharge lamps, like tungsten-quartz-halogen lamps, are usually calibrated as standards of spectral irradiance. In the USA tungsten strip lamps, tungsten-quartz-halogen lamps, and deuterium discharge lamps are available commercially with NIST-traceable calibration or can be directly calibrated by NIST (National Institute of Standards and Technology). Radiometric calibration of spectrometers in the vacuum ultraviolet from 115 nm to 330 nm in the ultraviolet is now much easier with the development of the argon miniarc (Bridges and Ott, 1977). This lamp is calibrated as a standard of spectral radiance or irradiance by NIST.

Accurate absolute emission spectroscopy requires many things including: a good radiometrically calibrated standard lamp, careful control of scattered light in the spectrometer, a well defined etendue (solid angle x area) for the optical coupling system, and knowledge of the angular distribution of the radiation from the discharge under study. The angular distribution of the radiation from a typical glow discharge is difficult to predict due to radiation trapping. A glow discharge is often not well described as a simple isotropic volume radiator nor as Lambertian surface radiator. Monte Carlo simulations of radiation trapping are increasingly useful for describing the angular distribution of the radiation (Doughty, 1995).

3. LASER INDUCED FLUORESCENCE SPECTROSCOPY

Laser induced fluorescence (LIF) with tunable dye lasers is now widely used to diagnose lighting discharges; it will be possible to cite only a few examples here. The high sensitivity provided by LIF enabled Battacharya (1990) to map the density of Ba and Ba^+ near the electrode of a Hg-Ar discharge. The loss of emission mix, Ba-Sr-Ca oxide, from the electrodes of fluorescent lamps is a common failure mechanism. Neutral Ba and Ba^+ have very convenient wavelengths, 553.5 nm and 455.4 nm respectively, for laser excitation. Fluorescence from the resonance level of Ba^+ is detectable at 614.2 nm, which is particularly convenient because it can be easily isolated from laser light scattered from the lamp wall or windows. Battacharya proposed using an absorption measurement to provide an absolute calibration of the LIF measurements. Detailed spatial (near the electrode) and temporal (during the ac period) maps of the Ba and Ba^+ densities from his experiment were compared with a model to gain a more quantitative understanding of the loss of emission mix from the electrode of a fluorescent lamp.

The excellent time resolution provided by pulsed LIF enabled van de Weijer and Cremers (1985a) to measure the trapped decay rate of the resonance level of Hg-Ar positive column discharge for a wide variety of conditions. A laser at 404.7 nm excited Hg atoms from the 6^3P_0 metastable level to the 7^3S_1 level, these atoms subsequently radiatively decayed to the 6^3P_0, 6^3P_1, and 6^3P_2 levels. The relaxation time of the perturbation of the 6^3P_1 population was measured by observing the 253.6 nm fluorescence. This relaxation time is the trapped decay rate of the resonance 6^3P_1 level. This experiment represents an important test of radiation trapping calculations for typical conditions in Hg-Ar lighting plasmas.

The very small spectral bandwidth provided by a continuous wave dye laser was essential for the detection of very small Stark broadening and shifts in an LIF experiment by Cornelissen and Burgmans (1982). Their experiment was performed in a Na-Ne discharge. They used Doppler-free two-photon excitation of the 3s-4d sodium transition from laser radiation at 578.3 nm and detected the 4p-3s cascade fluorescence at 330 nm. The use of a technique for high resolution Doppler-free spectroscopy enabled them to measure Stark shifts as small as 10 MHz. These Stark shifts and broadenings are due to Holtzmark fields in the plasma and thus are directly related to the local plasma density. Electron densities as small as 10^{12} cm^{-3} were measured with submillimeter spatial resolution.

A very high power Nd:YAG pumped dye laser was frequency doubled and Raman shifted into the vacuum ultraviolet in order to perform LIF on Hg^+ by Wamsley et al. (1991). A Hg-Ar discharge carrying 400 mA in a 35 mm diameter tube was used in the experiment. Wamsley et al. explored several schemes for LIF on Hg^+; they concluded that the simplest scheme which involved direct laser excitation and fluorescence detection at the 194.2 nm resonance line of Hg^+ was best. A fascinating nonlinear dependence of the negative glow ion density on discharge current was observed in this experiment. Wamsley et al. suggested that the nonlinear dependence of the plasma density in the negative glow might be due to variation in the ambipolar diffusion coefficient from changes in the temperature of the cold trapped negative glow electrons as the discharge current varied. The total ion production rate in the negative glow is very likely proportional to the discharge current and thus one would expect the ion density to be proportional to the current, unless the ambipolar diffusion coefficient is not constant.

The preceding four experiments illustrate some of the potential of LIF as a diagnostic of lighting discharges.

4. Laser Optogalvanic Spectroscopy

Optogalvanic effects are perturbations to the electrical characteristics of a discharge due to absorption of radiation at a wavelength corresponding to an atomic or molecular transition. Optogalvanic detection is particularly useful for observing transitions which do not fluoresce. Transitions to Rydberg levels do not fluoresce due to collisional quenching of the Rydberg atoms in a typical glow discharge. Rydberg atoms experience large and easily measured Stark effects. Lawler and Doughty (1994) have recently reviewed techniques for mapping electric fields in glow discharges using optogalvanic detection of Rydberg atoms and other methods. Techniques for mapping space charge fields have been extremely useful in microscopic studies of a variety of discharges; these techniques have not yet been extensively used to study lighting plasmas.

Although optogalvanic effects are most widely used as a detection method, they also provide insight on processes in a discharge. An interesting experiment by van de Weijer and Cremers (1985b) explored changes in the temporal characteristics of an optogalvanic effect due to varying the ballast resistor of a Hg discharge. It is simplest to think of an optogalvanic effect as a shift in the voltage-current characteristics of the discharge and then to use a load line analysis to determine if the shift results primarily in a voltage perturbation or in a voltage and current perturbation. A large ballast resistor forces the optogalvanic effects to become a perturbation to the discharge voltage and thus E/N (reduced field). A smaller ballast resistor results in optogalvanic effects which are significant perturbations to the discharge current. The ballast resistor couples the perturbation to the electron energy and density, which affects the temporal response.

Doughty and Lawler (1983) measured the absolute magnitude of optogalvanic effects in a Ne positive column for comparison to a model. It is not difficult, starting from a model of an unperturbed discharge, to apply first order perturbation theory and generate a model of the optogalvanic effect. Optogalvanic effects are almost always linearly proportional to the absorbed laser power. Saturation of optogalvanic effects is usually due to saturation or "bleaching" of the atomic transition while the discharge perturbation continues to be linear with absorbed laser power (Lawler, 1980). A direct comparison of absolute measurements of optogalvanic effects to prediction from a discharge model is a stringent test of the model. Although very sophisticated models of Hg-Ar discharges and other lighting discharges are available, the authors are unaware of any significant effort to test the models using optogalvanic effects.

5. Anomalous Dispersion Spectroscopy

The method of Rozhdestvenskii (1912) for using anomalous dispersion to form "hooks" in a spectrum and thus determine absolute column densities has been widely used as a diagnostic for lighting plasmas. The hook technique, although it is a close relative of direct absorption, is different in that it is useful for very large column densities. It is often impossible to determine accurate column densities from an absorption experiment in which the line center is opaque. The difficulty is usually due to a lack of knowledge of the wings of spectral lineshape which determine the total absorption signal or "equivalent width" in the limit of large column densities. The advantage of the hook technique is due to the fact that variation in the real part of the index of refraction in the vicinity of a spectral line are rather insensitive to variations in the spectral lineshape. Observation of the hooks typically requires a Mach-Zehnder or similar interferometer. Recent experiments using the hook technique have involved dye lasers. The dye lasers offer advantages such as better spatial and temporal resolution.

Absolute densities of 6^3P atoms in a Hg positive column discharge were measured by van de Weijer and Cremers (1982) using the hook technique with a dye laser. Densities were mapped for a range of wall temperatures and discharge currents. These density measurements were subsequently used to determine the cross sections for various 6^3P-6^3P collisional processes which populate highly excited levels of Hg.

More recently Bigio (1988) used the hook technique with a dye laser to measure absolute densities of 6^3P atoms in a Hg-Ar positive column discharge. He mapped absolute densities and radial variations for comparison to a model (Dakin and Bigio, 1988). He also tested the technique of saturated laser absorption in the Hg-Ar discharge.

6. Absorption Spectroscopy

Absorption spectroscopy has long been used to diagnose lighting plasmas. Absorption spectroscopy is suited to lower column densities than the hook technique described in the preceding section. It is possible in many cases to compute a curve-of-growth from the known spectral lineshape in order to quantify the relation between absorption equivalent widths and moderately large column densities. Both absorption and the hook technique provide absolute column densities unlike LIF.

In an important early experiment, Koedam and Kruithof (1962) used the line emission from one Hg-Ar discharge to measure densities of 6^3P atoms in a second Hg-Ar discharge. This approach has important advantages in terms of the simplicity of the experimental apparatus. Some significant care is required in the analysis of data from this type of line absorption experiment. Accurate quantitative density measurements are possible only if the detailed spectral lineshapes for both discharges are well known. The lineshapes must include radiative, collisional, and Doppler broadening, hyperfine structure, and any effects from radiation trapping in the first discharge.

Laser absorption experiments have several advantages including: superior spectral resolution, excellent spatial and temporal resolution, and great ease in discriminating laser light from discharge emission. Often very simple spatial filters can be used to separate the laser light from discharge light.

The saturated laser absorption method reported by Bigio (1988) offers the advantage of providing absolute column density measurements at rather large column densities without detailed knowledge of the spectral line profile. This technique requires a measurement of the total absorbed laser energy in the limit of a highly saturated atomic transition. The number of absorbed photons is approximately 50% of the absolute number of absorbing atoms in the laser beam path.

Wamsley et al. (1990) explored the use of a continuum from an arc lamp to perform absorption spectroscopy on excited Hg atoms in the 6^3P levels and ground state Hg^+ in a Hg-Ar discharge. Their experiment, like many others, used phase sensitive detection to discriminate against line emission from the Hg-Ar discharge and detect the continuum from the arc lamp with absorption features from the glow discharge. They used an echelle grating in their spectrometer to provide better spectral resolution. The absolute density measurements on Hg^+ in the negative glow, Faraday dark space, and positive column of a Hg-Ar discharge were one of the most interesting results from the experiment.

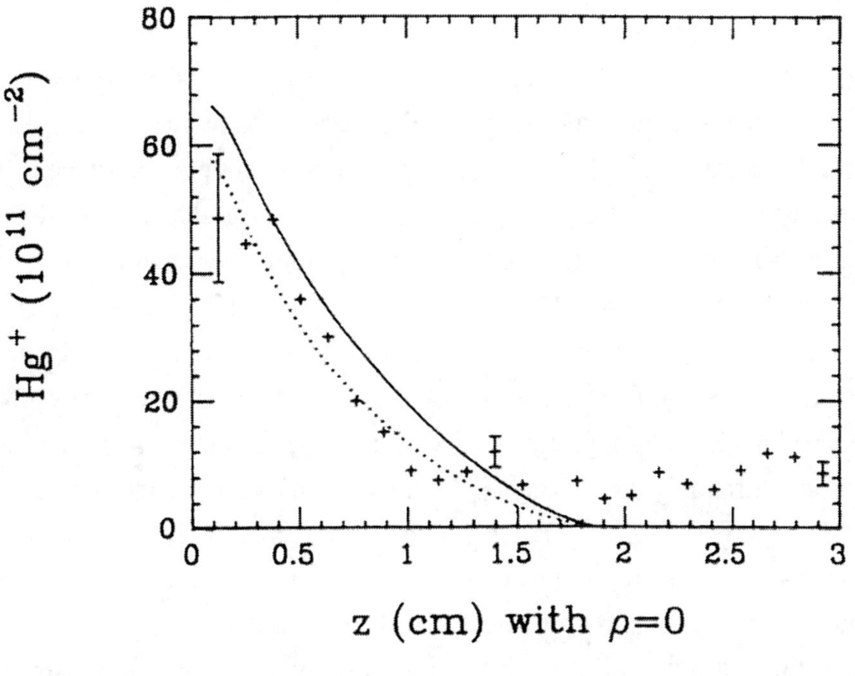

Figure 1 - The crosses are measurements of the Hg-ion column density taken through diameters of the lamp. The error bars represent the dominant random uncertainty in the measurements. The dotted and dashed lines show the Hg-ion density calculated using two different models.

Wamsley et al. (1993a) subsequently used absorption spectroscopy to map the absolute density of excited Ar atoms in the $3p^5 4s$ configuration and the density of ground state Hg^+ in the negative glow. These maps were compared to models of the negative glow. They found that Penning ionization of Hg atoms by excited Ar atoms played a major, possibly dominant, role in the ionization balance of the negative glow. Energetic electrons from the hot cathode first excited Ar atoms to the $3p^5 4s$ configuration. The populations of the metastable and resonance levels of the lowest excited $3p^5 4s$ configuration of Ar are tightly coupled in the negative glow by low energy electron collisions. All of these excited Ar atoms then behave as resonance atoms and they spread throughout the negative glow due to resonance radiation transport. Ultimately many of the excited Ar atoms Penning ionize Hg atoms.

Wamsley et al. (1993b) recently developed a technique for "High Sensitivity Absorption Spectroscopy" in glow discharge plasmas. The sensitivity of many single channel-sequentially scanned absorption experiments is limited by fluctuations in the continuum sources. Such fluctuations are mapped directly into noise in the spectrum. Use of a diode or CCD detector array in the focal plane of the spectrometer makes the absorption experiment insensitive to these fluctuations because all spectral elements are detected simultaneously. Photon statistical (Poisson) noise is the next important limit in the sensitivity of a typical absorption experiment. The detector array also provides a major advantage by simultaneously collecting good photon statistics in all spectral elements. Wamsley et al. (1993b) demonstrated a method for digital subtraction to discriminate against the line emission from the glow discharge and detect only the continuum emission from an arc lamp or synchrotron with absorption features from the glow discharge. Although use of detector arrays for emission spectroscopy has been common for many years, arrays have not been as widely used for absorption spectroscopy [see Fig. 2]. The advantages of arrays in absorption spectroscopy are very substantial. These simple, but powerful ideas enabled Childs et al. (1994) to detect fractional absorptions as small as 0.00001 in glow discharges used for chemical vapor deposition of diamond films. Such high sensitivity is essential for direct detection of many molecular radicals.

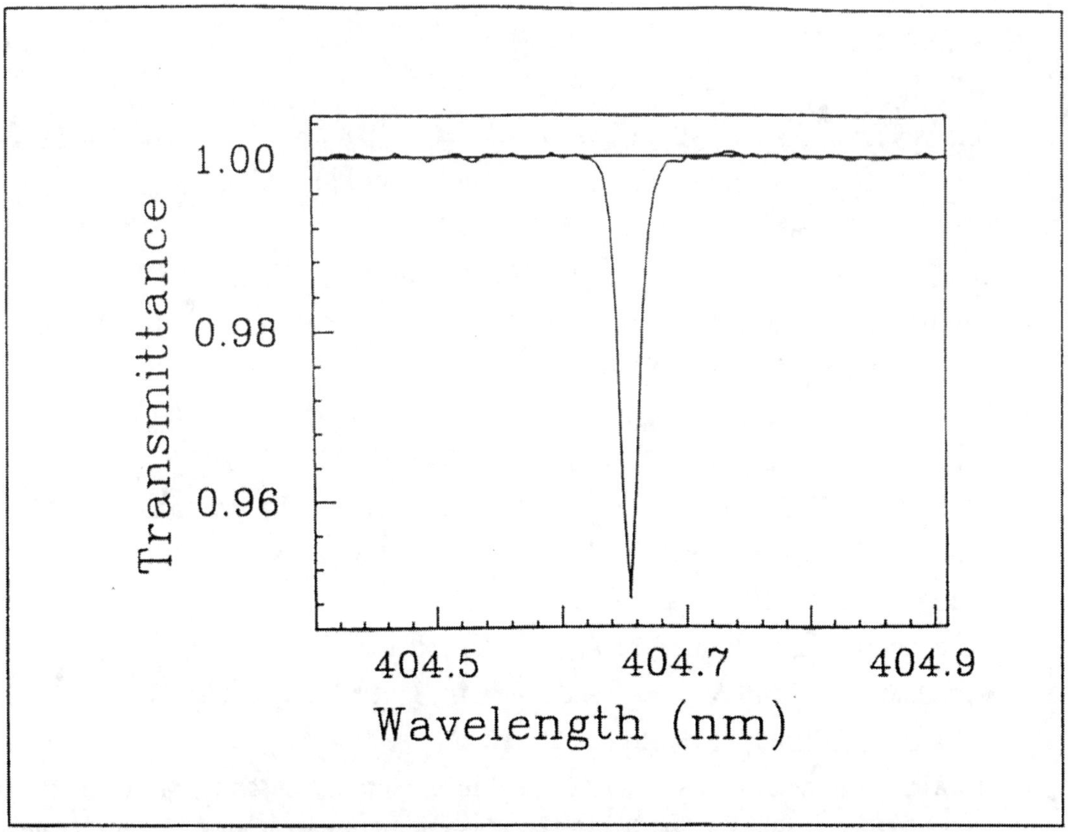

Figure 2 - Sample Absorption data recorded on the Hg 404.66 nm line using the gated, image-intensified, CCD detector array. The noise on the continuum is primarily shot noise.

7. Conclusion

Progress in spectroscopic diagnostics for glow discharges used in lighting and other applications has been quite substantial in recent years. New techniques using tunable lasers and advanced detector arrays have been developed. These techniques are now being exploited to provide a more quantitative understanding of these discharges at a microscopic level.

8. REFERENCES

Bhattacharya, A. K. 1990, in *Nonequilibrium Processes in Partially Ionized Gases*, ed. M. Capitelli and J. N. Bardsley (New York:Plenum), 669

Bigio, L. 1988, J. Appl. Phys. **63**(11), 5259

Bridges, J. M., and Ott, W. R. 1977, Appl. Optics **16**, 367

Childs, M. A., Menningen, K. L., Toyoda, H., Anderson, L. W. and Lawler, J. E. 1994, Europhys. Lett. **25**(9) 729

* Cornelissen, H. J. and Burgmans, A. L. J. 1982, Optics Comm. **41**(3) 187

Dakin, J. T. and Bigio, L. 1988, J. Appl. Phys. **63**(11), 5270

Dorleijn, J. W. F. and Jack, A. G. 1985, J. Illuminating Engineering Soc. **15**, 17

Doughty, D. A., and Lawler, J. E. 1983, Phys. Rev. A **28** (2) 773

Doughty, D. A., 1995, J.Appl.Phys. 77, 1393

Koedam, M. and Kruithof, A. A. 1962, Physica **28**, 80

Koedam, M., Kruithof, A. A., and Riemens, J. 1962, Physica **29**, 565

Lawler, J. E. 1980, Phys. Rev. A **22**, 1025

Lawler, J. E. and Doughty, D. A. 1994, in *Advances in Atomic, Molecular, and Optical Physics*, **34**, ed. Benjamin Bederson and Herbert Walther (New York:Academic Press), 171

Rozhdestvenskii, D. S. 1912, Ann. Phys. **39**, 307

van de Weijer, P. and Cremers, R. M. M. 1985a, J. Appl. Phys. **57**(3), 672

van de Weijer, P. and Cremers, R. M. M. 1985b, Optics Comm. **53**(2), 109

van de Weijer, P. and Cremers, R. M. M. 1982, J. Appl. Phys. **53**(3), 1401

Wamsley, R. C., Lawler, J. E., Ingold, J. H., Bigio, L., and Roberts, V. D. 1990, Appl. Phys. Lett. **57**(23) 2416

Wamsley, R. C., O'Brian, T. R., Mitsuhashi, K. and Lawler, J. E. 1991, Appl. Phys. Lett. **59**(23) 2947

Wamsley, R. C., Mitsuhashi, K. and Lawler, J. E. 1993a, Phys. Rev. E **47**(5), 3540

Wamsley, R. C., Mitsuhashi, K. and Lawler, J. E. 1993b, Rev. Sci. Instrum. **64**(1), 45

MEASUREMENT OF Ar LOWEST EXCITED STATE DENSITIES AND TEMPERATURES IN Ar AND Ar-Hg DISCHARGE

Kunio Yuasa*), kazue Yamashina**), Takeki Sakurai**)
*) Toshiba Lighting & Technology Corporation, 1-201-1 Funakoshi-cho Yokosuka-shi 237, Japan
**) Yamanashi University, 4-3-11 Takeda Koufu-shi Yamanashi-ken 400, Japan

Because of high stored energy and high population densities, the lowest excited states of rare-gas atoms have important roles in rare-gas discharges and also in fluorescent lamps especially near electrodes and in their low ambient temperature operations.
However, experimental estimations of them in such situations are still few[1),2)]. Here we report on our experimental results about not only lowest excited state densities of Ar, but also their temperatures.

Densities and temperatures of Ar lowest excited states $1s_i$ (i=2,3,4 and 5) are measured with single-mode AlGaAs semiconductor diode lasers. A typical temperature stability of the diode lasers is about 10^{-4}K, current stability is about 1μA, spectral width is less than 10MHz and short-time stability is less than 1MHz. Frequencies are swept over 5GHz around their absorption peak values, which are accomplished by changing their operating temperatures. Frequency calibrations, effect of saturation, effect of laser induced drift and overlap of the spectral lines are checked. Absorption and LIF(Laser Induced Fluorescence) technique are used. Absolute values of densities are estimated from the peak of absorption profiles, and temperatures are from fitting of the Voigt profiles.

Fig.1 shows dependence of the densities on the discharge current in a 30mm inner diameter Ar 400 Pa filled discharge lamp. Fig. 2 shows dependence of the temperatures on discharge current with the data of gas temperatures measured by thermo-couple in the same lamp.

Following results are obtained.

In a 30mm inner diameter Ar 400 Pa filled discharge lamp, density of $1s_3$, $1s_5$ at the center is $4x10^{16}$, $0.5x10^{16}$ [$/m^3$] respectively. Their distributions on axial and radius directions are obtained.

In Ar-Hg discharge lamp operated at 305K, Ar lowest excited state densities in positive column are less than 1/100 compared with them in Ar discharge lamp.

Temperatures of Ar lowest excited states are higher than gas temperature for 150~200 degrees.

1) K.Tachibana: Phys. Rev. 34, 1007 (1986)
2) R. C.Wamsley et.al.: 45th Gaseous Electronics Conference (1992)

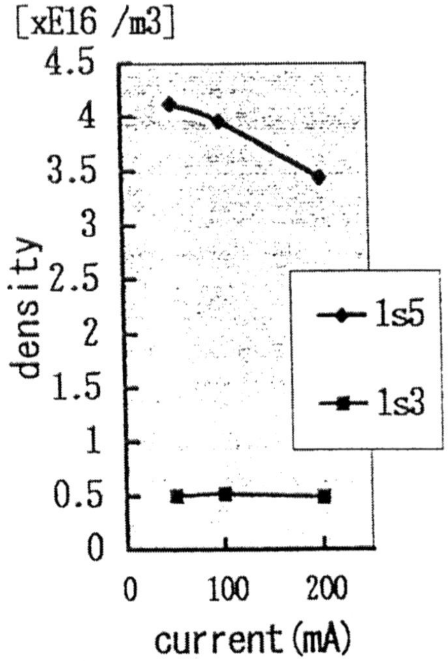

Fig.1 Density of 1s3, 1s5

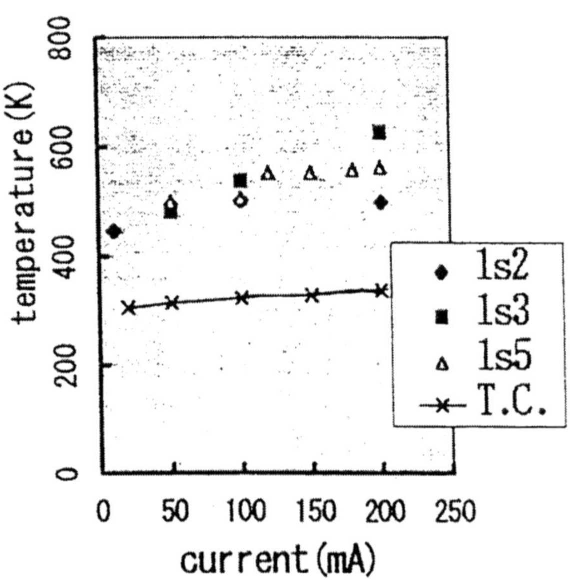

Fig.2 Temperatures of 1s2, 1s3 1s5 and thermo-couple

AN INVESTIGATION OF THERMODYNAMIC vs. KINETIC CONTROL IN OPERATING HALOGEN LAMPS

J.R.Woodward,* M.J.Abbott, D.D.Bruguier and R.Devonshire

High Temperature Science Laboratories, Chemistry Department, University of Sheffield, Sheffield, S3 7HF, UK

When modelling the spatial distribution of chemical species in reactive gas flows in cases where (large) temperature gradients exist it is convenient to assume that the chemical composition at any point in the system is defined by the temperature at that same point. This "local" temperature is assumed to be an equilibrium temperature and the resulting condition of the system is referred to as Local Thermodynamic Chemical Equilibrium (LTCE). If the temperature distribution in a system can be determined then species distributions can be predicted (N.B. to a first approximation only [1]) by minimisation of the Gibbs free energy at each point [2]. The necessary thermochemical data to carry out this procedure for many systems of practical interest is relatively secure and comprehensive compared to the kinetic data that would be necessary to describe the same system on the basis of a chemical kinetic model. The LTCE model may be viewed as a limiting case of a general kinetic model where the rate coefficients of all the reactions in the overall mechanism have been scaled up to ensure that equilibrium is maintained at all points in the system.

The growing importance of computer modelling in lamp design generally nas prompted us to carry out in the HTSL a comprehensive investigation of the validity and application of the LTCE assumption in halogen lamp systems. There is experimental evidence that LTCE conditions do not exist in operating halogen lamps [3]. The studies involve the direct measurement of physical property and multiple species distributions in operating lamps using spatially resolving laser spectroscopic techniques and comparison of the results with the predictions of both LTCE and kinetic computer models of the lamp chemistry. The computer models combine to different levels of approximation computational fluid dynamics (CFD) simulations of the bulk fluid's properties with thermochemical data and kinetic data of the species and reactions known / believed to be relevant.

For a number of important reasons the lamp type chosen for study was a GE DYS, 120V 600W halogen projector lamp. The cold-gas fill of this lamp is 600 torr comprising 95% N_2 and 5% HBr. CARS (coherent anti-Stokes Raman scattering) measurements of three species N_2, HBr and Br were carried out for numerous points within the 4.8cm^3 internal volume of the lamp along several tracks above, below, and through the filament at 3 filament operating temperatures (maximum values of 2030K; 2630K; and 2860K). Additionally, measurements were taken at points of special interest over a near continuous range of filament operating temperatures. The spatial resolution of each measurement is <1mm by 10s of μm.

A Gibbs free energy minimisation package MTDATA [4] shows the 3 principal species in the lamp over the full operating range to be N_2, HBr and Br, the same three species measured using CARS (N.B. the next most important species

H_2 could have been measured also by CARS had we wished to do so). The N_2-CARS spectrum provides a local temperature to ≅1%. The HBr-CARS spectrum provides information on the local HBr number density and an independent measure of the local temperature. The Br-CARS spectrum provides information on the local Br atom number density (one of the major general outcomes of this work has been to demonstrate that Br atom relative concentrations can be determined accurately down to 0.01%, well below the "textbook" CARS detection limit of 1%).

CFD simulations of the bulk fluid's behaviour were carried out using Fluent [5] after defining the complex lamp geometry with high fidelity using Ansys [6]. The 400,000 cell simulations provide a detailed picture of the temperature and velocity distribution in the lamp at each of the 3 selected filament operating temperatures.

As in previous studies by our group [7] excellent agreement was found between gas temperatures measured by N_2-CARS, T_{CARS} and those predicted by Fluent, T_{CFD}. Either temperature can be used to predict using MTDATA the relative concentrations of Br atom that would be present at the respective points within the lamp if LTCE conditions prevailed and if diffusion effects could be ignored [1]. We are interpreting the results we have obtained at several levels of approximation; the highest level at present being a 10 reaction kinetic model with explicit inclusion of diffusion effects [1,8].

To illustrate the detailed picture of halogen lamp chemistry emerging from our work we show in the Table below results obtained at different vertical heights above the centre of a filament operating at 2630K. $[Br]_{LTCE}$ is derived from T_{CARS} and MTDATA, i.e., assumes LTCE and no diffusion effects. The difference between $[Br]_{LTCE}$ and the measured Br atom relative concentration $[Br]_{CARS}$ is striking; a result that can be expressed in another way by working out the local temperature that would have to hold to achieve such a value of $[Br]_{CARS}$. The full analysis will appear in a series of publications in preparation.

Position /mm	T_{CFD} /K	T_{CARS} /K	$[Br]_{LTCE}$ /%	$[Br]_{CARS}$ /%	T_{LTCE} /K
0.25	2365	2326	1.88	3.60	2990
1.25	2184	2125	1.30	4.11	3290
2.25	1992	1875	0.679	4.53	3500
3.25	1850	1825	0.579	2.82	2650
4.25	1690	1750	0.446	2.20	2435
5.25	1545	1575	0.214	1.20	2090
5.75	1495	1425	0.094	0.40	1720

* Present Address: GE Lighting, Nela Park, Cleveland, Ohio 44112, USA.

References:

[1] P.Heeley, D.D.Bruguier and R.Devonshire (this volume); [2] see E.Schnedler in "High Temperature Lamp Chemistry", ed. E.G.Zubler, The Electrochemical Society, PV 85-2, p 95; [3] J.M.Ranish & L.E.Hoegler, LS:6, 52:P; [4] MTDATA, National Physical Laboratory, Teddington, UK; [5] Fluent Inc., Lebanon, NH, US; [6] Ansys, Strucom, Croydon, UK; [7] R.Devonshire *et al*, Chem.Phys.Lett., 129 (1986) 191; [8] M.J.Abbott, D.D.Bruguier, R.Devonshire, S.Hodson, P.Wild (to be published).

SPECTROSCOPIC STUDIES OF HIGH INTENSITY DISCHARGE LAMP

Makoto HAMAMOTO, Hidenori YAMASAKI, Takahiro KAI, Katsunori MURAOKA*, Taku SUMITOMO** and Shigeaki WADA**
Faculty of Engineering, Oita University, Oita 870-11 Japan
**Graduate School of Engineering Sciences, Kyushu University, Kasuga Fukuoka 816 Japan*
***Lighting Group, Matsusita Electric Works, Ltd., 1048 Kadoma Osaka 571 Japan*

Two types of spectroscopic measurements were performed in order to investigate a high intensity discharge lamp.

Spectral profiles of Na emission lines (568.266, 568.822nm), which are sensitive for the Stark effect, were measured to determine electron density distributions in a metal halide lamp (60 Hz AC 250 W ; metal halide additives [NaI, TlI, InI]) as a function of an AC voltage phase.

Stark-broadened and -shifted spectral profiles were measured using a high resolution spectrometer equipped with a photomultiplier. The spectrometer had a reciprocal dispersion of 0.030 nm/mm for the Na lines in the 39th diffraction order, and the photomultiplier had a slit of 100 μm wide. For the wavelength reference, emission lines from a Ne lamp were used.

Electron densities were obtained from comparison between the measured and the theoretical spectral profiles, and the results are shown in Figs. 1 and 2 as a function of an AC voltage phase at various radial and axial positions, respectively.

It is confirmed from the experimental results that (1) the electron density ranges from about 4 to 8 times 10^{21} m^{-3}, (2) the phase of electron density modulations is delayed about 30 degrees from those of AC voltage and current, (3) the radial distribution of electron density is rather flat around the arc axis ($r \leq 2$mm) where the density modulation is enhanced, (4) the electron density in the bottom of the lamp is higher than that in the top along the arc axis and (5) there is a strong

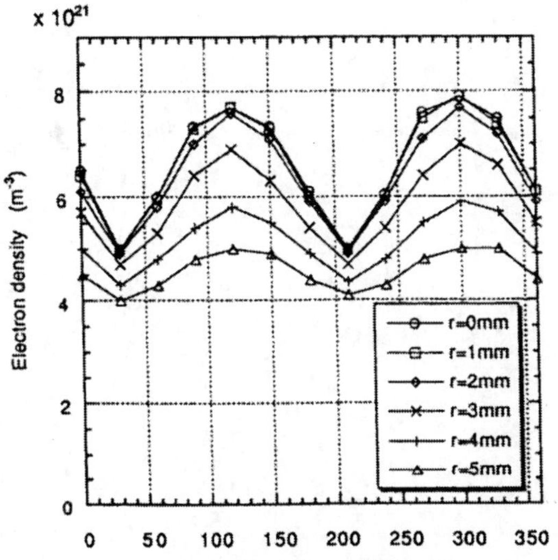

Fig. 1 Electron density variations in a metal halide lamp as a function of an AC voltage phase at various radial positions. [r=0mm:discharge axis]

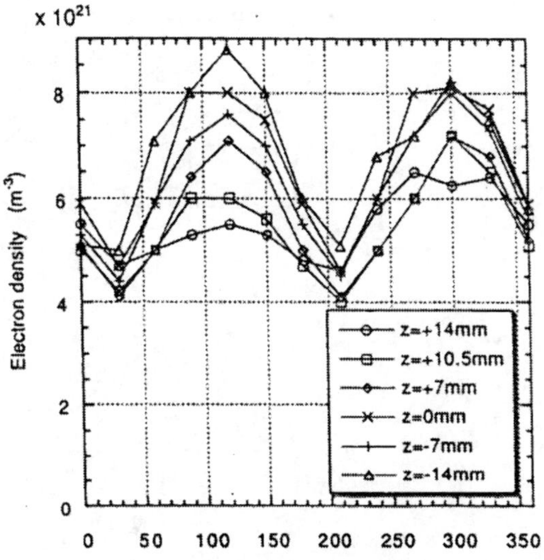

Fig. 2 Electron density variations in a metal halide lamp as a function of an AC voltage phase at various axial positions. [z=0mm:center between upper (z= + 15 mm) and lower (z= - 15mm) electrodes]

correlation between the variations of electron density and those of emission intensity of Na lines.

LIF measurements were also applied for three excited levels (6^3P_1, 6^3P_2, 6^1P_1) of mercury in three types of HID lamps (60 Hz AC 250 W ; (a) metal halide lamp [NaI, TlI, InI] same as the one used in the electron density measurements, (b) metal halide lamp [NaI, ScI_3] , (c) mercury lamp).

The tunable laser used for the measurements was a YAG-pumped dye laser, whose pulse were obtained synchronously with any phase of AC voltage of the lamp concerned. The observation volumes were chosen along the line of sight of the detection system (x-axis) and along the direction of the laser beam propagation (y-axis) in various cross sectional planes orthogonal to the arc axis (z-axis).

LIF signals, obtained for 6^3P_1 energy level of mercury in the metal halide lamp [NaI, TlI, InI] at 90 degrees of AC voltage phase, are shown in Figs. 3 and 4, as a function of reduced radius (r=8.5 mm at the lamp wall). Peaking of LIF signals in positive-x region near to the detection system and in negative-y region near to the laser source is clearly recognized in Figs. 3 and 4. This asymmetry of the radial distributions of LIF signals is explained by effect of the absorption of laser and fluorescence light.

Based on the suppositions that mercury is completely evaporated and local thermal equilibrium and pressure equilibrium exist, LIF signals were simulated in a following manner. An radial temperature distribution in the lamp was given and population densities of excited levels were calculated for the temperature, and then relative LIF signals were estimated for the population densities with considerations of the absorption of laser and fluorescence light. An example of the simulated data, which corresponded well with the experimental data shown in Figs. 3 and 4, showed that the temperature was about 4800 K and the population density of 6^3P_1 level was about 2×10^{20} m^{-3} at the arc axis.

The procedure mentioned above is shown to be effective for determining temperature and density distributions.

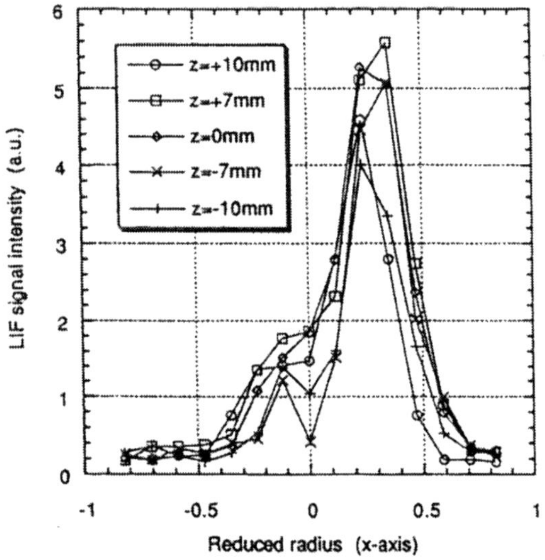

Fig. 3 LIF signal intensity for 6^3P_1 energy level of mercury measured along x-axis(the line of sight of detection system) at various axial positions in a metal halide lamp. [AC voltage phase:90 degrees]

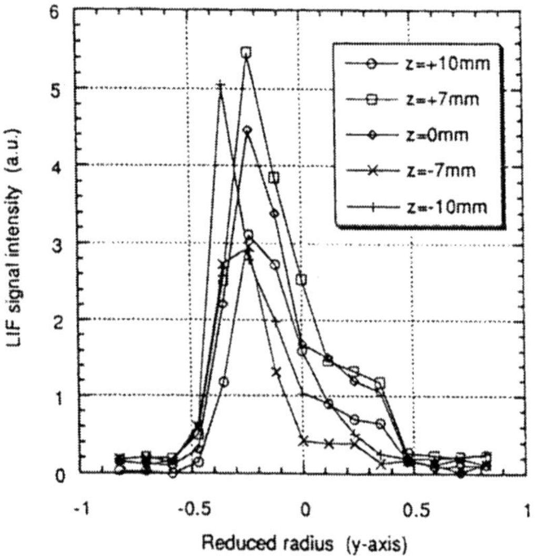

Fig. 4 LIF signal intensity for 6^3P_1 energy level of mercury measured along y-axis(the direction of laser beam propagation) at various axial positions in a metal halide lamp. [AC voltage phase:90 degrees]

WALL EFFECT ON OPTICAL EMISSION SPECTROSCOPY

I.Arikata, T.Kubota and K.Amakawa
Himeji Institute of Technology, Department of Electrical Engineering
2167 Shosha, Himeji, 671−22 Japan

It is am important task to study the effect of the wall on the optical emission of spectroscopy(OES) in the field of light sources. In this paper, the sheath width estimated from OES, in the vicinity of the walls, is compared to that calculated from the parameters of an argon plasma.

The experimental apparatus is shown in Fig.1. The chamber with a quartz window was made of stainless steel(50 mm in the inside diameter by 200 mm in length). A turbo−molecular pump (150 l/s) was used to evacuate the chamber. An oxide coated coiled coil tungsten filament was used as the cathode. A collector(20x12x0.5 mm^3 ,SUS304) was located with the inter−electrode spacing of 8.1 mm. The argon plasma was produced by the thermionic cathode discharge. The plasma parameters such as electron temperature, electron density, potentials were measured by a Langmuir probe. A sheath forming plate made of stainless steel(35x113x1 mm^3) covered with glass of a half area was located in the chamber. The emission spectra from the plasma were intro−duced to a monochrometer (Shimadzu :GE−100, Ebert type, the focal length :1 m).

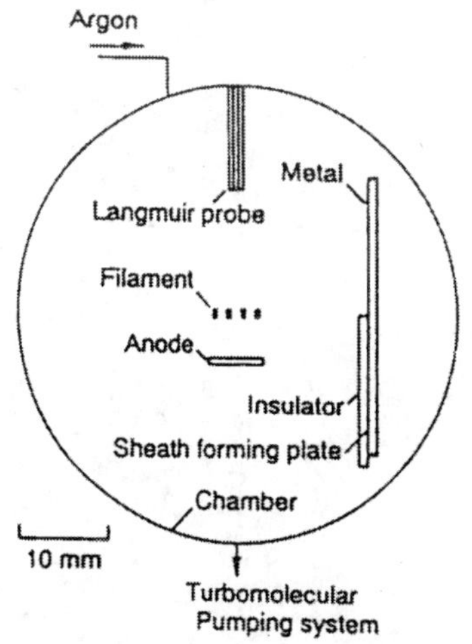

Fig.1 Experimental apparatus.

In this paper, spectral intensities of Ar_{II} : 480.6 nm(upper level:(4p) $^4P_{5/2}$:34.975eV), and Ar_{I} :470.23 nm(upper level:5p(1/2): 14.46eV) were used for diagnostics.

The normalized Vp−Ip curve at the space potential is shown in Fig.2, which denotes that the plasma has two electron−tempera−tures (Tes:slow, Tef:fast). Figure 2 shows that as P_{Ar} increases, the spectral intensity of Ar_{II} decreases drastically with Tef. Otherwise the intensity of Ar_{I} varied with the electron density n_e.

In the vicinity of the wall, because of the retarding field for electrons, slow electrons are repelled and fast electrons remain, therefore we can estimate the sheath width d_{sOES} from the ratio of the spectral intensities Ar_{II}/Ar_{I}.

The sheath width d_{sp} was calculated from plasma parameters by eq(1).

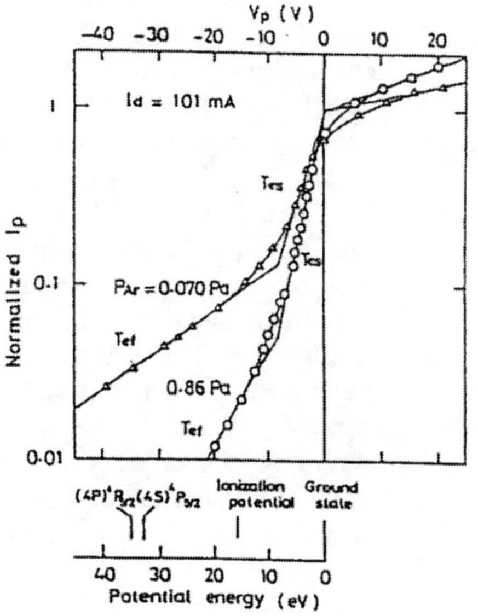

Fig.2 Normalized Vp−Ip curves and potential energy.

$$0.965\ \xi^2 = (\eta - 1/2)^{3/2}, \qquad (1)\ [1]$$

where $\xi = d_s/\lambda_D$, $\eta = eVp/kTe$, λ_D: Debye length: $(\varepsilon_0 kTe/n_e e^2)^{1/2}$.

The spacial variation of Ar_{II}/Ar_{I} is shown in Fig.3, which shows that we can estimate the sheath width $d_{sOES} ≒ 0.70$ mm. Otherwise the value of $d_{sp} ≒ 0.67$ mm was obtained at the same situation as Fig.3. Then we get $d_{sOES}/d_{sp} ≒ 1.04$. Thus we can estimate the sheath width by OES within an error of about 5%.

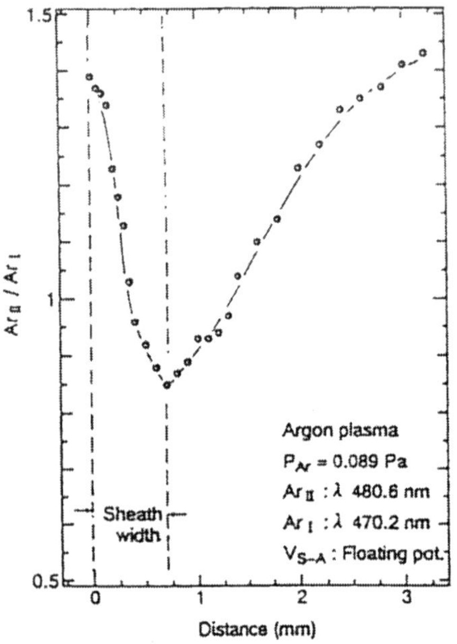

Fig.3 Ar_{II}/Ar_{I} as a function of the distance d from the sheath forming plate.

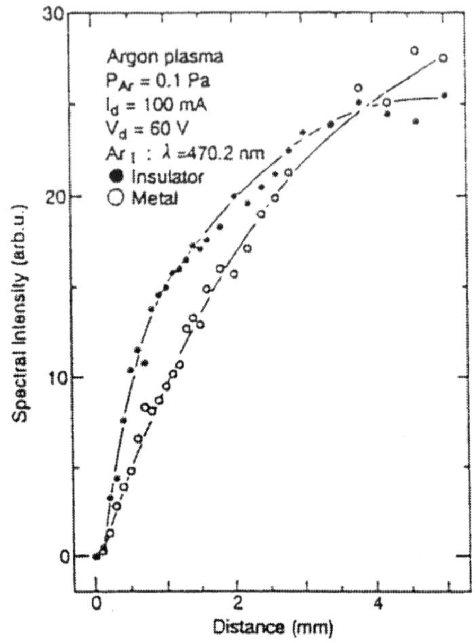

Fig.4 Intensity of Ar_{I} as a function of d.

The distsnce dependence from the wall (metal and insulator) of Ar_{I} is shown in Fig.4, which indicates that sheath width near the insulator is thinner than that near the metal from each decreasing point of Ar_{I} in the vicinity of the walls.

The intensities of Ar_{I} and Ar_{II} as a function of V_{S-A} (S:Sheath forming plate, A:Anode:refference electrode) in the plasma(distance from the plate d=2 mm) is shown in Fig.5, which denotes that each curve is devided into three regions: region I ($V_S > V_A$), region II ($V_W < V_S < V_A$), region III ($V_S < V_W$). The characteristics of each region can be explained from the fact that the intensity of Ar_{I} proportions to n_e and the intensity of Ar_{II} is affected by Tef.

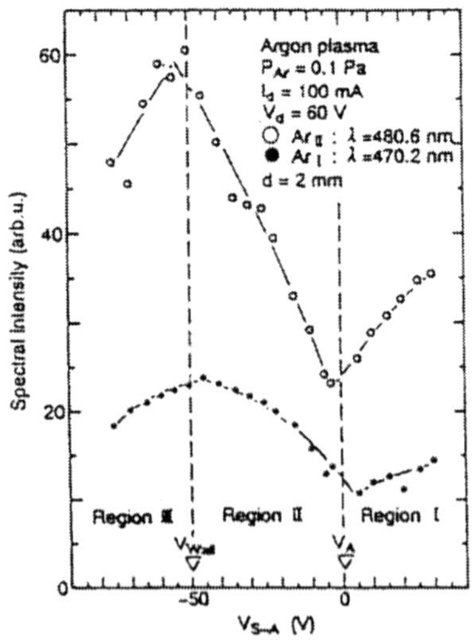

Fig.5 V_{S-A} dependence of Ar_{I} and Ar_{II}.

1) I.Arikata et al.:Proc.Jpn.Symp.Plasma Chem.,vol.1, 169(1988).

A VERSATILE MANUFACTURING TEST CELL FOR LOW PRESSURE DISCHARGE LAMPS

W.Yu, G.Gregory, P.Ingram and R.Devonshire

High Temperature Science Laboratories, Chemistry Department, University of Sheffield, Sheffield, S3 7HF, UK.

Computer based models play an important and increasing role in lamp design. They can contribute to the prototyping process by guiding and interpreting the empirical stages of lamp development, and to predicting the effects of iterative changes to existing lamp designs. The versatility and productivity of such models is not easily matched by the traditional lamp prototyping process which involves the laboratory (if not laborious!) production of small numbers of lamps to explore the design variables. The process itself, involving as it does individual lamps, introduces new variables. Can this important practical process, unlikely to be supplanted entirely by computer models, be made more efficient?

In the HTSL we have developed a novel chamber which we call a Manufacturing Test Cell (MTC). As presently configured it enables us to explore systematically the effect on the performance of a low pressure discharge lamp of making changes in a range of important lamp design features for a particular set of lamp components. The lamp currently of interest is the deuterium lamp used widely as a source of continuum UV radiation. These lamps are three electrode (anode, cathode and cup electrodes), hot cathode, glass / quartz envelope, devices filled with deuterium gas at low pressure (10 mbar) and represented schematically in Figure 1. A computer based model of the discharge is being developed in the HTSL (Harris et al (1995)).

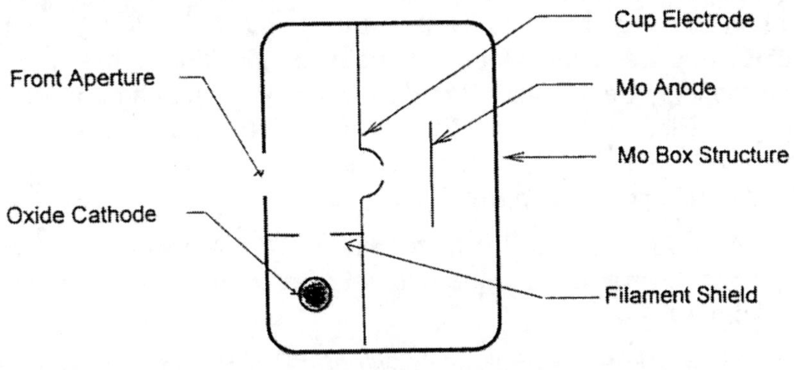

Fig.1: Schematic of the cross-section of a deuterium lamp

The cathode is typically a triple oxide type cathode. Movement of the plasma attachment area on the cathode's surface during lamp operation is believed to be a major contributor to fluctuations and instabilities in the radiant output (Page (1992)). Lamp performance is sensitive to the spacing of the electrodes, gas fill pressure, etc., and for these reasons it was a suitable candidate for investigations based on the MTC.

The MTC has UHV chamber performance (10^{-10} mbar). The lamp components are configured and, where necessary, processed in the UHV chamber which is then filled with gas / gas mixtures and the device is then operated as a normal lamp. Lamp components can be aligned, positioned, and rotated, gas pressures changed and monitored, gas composition changed, etc., during operation of the same basic device providing clear, revealing insights into the critical aspects of the design.

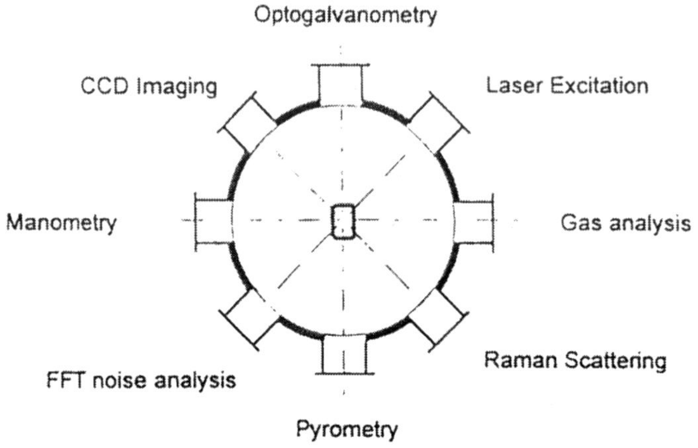

Fig. 2: Schematic of the main MTC chamber and several of the available diagnostic techniques

The MTC is equipped with a wide range of measurement techniques to monitor all stages of the assembly, processing and operation of the lamp devices. In addition to the conventional electrical and optical characterisation of the device's operation, and the ability to fill the devices with gas mixtures of accurately known composition (down to "impurity" levels), a number of advanced diagnostic techniques are available. These can characterise, for example: the radiant output level and its noise components; the radiant output spectrum; the distribution and condition of plasma constituents; the gas composition with high dynamic range; the time dependent behaviour of the plasma-cathode attachment areas; the distribution of work function over the cathode's surface; the temperature profile of the lamp structure and lamp components; the nature of deposits and films formed on surfaces during device operation.

We believe the work based on the MTC will provide an unprecedented insight into the operation of the deuterium lamp and a sound basis for lamp design improvements.

The authors gratefully acknowledge EPSRC and Cathodeon Ltd. for supporting this work.

References:

A.J.Page, "A study of electrical characteristics of deuterium lamp cathodes", LS:6, Budapest, 1992, 65:L.

G.R.Harris, A.Hartley and R.Devonshire, "Computer model of the gas discharge in a deuterium lamp", (this volume).

DEVELOPMENT OF SPECTROMETER FOR NIR REGION BY NEW METHOD

Authors T. Okura[1] Y. Yamaguchi[2] T. Ooka[3] T. Inoue[4] K. Uchida[4] T. Inoue[1]
Institution and address
1. OPTO RESEARCH CORPORATION HACHIOJI TOKYO JAPAN
2. GEOLOGICAL SURVEY OF JAPAN TSUKUBA IBARAKI JAPAN
3. METAL MINING AGENCY OF JAPAN MINATO-KU TOKYO JAPAN
4. DOWA MINING CO., LTD CHIYODA-KU TOKYO JAPAN

Recently request for measuring NIR light has become greater. Detector responsivity is not so good for that wavelength region as UV-VIS's that measuring speed and stability used to be limited. However NIR reflection spectrum has as much information as UV-VIS. By NIR reflection spectrum quality check or sample identification is well done. In the fie f mining they can process the work very quickly if identifications of altered minerals are possible instantly in the fields. And these identifications are possible by NIR spectrum. But they hadn't small, fast portable spectrometer for outside.

A project to develop such a new instrument has been carried out since 1989 and final model is completed. We are able to develop a new NIR spectrometer which will be a helpful instrument in the field of mining. We've developed a new technique named "multiband method".
By this technique we are able to measure from 1.3 to 2.5 micron within 10 seconds. Optical system is described by fig. 1. as follows.

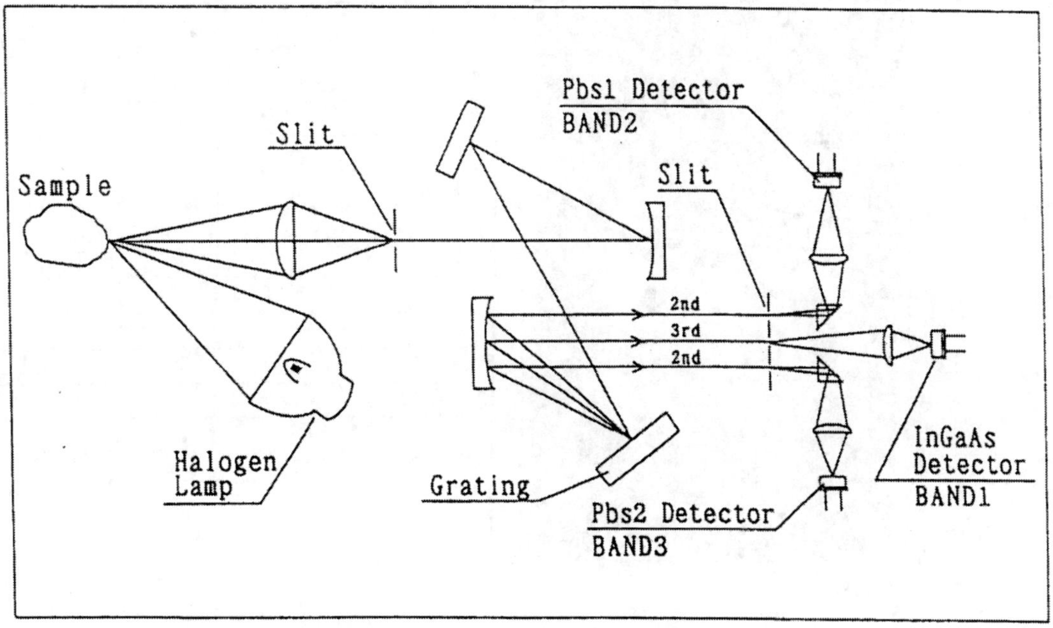

Fig. 1 Optical System of Instrument

Sample is irradiated by halogen lamp. Reflected light is collected to grating type monochrometer. In the monochrometer incident light is diffracted according to their wavelength. Usually we use only the 1st. order diffracted light and single detector. But in our new "multiband method", we use higher order diffracted lights and 3 detectors for each 3 bands as shown in Table 1. In this method we can select the order of each 3 band to come near point at exit slit. So all of them can be intensified by grating blaze effect if we select appropriate blaze angle. And when we use higher order diffracted light high dispersion will be realized. High dispersion will give high optical entrance aperture. On top of it we are able to use effective detector for each 3 bands. All detectors give signals simultaneously when grating is scanned.

Table 1 Wavelength and Order of 3 Bands

	Detector	Wavelength μm	Diffraction Order
1	InGaAs	1. 3-1. 7	3rd
2	PbS 1	1. 7-2. 2	2nd
3	PbS 2	2. 2-2. 5	2nd

For these reasons, high optical throughput is possible and high performance is attained. Fig 2. is the photograph of the instrument named "POSAM" we developed. And Fig 3. is the reflection spectrum of kaolinite which shows high performance of our system.

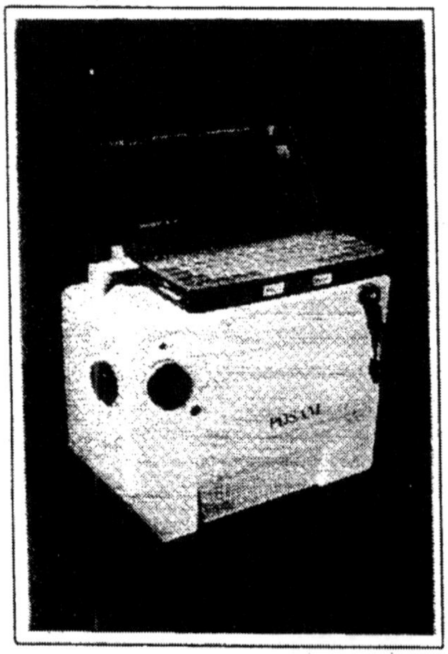

Fig. 2 Photo of Instrument

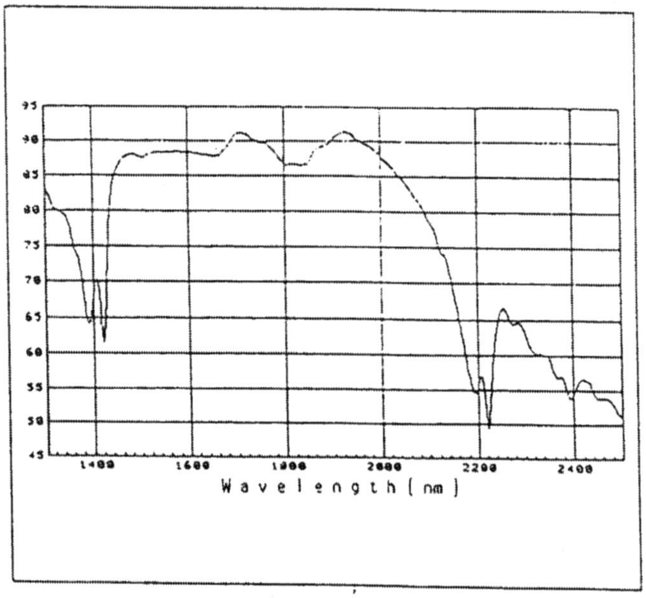

Fig. 3 Spectrum of Kaolinite

MEASUREMENT OF $Hg6p\,^3P_{0,1,2}$ STATE DENSITIES IN LOW-PRESSURE Ar-Hg DISCHARGE PLASMA USED FOR LIQUID CRYSTAL DISPLAY BACK-LIGHTING

Miki GOTO and Toshihiko ARAI
Department of Electrical and Electronic Engineering, Kanagawa Institute of Technology, 1030 Shimo-Ogino, Atsugi 243-02, Japan

1. Introduction

The positive column of a low-pressure Ar-Hg discharge has been applied as a fluorescent light source for illumination. Many studies on the diagnostics and fundamental mechanisms have been carried out on both the classical fluorescent lamp (d=36 mm)[1-3] and the compact fluorescent lamp (d=12 mm).[4-6]

On the other hand, a lamp of extremely narrow diameter (usually below 6mm) has been recently developed for liquid crystal display (LCD) back-ligting and its importance is undoubtedly increasing. Some characteristics or mechanisms of the narrow-diameter lamp may be similar to those of the 36mm one; however the similarity rule does not hold between them due to the contributions from a stepwise ionization process.[3] Therefore, in order to clarify the excitation mechanism in the narrow-diameter lamp quantitatively, various parameters must be measured directly and some analysis must be done.

The $Hg6p\,^3P_{0,1,2}$ state densities are important parameters for the purpose of clarifying the excitation mechanism quantitatively. In this work, using the modified absorption method[7], we have measured the dependences of the $Hg6p\,^3P_{0,1,2}$ state densities in the Ar-Hg discharge of the 4mm bore tube on bath temperature.

2. Experiment

The $Hg6p\,^3P_{0,1,2}$ state densities were measured with the modified absorption method. The principle of this method is described in detail in Ref.7. The discharge tube was made of quartz glass of 4mm inner diameter and has one anode; A, and two cathodes; K_1 and K_2. The discharge can be obtained at A-K_1 (ℓ =3cm) and A-K_2 (2ℓ =6cm) by changing the switch SW. The Ar filling pressure is 3Torr. The tube is placed in a water bath to control Hg vapor pressure by changing the water temperature.

First, the discharge is obtained between an anode A and a cathode K_1 and the end-light intensity I_1 of the Hg(I) line ($7s\,^3S_1 - 6p\,^3P_{0,1,2}$) from the plasma I is measured with monochromator. Next, the discharge is switched from A-K_1 to A-K_2, and the end-light intensity I_2 of the line from the plasma I+II is measured. Then, the ratio of the two light intensities is given by

$$I_2/I_1-1=G(k_0\ell)$$
$$=\{\int[1-\exp(-k_0\ell f_\nu)]\exp(-k_0\ell f_\nu)d\nu\}\times\{\int[1-\exp(-k_0\ell f_\nu)]d\nu\}^{-1}. \quad (1)$$

Here, k_0 is the absorption coefficient at the line center of the Hg^{202} isotope component, ℓ is the plasma length, and f_ν is the function giving the profile of the 404.7, 435.8 or 546.1nm line. Since the transmissivity of the window is not included in Eq.(1), the absorption coefficient of ultraviolet line can also be measured exactly with this method. Then $k_0\ell$ is known from the measured I_2/I_1 and Eq.(1), the $Hg6p\,^3P_{0,1,2}$ state densities can be obtained.

3. Results and Discussion

Figure 1 shows the bath temperature dependences of the $Hg6p\,^3P_{0,1,2}$ state densities measured at the discharge current of 20mA. They increase with the bath temperature and tend to be saturated at the bath temperature above 70℃.

Figure 2 shows the bath temperature dependence of the Hg^+ ground state ion density $n(Hg^+)(6s)$ reported in Ref. 8. This trend of $n(Hg^+)$ can be interpreted qualitatively by considering the variations of the $Hg6p\,^3P_{0,1,2}$ state densities which are supposed to contribute to the formation of Hg^+ ions, and the variations of the electron temperature and electron density. The electron temperature decreases to less than 2eV with the bath temperature while the electron density has a minimum around 50℃ and then turns out to increase.[9] Then the number of electrons of a higher energy than 5.5eV decreases above 50℃. Since the stepwise ionization through the $Hg6p\,^3P_{0,1,2}$ states by electron impact and the direct and indirect ionizations caused by the collisions of two $6p\,^3P$ atoms are expected to be important processes at higher temperature 50℃ and the threshold energy for the ionization process by the electron is about 5.5eV, it is understandable that $n(Hg^+)$ shows a trend to be saturated.

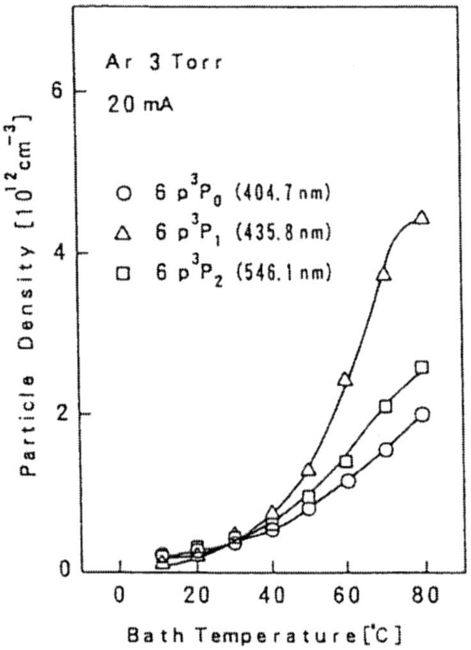

Fig. 1. Bath temperature dependences of $Hg6p\,^3P_{0,1,2}$ state densities.

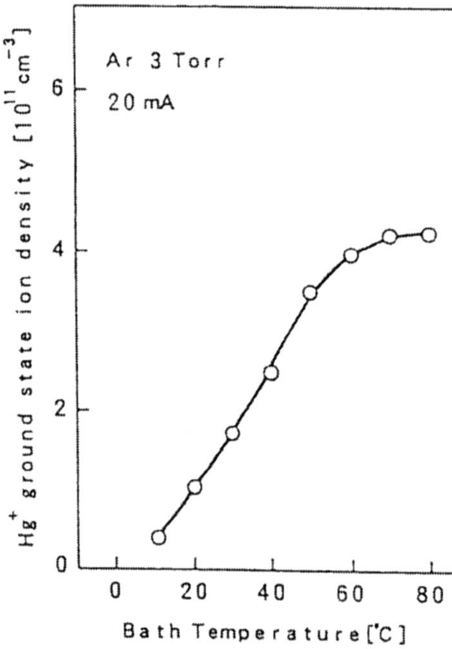

Fig. 2. Bath temperature dependence of Hg^+ ground state ion density.

1) For example, W. Verweij: Philips Res. Rep. Suppl. 2 (1961) 1.
2) M. Koedam and A. A. Kruithof: Physica 28 (1962) 80.
3) L. Vriens, R. A. J. Keijser and F. A. S. Ligthart: J. Appl. Phys. 49 (1978) 3807.
4) T. Lin, T. Goto, T. Arai and S. Murayama: J. Appl. Phys. 66 (1989) 2779.
5) T. Lin, T. Goto, T. Arai and S. Murayama: J. Appl. Phys. 67 (1990) 4012.
6) T. Lin, T. Goto, T. Arai and S. Murayama: J. Appl. Phys. 69 (1991) 4201.
7) T. Goto, M. Mori and S. Hattori: Appl. Phys. Lett. 29 (1976) 358.
8) M. Goto and T. Arai: Proceedings of the 2nd Asia-Pacific Conference on Plasma Science & Technology, Korea, (1994) 248.
9) M. Goto, T. Sakai and T. Arai: Jpn. J. Appl. phys. 33 (1994) L896.

CADMIUM ION LAMPS WITH HIGH RADIANT EFFICIENCY

Akiyasu Yamaguchi, Yukio Yasuda and Tatsumi Hiramoto
USHIO Inc. , 1194 Sazuchi, Bessho-cho, Himeji, Japan.

Introduction

Since early in this century, characteristics of absorption or emission of cadmium lamps have been studied[1)2)]. Recently, it was suggested that the cadmium ion lamps could be used for industrial applications, such as UV-surface modification, photo-CVD, ashing, UV-curing or photolithography[3)], since cadmium ion lamps have strong emission line in the short wavelength region of UV-C (far deep UV). Conventional cadmium lamps could not be used, however, because their radiant intensity in this wavelength region was too weak for practical use. Therefore, it becomes important to increase both radiant intensity and efficiency in far deep UV. Many investigations have been done in order to realize a lamp with high radiant intensity in the far deep UV region. A cadmium lamp emits the resonance line of 228.8nm. However, since the line becomes weak at high pressures by self-absorption, the maximum intensity is not high.

In this paper, cadmium ion lamps with both an intensively high radiance and a high radiant efficiency is reported.

Lamp Characteristics and Discussions

Spectral irradiance was measured by spectrometer with a Czerny-Turner mounting calibrated by a standard deuterium lamp and a standard halogen lamp. The bulb of cadmium ion lamps was made of fused quartz and the lamps were operated at about 2kW. The radiant efficiency η can be defined with the measured spectral irradiance and input power as

$$\eta = \frac{4\pi r^2 \int_{\lambda_1}^{\lambda_2} I_\lambda d\lambda}{J_L \times V_L} \times 100 \qquad (1)$$

where

J_L, V_L ;lamp current, lamp voltage, respectively

r ;distance from the lamp

I_λ ;measured spectral irradiance at wavelength λ

λ_1, λ_2 ; lower and upper limits of the wavelength of an integration, respectively.

One of spectral distribution of the cadmium ion lamps is shown in Fig.1 as an example. In general, the spectral distribution depends upon the ionic density and the arc temperature. The ionic density is determined from Saha's equation. As a result the spectral distribution varies with the arc temperature and the filling amount of cadmium in the lamp. The resonance line of 214.4nm of the cadmium ion is dominant as shown in Fig.1. However, the atomic resonance line of 228.8nm is absolutely self-absorbed except in the far wings. An ionic line of 219.5nm is observed. A line of 216.8nm may be a satellite line of the ionic resonance line. Radiant efficiency was obtained by Eq.(1) integrated from 210nm to 225nm (hatched area in Fig.1). Figure 2 shows relative radiant efficiency as a function of the filled gas pressure of three kinds of the rare gas as a buffer gas at room temperature. Figure 2 indicates that the radiant efficiency increases with the rise in ionization energy of the rare gas. In case of argon as a buffer gas, the radiant efficiency increases as the pressure rises.

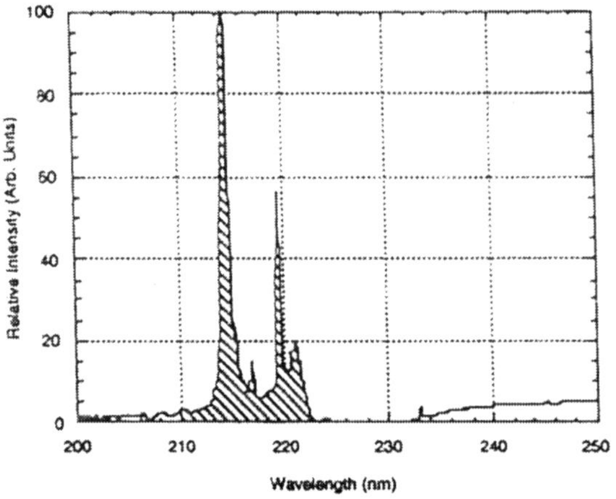

Fig.1.
Measured spectrum distribution of Cd-ion lamp in the wavelength range of 200nm to 250nm.The resonance line of 228.8nm of Cd-atom is self-absorbed except in the far wings.

In general, the radiant efficiency is high at high-arc temperatures. Moreover, the efficiency increases with the rise in the ionization energy of the buffer gas, since the cadmium ionic density becomes higher with the increase in the ionization energy of the buffer gas. From Saha's equation of the mixture of cadmium with a rare gas, it is understood that the cadmium ionic density increases with the rise in ionization energy of the buffer gas even at a constant temperature.

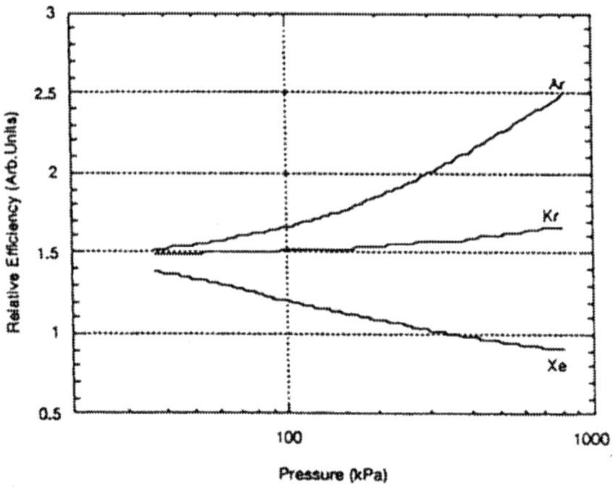

Fig.2.
Relative efficiency v.s. pressure of a buffer gas.

It is found that using a rare gas with a high ionization energy in high pressures makes radiant efficiency high. When the argon gas is used as a buffer gas, the increasing gas pressure causes high radiant efficiency. Thus, it can realize the radiant efficiency of 1.5% or more. Higher radiant efficiency may be expected with a synthetic-quartz bulb.

In general, ionic lines have weak self-absorption, because the density of ionic absorbing species steeply decreases with the reduction in temperature in the periphery of the arc.

Conclusion

The resonance line of cadmium ions (214.4nm) can be very strong and the radiant efficiency becomes high. Ionic resonance lamps emit high-intensity radiation if the operation condition for the atomic density and the arc temperature is optimized. The cadmium ionic lamps can be applied to a wide range of applications including photo-CVD and photolithography.

References

1) e.g. Von S. Mrozowski, Z. Phys. *62*, 314 (1930)
2) e.g. F. Bates, Phil. Mag. *39*, 353 (1920)
3) e.g. T. Hiramoto *et. al.*, HEI-5-130103 (1993)

HIGH POWER CONTINUOUS VUV RADIATION OF D_2-Hg LAMP

Makoto KUBO[(A)], Ichiro TAKAHASHI[(B)]

Masaharu AONO[(C)], Ryohei ITATANI[(D)]

(A) Dept. of Electronic Science & Engng.,Kyoto University
Yoshida-honmachi,Sakyo-ku,Kyoto 606-01,Japan

(B) Dept. of Electrical Engng.,Maizuru College of Technology
234 Shiraya,Maizuru 624,Japan

(C) Dept. of Electrical & Electronic Engng.,Ehime University
3 Bunkyo-cho,Matsuyama 790,Japan

(D) Niihama College of Technology
7-1 Yakumo-cho,Niihama 792,Japan

1. Introduction

Semiconductor production such as thin film deposition,impurity doping,etching and so on requires low temperature chemical reaction processes. Currently , photo-CVD (Chemical Vapor Deposition) attracts a great attention on its low temperature process. At present , mercury lamp,rare gas lamp and excimer laser are used as photo reaction light sources , but all of them have merely line spectrum.

In this paper we describe the pulse discharge low-pressure lamp the that can be used as a source of vacuum ultraviolet(VUV) radiation in the wavelength region of 115-200nm. This lamp composes the special liquid metal cathode[1],[2]. It was discovered in the study of variable luminous color discharge lamp controlled over the electron energy in a positive column by using pulse discharge method[3],[4]. This cathode has the characteristic of high-speed glow-arc transition in every pulse period. Additionally it is very strong for pulsed ion spattering.

2. Experiments and Results

The main tube of this lamp was made from PYREX-glass(d=30mm,gap length=50mm) and equipped by the branch tube with MgF_2 window for penetration of VUV radiation.

In this case,mercury(Hg) was used as the special liquid metal cathode. The lamp was filled in by deuterium gas(D_2) under the pressure of 5 torr and Hg under room temperature. The applied pulse voltage was 2KV(i≅2KA, pulse width≅10μsec). Measurement equipment consisted of vacuum monochromator with the VUV photomultiplier tube. Measurement system and the lamp were pumped out by the same vacuum pumping system. Experimental radiation spectrum is shown in Figurer 1.

3. Discussion and Conclusion

The continuous spectra of DC type D_2 lamp and D_2-Hg lamp and the line spectrum of Hg lamp (W=40W) were examined and compared. Spectrum intensity of continuous radiation component of D_2-Hg lamp(Fig.1) is much stronger than that of DC type D_2 lamp in the region of 115-155nm.

For the region of 155-200nm the corresponding spectrum components are approximately the same for both lamps. The spectrum intensity of radiation component of D_2-Hg lamp(Fig.1) is 5.5 times lager than the line spectrum intensity peak of Hg lamp on the same λ=185nm.

The peak of maximum spectrum intensity for the first lamp is 70 times lager than that of the second lamp on λ=185nm correspondingly. We have developed high intensity VUV output pulse discharged D_2-Hglamp. It can be used in photo-CVD technique instead of Hg lamp.

4. Acknowledgment

The authors wish to thank N. TAGUTCHI and H. KUROKAWA for experimental assistance. This work is supported by The Grant for Research and Education from The Illuminating Engineering Institute of Japan (current year).

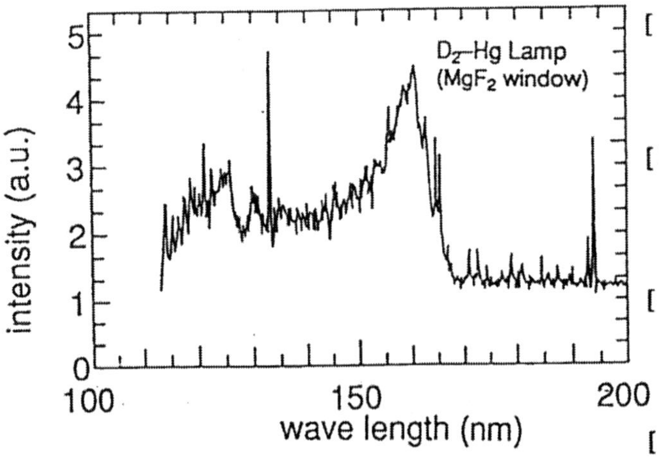

Fig.1 VUV radiation spectrum of D_2-Hg lamp

References

[1] M.KUBO, N.TAGUCHI R.ITATANI: Transaction I.E.E.Japan (1987) G13-3

[2] I.TAKAHASHI, M.KUBO R.ITATANI : Transaction I.E.E.Japan (1994) G1-11

[3] M.AONO, H.KUROKAWA, M.KUBO R.ITATANI : J.Light & Vis. Env. Vol.13(1989) p.1-9

[4] R.ITATANI, M.KUBO, M.AONO M.RYOKO : LS-6 Budapest, (1992) p.85-86

MICRO HOLLOW CATHODE ARRAY DISCHARGE

W.W. Byszewski*, K. H. Schoenbach**, R. Verhappen**, and F.E. Peterkin**
*OSRAM SYLVANIA INC., Danvers, MA
**Old Dominion University, Norfolk, VA 23529

Hollow cathode glow discharges develop the cathode fall and the negative glow in the cathode cavity. The cathode hole diameter must be on order of the mean free path for collisional ionization. Increasing the gas pressure, p, in a hollow cathode discharge, therefore, requires a reduction in the cathode hole diameter,. D. Experiments in argon with cathode holes of submillimeter diameter have demonstrated hollow cathode operation for pD on the order of 1 Torr cm. This scaling law is similar to the one for large (> mm) cathode hole diameters (1).

The current-voltage characteristics of micro hollow cathode discharges were found to have a positive slope in a voltage range up to a power level where heating of the cathode leads to thermal emission of electrons. In the range of the positive slope two discharge modes were observed. For low voltages, close to the breakdown voltage, the current increases exponentially. The slopes of the current voltage characteristics become steeper with increasing pressure. Increase of the voltage, at constant pressure, causes a discontinuity in the current-voltage characteristics. Above this point the current rises again exponentially but with a different slope. The transition into this mode is assumed to be due to a change in gas composition caused by an increasing rate of sputtering of the cathode material. Spectral measurements of the discharge radiation, which showed the presence of electrode material in the second mode, support this assumption.

In the range of positive current-voltage characteristics, hollow cathode discharges can be operated in parallel without ballast resistors. An array of hollow cathode discharges is shown in Fig. 1. The discharge geometry consists of a nickel mesh with rectangular openings of 0.2 mm. Each opening serves as a cathode hole. A flat anode was placed approximately 0.1 mm apart from the mesh. The pressure was 50 Torr in argon. The applied voltage was 430 V. The discharges were operated in a pulsed mode (300 microseconds). Experiments with arrays where the cathode hole diameter was 0.7 mm showed that the discharges could also be operated with rectified ac-voltage and even dc. The discharges in any mode of operation are stable (reproducable) and do not influence each other.

Fig. 1. An array of micro hollow cathode discharges in argon at 50 Torr. The dimension (opening) of the individual holes is 0.2 mm.

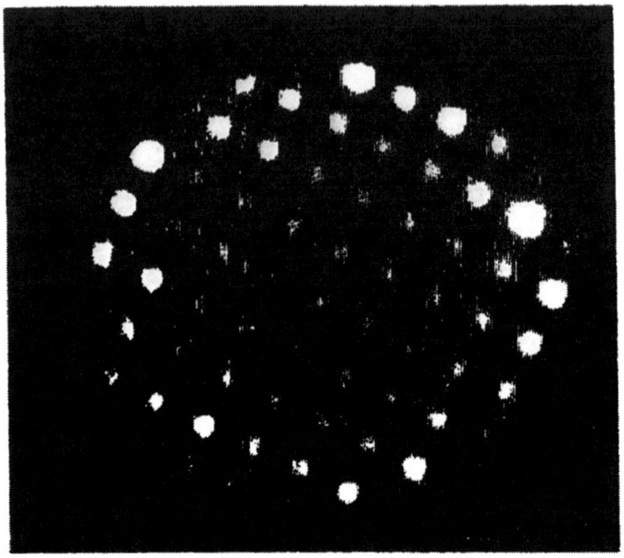

Results of Monte-Carlo calculations (2), indicate that the discharge in the cathode hole forms a nonthermal plasma. The electrons originating from the cathode wall through ion impact and accelerated towards the axis can be considered as a radial electron-beam which can be used to dissociate, excite, and ionize the gas with high efficiency. It therefore offers the possibility to use the discharge array as a flat panel UV-light source, or as a flat light source in the visible, where the emitted UV photons are used to excite a phosphor. The thickness of these light sources is on the order of one millimeter. Another application is the use of micro hollow cathode discharge arrays as cathodes (electron sources) in fluorescent lamps.

References

(1) J.W. Gewartowski, and H.A. Watson, "Principles of Electron Tubes," Princeton, NJ, Van Nostrand, 1965, p. 56.
(2) K.H. Schoenbach, "The Effect of Magnetic Fields on Hollow Cathiode Discharges," Proc. XXI Int. Conf. Phenomena in Ionized Gases, Bochum, Germany, September 1993, Vol. III,p. 287.

NEON-DISCHARGE BASED MOTORCAR STOPLIGHTS

F.A.S. Ligthart and J. Geboers*
Philips Lighting B.V.
Central Development Lighting
P.O. Box 80020
5600 JM Eindhoven
The Netherlands
*) Development Philips Automotive Lighting, Chartres, France

A Neon-discharge based motorcar stoplight system was developed which makes use of narrow diameter (about 4 mm outer diameter) discharge tubes of about 40 cm length. Presently, the main application of the Neon system is the Central High Mounted Stoplight (CHMSL) or third stoplight. In a later phase the Neon system may also be applied as a low mounted stoplight and as a rearlight replacing incandescent lamps and LED's, which are already on the market as third stoplights.

The main characteristic of the Neon lamp is its **feature of instant light** (100% light within 1 millisecond) while the standard incandescent lamp shows a delay of about 150 milliseconds. In Fig.1 the human response or reaction times are shown for the different types of stoplights. The reaction time appears to be about 150 milliseconds shorter for Neon lamps and LED's as compared to standard incandescent lamps.

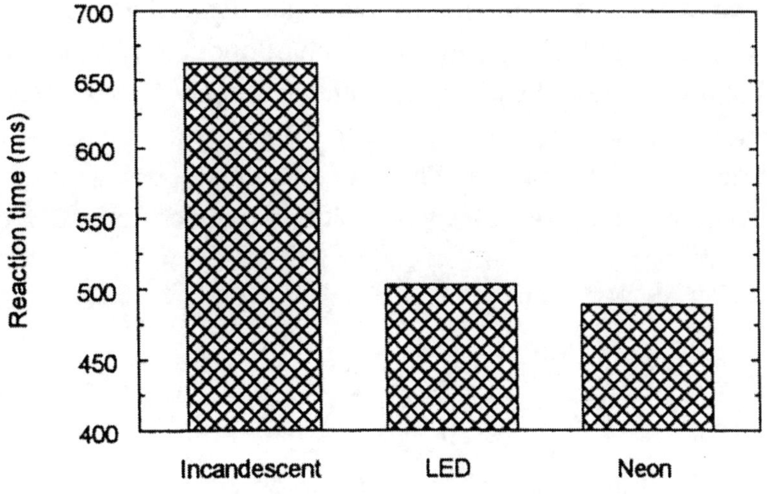

Fig. 1: Mean reaction time by lamp type

Thanks to the narrow tube and the high luminance the **overall height of the Neon system**, including the reflector, can be reduced to 10 mm. Furthermore, the long linear light source of nearly **uniform luminance** provides a light-bar effect. It is this effect which makes the Neon Stoplight so effective and appealing. The competing LED and incandescent systems lack this uniform appearance.

System characteristics:

To obtain a high luminous central intensity (70 cd) at minimum system power an optimization was carried out of the different parameters that influence the luminous intensity i.e. the electric current, the lamp dimensions (including length), the Neon pressure and of course the optical reflector. A lamp with a narrow diameter (3.0 mm internal), 40 cm length and a relatively low electric current (8 mA) appeared to be an attractive choice because of its high luminance, its low ballast losses and its low bulb temperature. The robust hollow cathode helps to lower the cathode fall and to extend the lamp-life.

The resulting life-time of the lamp easily exceeds 2000 hours and this is independent of the switching cycle. This long life-time makes possible sealed-for-life operation because even extreme drivers which brake 25% of the time during driving are taken into account.

A dedicated electronic circuit was developed which assures reliable ignition and lamp operation at about 8 mA lamp direct current under all automotive conditions (temperature, humidity etc.). Thanks to the direct current operation of the lamp the electro-magnetic interference can easily be kept below the limits required by the automotive industry. The light output is kept stable between 9 V and 16 V battery voltage. Using this circuit the lamp produces a luminous flux of about 90 lumen at about 7 Watt lamp power. Under nominal operation the maximum lamp luminance is 2.0 cd/cm^2 which leads to a luminous intensity of about 70 cd when combined with a dedicated optical reflector. This luminous intensity by far exceeds the minimum luminous intensity of 25 cd which is required for a CHMSL.

CARBON-EMITTERS -- A NEW ASPECT OF INFRA-RED TECHNOLOGY

K.J.Dietz, K. Schmitz, C. Zou
Heraeus Noblelight GmbH, Heraeusstraße 12-14, D-63450 Hanau, Germany

A.C. Dexter, A. Heaton, W. Jones
EA Technology Ltd., Capenhurst, Chester, CH1 6ES, Great Britain

With carbon fibres as the radiating filament Heraeus Noblelight and EA Technology have developed a new generation of infra-red emitters for applications in heating and drying processes. This emitters combine the advantages of different kinds of standard emitters: the spectral radiation of a medium wave emitter and the very short response time of a short wave emitter. So these IR-sources are useful in drying solvents containing water and waterbased lacquers because of the optimal fit of spectral output and absorptionlines of water. In general all heating processes with high power densities require a short response time and good radiation absorption in the used materials.

In carbon emitters carbon fibres joined to a filament replace the commonly used resistance wires. Because of the small thermal mass of the carbon filament the response time of the emitter is less than 3 seconds, compared with a typical timescale of minutes for commercial lamps. The temperature can be adjusted between 1000 °C and 1200 °C, so that the maximum of the spectral power distribution is in the range of $\lambda_{max} = 2$ µm. High power densities (up to 80 kW/m^2), variable emitter power, homogenious temperature distribution, high radiation efficiency and long life time are typical attributes of carbon emitters.

THE INFRA-RED SUPPRESSION IN INCANDESCENT LIGHT FROM SUBMICRON HOLES

Masaru SUGIMOTO, Toru FUJIOKA, Takahiro INOUE, Hiroshi FUKUSHIMA, Yousuke MIZUYAMA, Shin UKEGAWA, Tomoaki MATSUSHIMA and Makoto TOHO
Matsushita Electric Works, Ltd.
1048 Kadoma, osaka 571, JAPAN

INTRODUCTION

In general efficacies of incandescent lamps (IL's) are very low, while they have many advantages such that they are still widely used. As well known IL's waste energy due to the emission of infra-red light (IR) at the expense of visible light.

An idea to save energy; i.e. submicron holes can suppress the IR; has been proposed by J.F.Waymouth[1] and a few works to check it has been published[2][3][4]. Through them we noticed that the idea being able to checked by three steps as below.

1. Although the emission of the both visible light and IR increase by the submicron holes, the increment of IR is smaller than that of visible light. Thus the efficacy of IL will increase.

2. If IR emission from the submicron holes is absolutely less than that from flat surface then the efficacy will increase very much.

3. If the submicron holes are formed in very high density such that the spaces between holes are as narrow as the skin depth then little IR will be emitted from the surface without holes too and the efficacy will drastically increase.

We tried to check the idea experimentally and got evidence of the step 1, however, step 2 and 3 did not become so clear. We will discuss here the step 2 and 3.

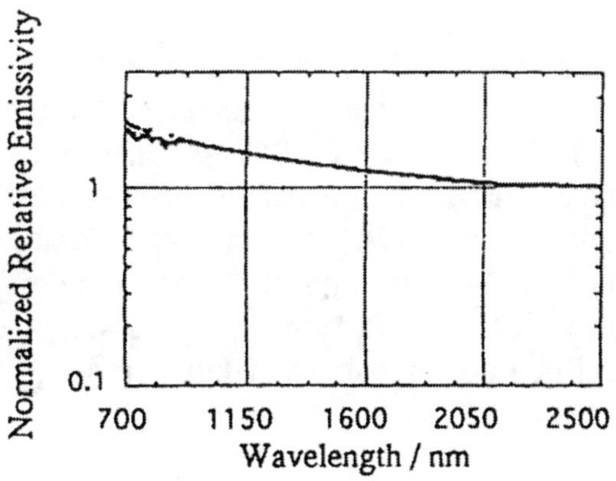

Fig.1 Ratio of the spectra of Incandescent light from porous surface to that from flat surface.

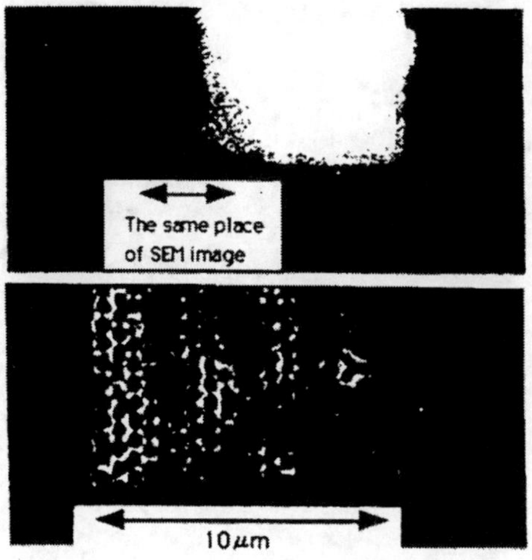

Fig.2 IR image and SEM image.

EXPERIMENT and RESULT

We fabricated samples covered with submicron holes by several ways. Some of them were measured spectra of incandescent light heated at 1400K. We also measured spectra of flat surface at the same temperature and compared both of them as shown in the Fig.1.

Samples with holes formed regularly and in high density were measured the two-dimensional distribution of IR emission at about 700K as shown in the Fig.2. The IR emission from the good fabricated porous region is not higher rather equals to that from the flat surface.

During heating the fabricated surface was destroyed by atomic diffusion and evaporation. The critical deformation temperature was expected to depend on the crystal structure of the sample. Since low density compound such as deposited film was expected to tend to be deformed, samples made from single crystal tungsten was heated at 2100K in the hydrogen. However the surface was deformed by an hour.

CONCLUSION

From the Fig.1 we can clearly conclude that the IR ratio is decreased by the submicron holes (step 1 is true). From the Fig.2 the possibility of suppression of IR below the normal level is suspected. Although we could not check the step 3 experimentally, we consist that such a small structure cannot keep the shape at the high temperature for incandescent. These mean that an IL whose efficacy is about 2.5 times of the conventional one may be available.

ACKNOWLEDGMENT

The authors would like to acknowledge to Dr.H.Masuda at Tokyo Metropolitan University, Dr.N.Watanabe and Y.B.Chong at Applied science laboratory and Dr.J.Itoh, Dr.F.Kanayama and the other members in process fundamental section and micro-beam section of Electrotechnical laboratory for their help in preparation of the samples. Thanks to Nippon Barnes Co.,Ltd for IR close-up lens. Thanks to Dr.J.F.Waymouth for extremely valuable discussions of the infra-red suppression effect.

REFERENCE

[1]Waymouth,J.F., J.Illum.Engng.Inst.Jpn., 74 no.12 (1990) pp.800-805

[2]Sugimoto,M., Fujioka,T., Inoue,T. Fukushima,H., Mizuyama,Y., Ukegawa,S., Matsushima.,T., Toho,.M., J.Light & Vis. Env., 18 no.2 (1994) pp.5-9; Proceedings of 1994 Annual Conference of the Illuminating Engineering Institute of Japan, p.125

[3]Seki,S.,Takisawa,M.,Kawabe,K.,Ookawa,T., Hirono,M. and Sekine,M. Proceeding of 1994 Annual Conference of the Illuminating Engineering Institute of Japan, p.126

[4]Sibata,T., Sakuraoka,T., Seki,M., Ookawa,T. and Sekine,M., Proceedings of 1995 Annual Conference of the Illuminating Engineering Institute of Japan, p.47

MECHANICAL ASPECTS OF COIL GEOMETRY

B. van Bakel and E. Gerritsen
Philips Lighting B.V., Eindhoven, The Netherlands

In recent years much effort has been devoted to the effect of grain boundary structure in tungsten coils on their mechanical stability. This structure should provide a dragging force against sliding of the grain boundaries at high temperature.

This paper, on the other hand, considers the driving forces for the sliding process. These forces are determined by the geometrical coiling parameters and the amount of prestretch applied to the coil during mounting in the lamp.
The magnitude and type of these forces and their effects on elastic deformation and creep have been analyzed by numerical as well as experimental methods.

Prestretch during mounting is applied in order to reduce the initial (elastic) sag of the coil. Without prestretch this elastic sag increases with coil length. The reduction of elastic sag by application of prestretch is seen to be more effective for longer coils, having smaller mandrel to wire ratios (**figure 1**).
The creep rate of a tungsten wire is fully determined by the grain structure and the stresses driving the creep process. Nevertheless, when coils of different geometry are operated under conditions of equal grain structure and driving stresses, their sag rate will still be different. Under these conditions the initial sag rate, expressed in microns per hour, is calculated to be lower for short coils, as can be anticipated from the slope of the curves in **figure 1**.

Coil sag was determined at room temperature and at operating temperature. It appeared that sag at room temperature can be converted to that at operating temperature by taking into account the thermal expansion and elastic modulus decrease of tungsten (**figure 2**).

The resonance frequency of a coil and its relation to coil sag could well be described by numerical means (**figure 3**).

Also the geometry of the coil ends is calculated to be of major influence on coil stability (**figure 4**). The amount of sag is seen to depend on filament position (legs up or down) when the legs are not centered on the filament axis. Sag is drastically reduced by applying stiff lead-in poles.

Figure 1
The calculated effect of prestretch on elastic sag of long (15 mm) and short (9 mm) coils.

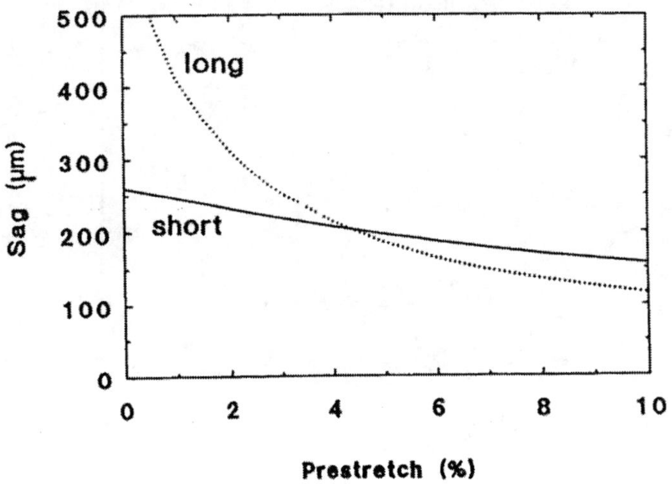

Figure 2
Sag at room temperature (C) and at operating temperature (H) for coils with different amounts of prestretch. The solid curves are calculated.

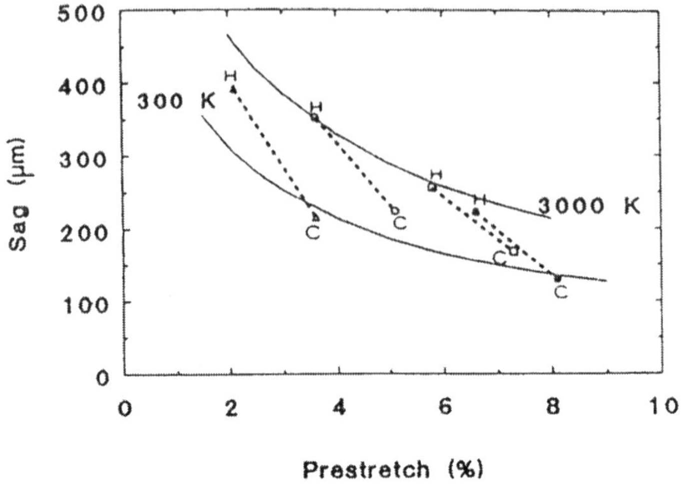

Figure 3
Resonance frequency of 15 coils with various amounts of prestretch. All coils have been analyzed at both 300 K and 3000 K.
The solid curve is calculated.

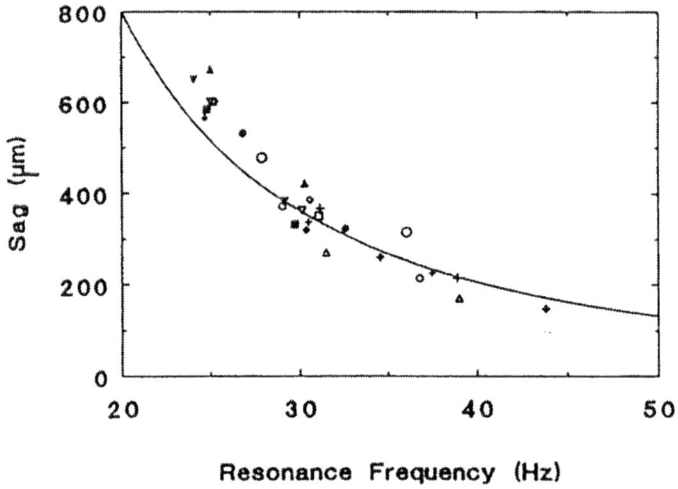

Figure 4
The calculated effect of single coiled legs on elastic sag.

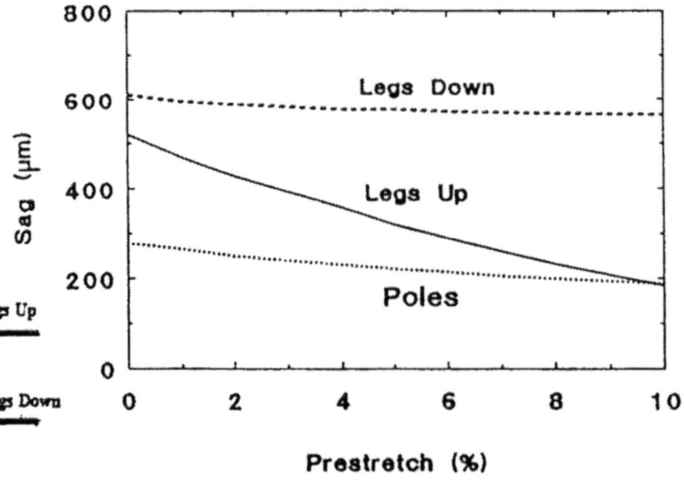

TENSILE STRENGTH OF W-WIRE AT HIGH TEMPERATURES

Hideyuki Ikegami , Tatsushi Igarashi , Tatsumi Hiramoto

USHIO Inc. Research and Development ,1194 Sazuchi , Bessho-cho Himeji-City ,
Hyogo-pref.671-02 Japan

1 . Introduction

In general , the lamp life of incandescent lamps depends upon a creep property of the W-filament. [1] The halogen lamps for an exposure light of copier machines are operated flashingly , not continuously. The life of these lamps is mainly controlled by the thermal stress due to rapid variation in the temperature of the filament , not by the creep. A W-wire with high tensile strength at high temperatures is a key-material for the halogen lamps with long lives in the case of the exposure lamps in the copier machines. The W-wire for halogen lamps having long lives even at the flash operation has been investigated in this research. At first , an apparatus was designed and constructed in order to measure the tensile strength at high temperatures. In this paper this apparatus is explained and then a relationship between the life of the halogen lamps for copier machines and the tensile strength measured at high temperatures in the apparatus is shown.

2. Experimental

(a) Apparatus

A schematic diagram of an apparatus for measuring a tensile strength of the W-wires at high temperatures is shown in Fig.1. The W-wires of 99 to 173 μ m in diameter and 100 mm long can be hold with W-holders. The apparatus is evacuated by a turbo-molecular pump (T.M.P) below 10^{-3} Pa. Atmosphere such as oxygen is also available. After the W-wires were recrystallized by resistively heating them at 2727 ℃ , the tensile strength was measured at 2227℃ , 2493℃ , 2727℃ , 3004℃. Many W-wires manufactured in

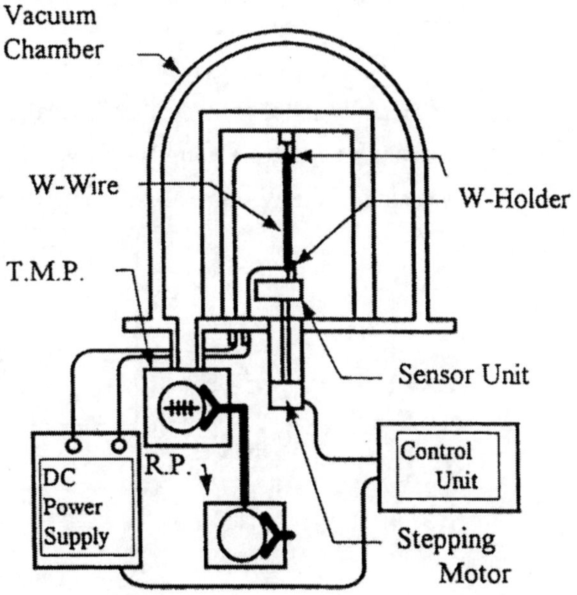

Fig.1 Schematic diagram of an apparatus for measuring tensile strength at high temperatures

various processes were tested. One W-holder with the sensor unit shown in Fig.1 moves down at a speed of 1 mm / min.

(b) Lamp

The wires in a lot had been grouped at random into two sub-groups. One was for measuring the tensile strength at high temperatures and the other was for the filaments of the halogen lamps with standard specifications for the exposure. An example of the lamps is shown in Fig. 2. These lamps were operated under the following standard condition : 1.3 sec ON - 0.2 sec OFF , this ON-OFF is repeated 250 times , then a pause for 300 sec , and this cycle is repeated. The input power during ON-time is 260 W at 85 V AC.

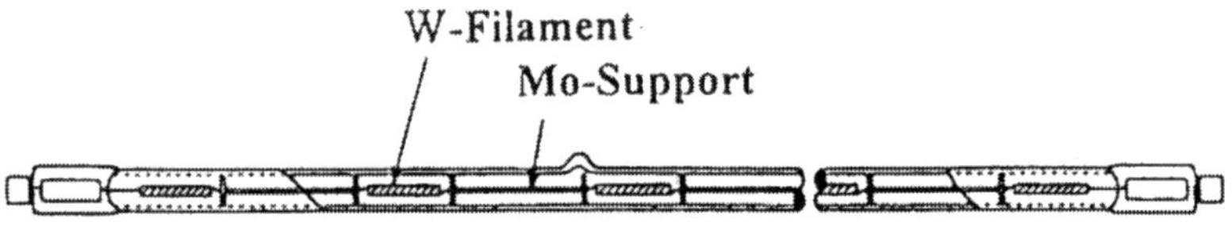

Fig.2 Schematic diagram of a halogen lamp used for the exposure in a copier machine

4. Result & Conclusion

Fig. 3 shows the relationship between the tensile strength of the W-wires at 2727 ℃ and the lamp life. It was found that the lamp life has a direct correlation to the tensile strength at 2727 ℃ of the W-wires belonging to the same manufacturing lot of the lamp filaments. As a result the lamp life can be estimated from the tensile strength at 2727 ℃. It can be concluded that the life of the halogen lamps flashingly operated is dominantly influenced by the thermal stress of the filament wires. The halogen lamps flashingly operated with long lives has been developed with increasing the tensile strength of the W-wires at high temperatures.

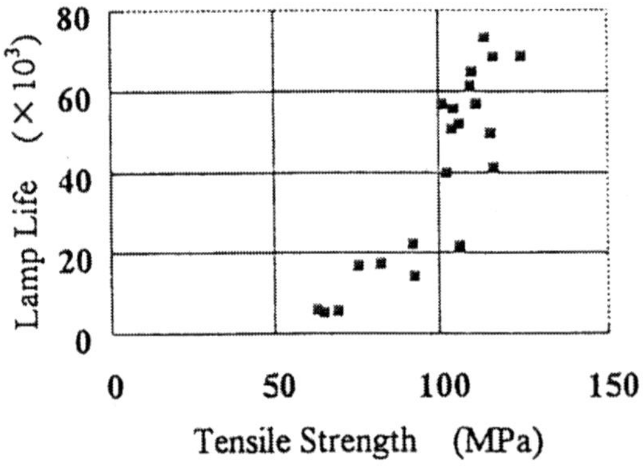

Fig.3 The relationship between lamp life and tensile strengthat 2727℃

Reference

1 . M.F.Ashuby : Rate Processes in Plastic Deformation of Materials ; ASM , Metalspark , p.70 , 1975

CHOICE OF SUITABLE FILLING PRESSURES FOR (HALOGEN) INCANDESCENT LAMPS: THEORETICAL AND EXPERIMENTAL INVESTIGATIONS

A. Bodmer, M. Damm, and R. Minder
OSRAM GmbH, Entwicklung Lampentechnologie, Industriestraße, D-89542 Herbrechtingen

In lamp technology frequently there is found that at constant efficacy and power life of incandescent as well as halogen incandescent lamps increases almost linearly with filling pressure. Therefore one employs the largest possible cold filling pressure which is determined by safety and cost aspects. Here it is shown for incandescent lamps as well as halogen incandescent lamps that the use of the largest possible filling pressure isn`t always a good choice.

First there were performed simulations of heat and mass transport for an incandescent lamp (stop lamp #7506, 21W, 12V) using the CFD-code FLUENT (CFD = Computational Fluid Dynamics). At relative low pressures the heat transported from the coil to the bulb walls is almost only due to molecular heat conduction and is therefore almost independent on pressure. Above a "critical" pressure the heat flux from the coil to the bulb walls is increasing with increasing filling pressure due to the convective heat transport (→ reduction of efficacy). The following figure shows the calculated heats transferred from the coil to the filling gas in dependence on operating pressure:

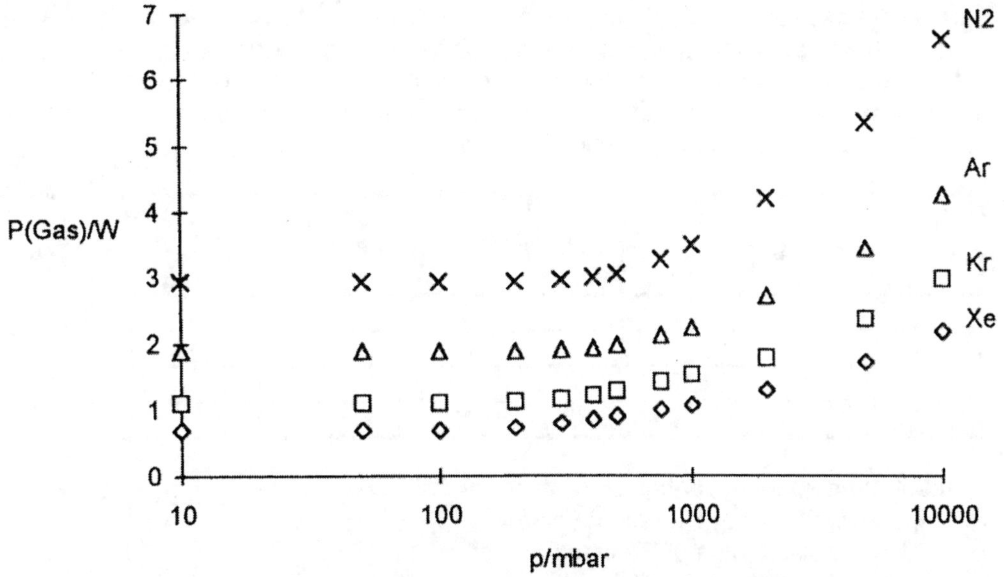

Analogous considerations apply to tungsten mass transport. At low pressures the tungsten is transported almost only by diffusion from the coil to the bulb walls. Due to the $D \propto 1/p$ dependency of the diffusion coefficient the mass transport rate decreases with increasing filling pressure (→ enlargement of life). This decrease of tungsten mass flow with increasing pressure is weakened stronger than predicted by the $D \propto 1/p$ dependency above the "critical" pressure due to the convective mass transport by the flow. Based on simulation of heat and tungsten transport one obtains the following results for the standard comparison efficiency

(SCE) in relation to the SCE of the "standard" lamp #7506 (88% Ar, 12%N_2, operating pressure ~ 1 bar) in dependence on operating pressure:

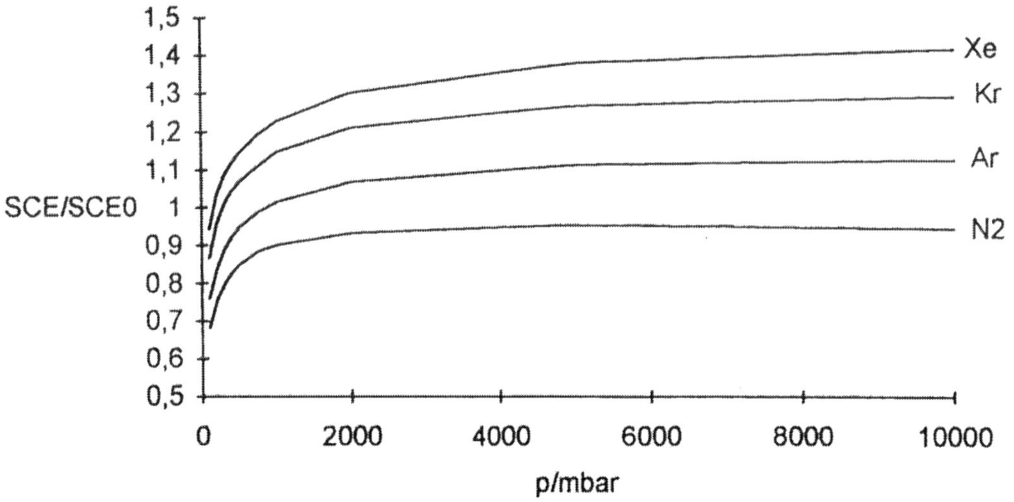

One realizes that enlargement of filling pressure doesn`t improve the SCE in general, for nitrogen there is predicted the existence of an optimal filling pressure with regard to SCE.

The considerations concerning heat transport can be applied to halogen incandescent lamps also whereas with regard to mass transport many modifications (not well understood presently) have to be brought in due to the complex chemistry of the halogen cycle. Experiments show clearly that it is possible to build halogen incandescent lamps with much smaller filling pressures than employed in the past because the decrease of SCE with decreasing filling pressure is smaller than expected. For example one obtained the following results for the miniwatt lamp #64405 12V 5W:

Filling gas	Filling pressure	SCE [%]
Kr	1 bar	104,2
Kr	13 bar	105,3
Xe	1 bar	108,1
Xe	13 bar	122,3

Furthermore some advantagous features of the low pressure version were observed compared with the high pressure version:

- Smaller distribution of life times
- Larger luminous output towards the end of life

Both features can be explained by the reduced rates of tungsten transport processes along the surface of the filament (reduced development of facets) which are due to the reduced temperature of the coil. The reduced temperature of the coil at constant efficiacy and power results from the reduced heat flux from the coil to the filling gas.

A further advantage of the low pressure version is the reduced difference between the largest and the smallest temperature along the bulb walls.

INVESTIGATION ON THE QUALITY OF TUNGSTEN HALOGEN LAMPS WITH DIFFERENT FILLED GAS COMPOSITION

Chen Dahua
Institute for Light Sources,
Fudan University
Shanghai, China

Zhan Yunxiang
Moganshan Electric Lighting Co.
Zhejiang Province, China

1. Introduction Single ended JD type tungsten halogen lamps for mains voltage (220 V) are rapidly being developed in new design version as well as application field. Nowadays, these products are widely accepted in the market. The main filling gas composition of these type lamps are Kr, N_2 and halides mixture (HBr + CH_2Br_2). The principle of halogen cycle generated in tungsten halogen lamps are well known and are employed for various version of lamps. However, the detailed chemical processes which give the influence on the lamp quality and life are yet clear.

2. Experiments Experiments are carried out by changing the filling gas composition and pressure in the range of Kr(1.95 - 2.34 atm.) N_2 (39.90 - 65.17 KPa). The mixing ratio of methylene bromide (CH_2Br_2) and hydrogen bromide is also studied. All the experiments are made on the JD 220V 75 W lamps illustrated in the drawing.

3. Results and discussion Experimental results are shown in table 1, 2 and 3. Figures are based on the average of 5 test lamps respectively.

Table 1. The efficacy and life of JD 220V 75 W lamps as the function of Kr filling pressure
(N_2= 59.85 KPa, and CH_2Br_2= 200 Pa)

Kr.(atm.)	1.95	2.01	2.11	2.18	2.26	2.34
efficacy (lm/W)	7.23	7.36	7.68	7.70	7.38	8.53
life (hrs)	1108	1100	1050	1065	985	980

Table 2. The efficacy and life of JD 220V 75 W lamps as the function of N_2 filling pressure
(Kr = 1.95 atm. and CH_2Br_2= 200 Pa)

N (KPa)	39.90	45.22	50.54	55.86	61.18	65.17
efficacy (lm/W)	7.62	7.58	7.57	7.59	7.36	7.68
life (hrs)	940	948	960	1100	1210	1315

Table 3. The efficacy and life of JD 220V 75 W lamps as the function of the mixing ratio of HBr and CH_2Br_2, where the total filling pressure is kept as 106Pa.
(Kr = 1.95 atm. and N_2= 59.85 KPa)

HBr (%)	0	20	40	60	80	100
efficacy (lm/W)	8.93	8.90	9.35	8.57	8.30	9.72
life (hrs)	1205	1100	1020	914	915	910

The conclusions derived from the experiments are;

(1) The efficacy of 100% HBr filled lamps is 1 – 2 lm/W higher than those of HBr – CH_2Br_2 mixture gas filled lamps.

(2) With increase in the Kr pressure, efficacy becomes higher whereas the lamp life becomes shorter.

(3) With increase in the nitrogen filling pressure, lamp life becomes longer whereas efficacy becomes lower.

(4) The regenerative transport cycle in JD220V75W halogen lamps is able to be taken place in good order when the filling pressure of halogen compounds mixture is kept within 106 to 133Pa during lamp life.

This is one of the attempt to clarify the factors which have influence to lamp efficacy and lamp life, especially mains voltage (220V) low wattage JD lamps. Next step might be the study on the relationship inbetween the filling and coil design.

References

1. L. Nagy, E. Biro and J. Balla   J. Chromatography 520, 213 (1990)
2. L. Bgio, B. Kubicki and P. J. Codella   Appl. Spect.,45, 819 (1991)
3. G. Dittmer and U. Niemann   J. Philips Research Vo.l 42 No.1   (1987)

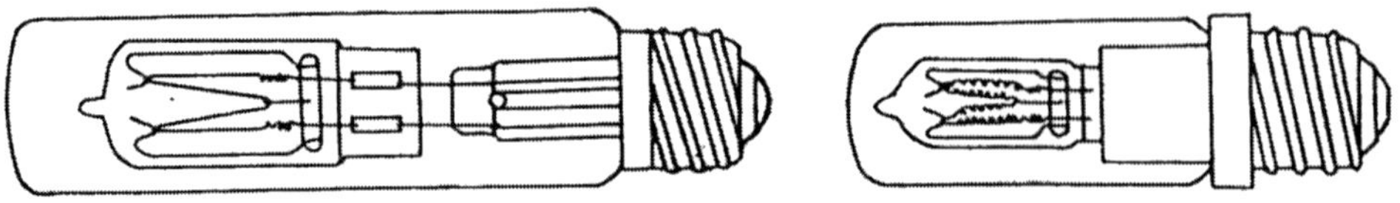

The Structure of Single Ended JD 220V 75W Halogen Lamps

INVESTIGATION OF EXHAUSTING PROCESS WITH INJECTING OXYGEN IN MANUFACTURING LAMPS

Pan Zhong qiao
Haining Illuminating Appliance Inc.
Haining, Zhejiang Province, P.R. of China

To inject small amount of dry oxygen during exhausting process in lamp manufacturing is an effective method for eliminating the air involved inside the bulbs and for decreasing the ratio of releasing residual gases.
This technique has been applied to the exhausting process of flash lamp manufacturing. The result shows that the starting voltage decreases by 13.2% and the times of flashing increase by 78%. This technique has also been applied in manufacturing vacuum and gas-filled incandescent lamps. Under the production condition, the lifetime of various lamps is improved by at least 56% and the total luminous flux becomes stable and its spread becomes narrower.
When this technique is applied for the outer bulb exhausting process of high pressure mercury lamps, the luminous flux increased by 5.3% and the lifetime of self-ballasted high pressure mercury lamps increased by 38%. The quality of tubular fluorescent lamps (standard 40W) is also able to improve by means of this technique. The initial luminous flux raised by 1.4% and lifetime increased by 11.0%. The maintenance after 100 hours burning improved by 20%.

It is well known that oxygen is a very active oxidizing element for various materials. The components inside bulb have complicated geometry and the surface of these components is covered by the layers consisting oxides, absorbed gases, liquid state molecules and other pollutants. The injected dry oxygen works effectively to oxidize the pollutants and serves to easy removal of them.

As oxygen, water vapour and carbon oxide are all covalent molecules, they can easily combine each other. The compound of oxygen and water vapour molecules is $O_2 \cdot H_2O$ of which structure can be shown as ;

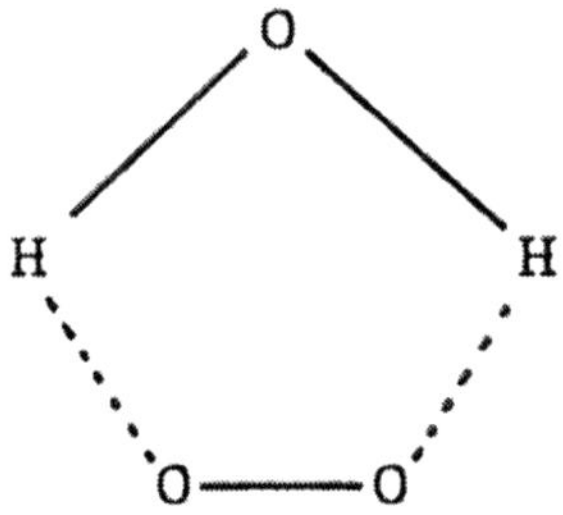

The reactions can be explained as follows;

$$4O_2 + 4H_2O \longrightarrow 4O_2 \cdot H_2O\uparrow$$

$$3O_2 + 2H_2 \longrightarrow 2O_2 \cdot H_2O\uparrow$$

$$2O_2 + 2C\cdot \longrightarrow 2CO\uparrow + O_2 \longrightarrow 2CO_2\uparrow$$

$$5O_2 + 2H_2S \longrightarrow 2SO_2\uparrow + 2O_2 \cdot H_2O\uparrow$$

$$C_6H_8O_3(CNO_2)_2 \xrightarrow{\Delta\ O_2} O_2 \cdot H_2O\uparrow + CO\uparrow + NO_2\uparrow + CO_2\uparrow$$

$$\xrightarrow{\Delta\ Vacuum} 4H_2O + CO\uparrow + 2NO_2 + 5C$$

$$4O_2 + CH_4 \longrightarrow CO_2\uparrow + 2O_2 \cdot H_2O\uparrow$$

Obviously the injection of a small amount of dry oxygen facilitate the exhausting ability and results the higher vacuum level of inside bulb.

A 26-INCH FULL COLOR PLASMA DISPLAY PANEL: DESIGN AND LUMINOUS CHARACTERISTICS

N.Kosugi*, K.Hamada*, K.Takahashi*, N.Isobe*, S.Fujiwara*, K.Wani*, M.Seki**, H.Murakami**

*Matsushita Electronics Corporation, 1-1 Saiwai-cho, Takatsuki, Osaka 569 JAPAN

**Science and Technical Research Laboratories of NHK, 1-10-11 Kinuta, Setagaya-ku, Tokyo 157, Japan

1. Introduction

A dc color plasma display panel (PDP) operated with pulse-memory scheme[1] is suitable for large area flat display such as, for example, HDTV monitor, because of its simple structure, sufficient gray levels and scaling capability.

Recently, the authors have demonstrated a 26-in. experimental panel with resistor-in-cell structure[2]. In this new structure, the discharge current per cell is controlled as low as possible with built-in resistors. As a result, panel lifetime of over 30,000 hours was achieved.

In this study, we optimized the gas mixture and the gas pressure in a pulse-memory dc-PDP with resistor-in-cell structure, in order to achieve high efficiency, and consequently, high brightness.

2. Panel Design

Figure 1 shows the structure of the display cell with built-in resistors. The specifications of the panel is shown in Table I. The resistor is inserted between the anode electrode (for display or auxiliary anode) and the bus line, which are formed on the back plate. On the front plate, the aluminum cathode lines are formed[3].

All these patterns are made of thick film paste. While most of thick film patterns were screen printed, we employed a photo-etching process for patterning the anode bus lines and the anode electrodes. This brought about an accurate effective length of resistors, and the variation of the resistor was within 10%.

We also coated the side walls as well as the bottom surface of the cell with phosphor paste.

3. Performance

One of the drawbacks of the resistor-in-cell structure is limited luminance due to low discharge current per pulse. On the other hand, gas mixture and pressure can be selected from wide operating range, because the built-in resistor stabilizes the current in any discharge conditions.

We fixed the gas mixture to the He-Xe mixture, which does not emit strong visible light. The total pressure and Xe content was changed and the brightness of the panel was measured. Figure 2 shows the experimental result. When the total gas pressure is increased, the brightness goes up monotonously. The brightness also increases along with Xe content.

However, the sustaining voltage of the discharge cell also increases as shown in Fig.3. From the view point of driving ICs, pd (p:gas pressure, d:gap distance) was chosen to be 8.0 [Pa·m]. With this condition, white luminance of 150 cd/m^2 was obtained, when the panel was operated with 256-gray scale digital video data.

4. Conclusion

A experimental 26-in. dc pulse memory PDP with white luminance of 150 cd/m^2 was fabricated. The lifetime of the panel is expected to be 30,000 hours. As a next step, we started a development of mass production technology for large-area (~40-in.) practical panel.

References

1. H.Murakami:LS7 Invited paper (1995).
2. K.Wani:SID'94 Digest, 33.1 (1994), M.Ushirozawa: SID'94 Digest, 33.2 (1994).
3. H.Kono:SID'95 Digest, 37.5 (1995).

Table I. Specification of 26-in.PDP.

Screen size (mm^2)		582×333 (26-in.)
Aspect Ratio		16:9
Cell pitch (mm^2)		0.65×0.65
Number of cells		896×512
Gray levels (Number of colors)		256 (16.8 million)
Cell arrangement		R G / G B
Phosphors	R :	Y_2O_3: Eu
	G :	Zn_2SiO_4: Mn
	B :	$BaMgAl_{16}O_{27}$: Eu

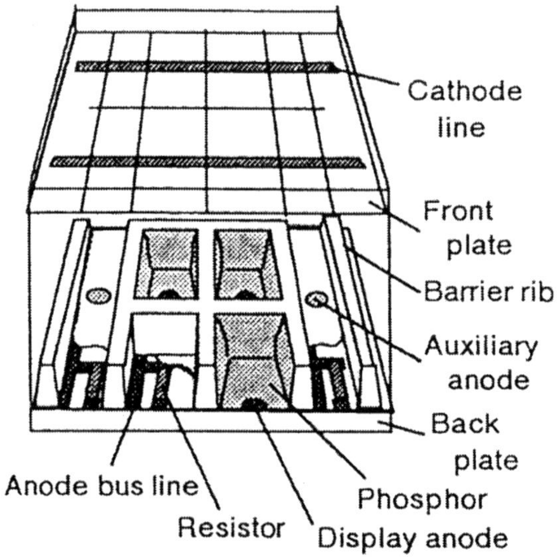

Fig.1 Structure of the discharge cell.

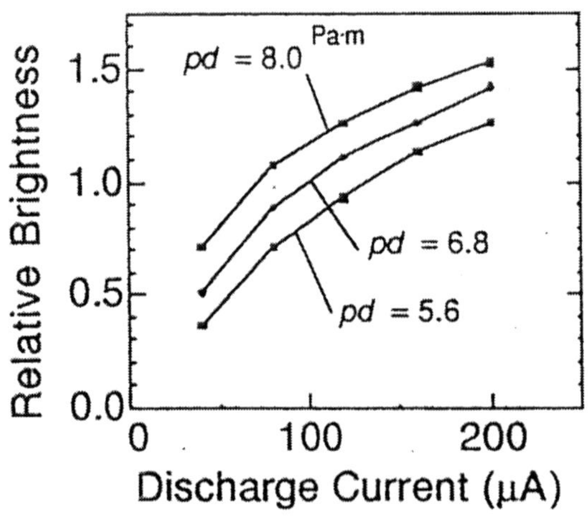

Fig.2 Relative brightness.

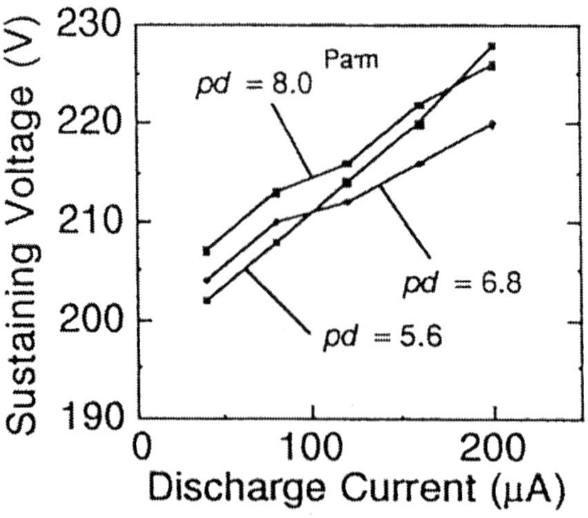

Fig.3 Sustaining voltage.

THE PROGRESS OF AND PROSPECTS FOR THE PLASMA DISPLAY

Hiroshi Murakami
NHK Science and Technical Research Laboratories
1-10-11 Kinuta, Setagaya-ku, Tokyo 157

1 INTRODUCTION

Plasma display panels (PDPs) were developed mainly for use in office automation. In the 1980's, the era of the personal computer was in full swing and the lap-top had begun to emerge as a major market force. The 640 x 400-dot orange-yellow PDPs were introduced to the PC display market by various manufactures as very thin display units. However, the full-color active-matrix LCDs, which were commercialized in the late 1980's, have since expanded their share of the PC display market and left the orange monocolor PDPs behind. A few orange-yellow PDPs are used as small displays for devices such as train ticket vending machines, ticket examining machines, and cash resisters.

Research and development for the color PDP has been major theme since the 1980's. It is not easy for a full-color PDP for graphic display to compete with the CRT and LCD because of its limited resolution. This is due to the occupation of areas by phosphor, electrodes, and so on. Full-color PDPs with somewhat inferior resolution for their screen size should, however, be used in fields where the efficient utilization of space is a major consideration. An example is the 21-inch-diagonal full-color AC PDP which was recently commercialized[1] and introduced to display terminals at the New York stock exchange.

Figure 1 shows the range in which the PDP will be available in the near future. The PDP will be used for flat displays twenty to sixty inches in the diagonal, with the display field between the LCD and projection. The field in which the PDP is especially superior to the CRT and other flat displays is that of 40-inch-plus large-screen displays. A compact large-screen Hi-Vision(HDTV) display is now demanded for the diffusion of the HDTV system. Flat displays, lighter and thinner than CRTs, are considered indispensable for bringing Hi-Vision receivers into general home-use, because a display of at least fifty inches in width is required to make the pictures seen truly realistic. The PDP is the most promising option for realizing such large-screen flat Hi-Vision receivers. Two kinds of experimental 40-inch Hi-Vision PDPs have already been fabricated[2][3].

From the foregoing, it appears that research on the color PDP may focus on the development of medium-sized graphic displays of relatively low resolution and large Hi-Vision receivers. This paper consider the latter. It describes the present state and future prospects of PDP research aimed at realizing the large flat TV.

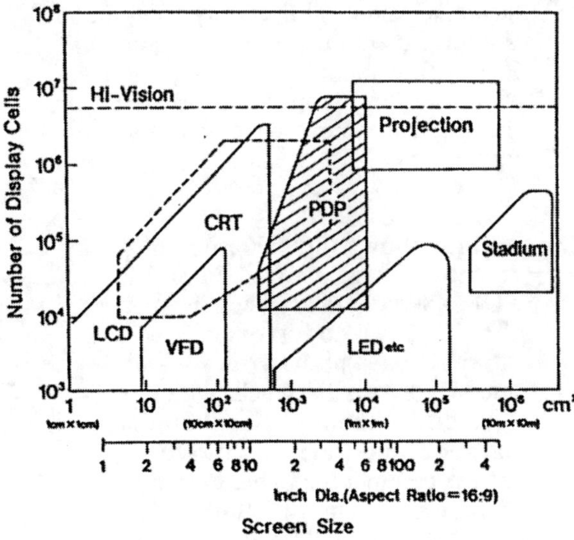

Fig.1 Range of PDP in the future

2 FEATURES OF THE PDP

The PDP is categorized to two main types: DC and AC. Figs.2(a),(b) are schematic illustrations of the operaional principles of the color DC and AC PDPs, respectively. Both the cathode and the anode in the DC PDP are exposed in the part filled with discharge gas. Two electrodes in the AC PDP are covered with a dielectric layer and an MgO layer. Vacuum ultraviolet (VUV) light is generated by the ignition of gas-discharge between two electrodes. The VUV light excites the phosphor layer on the

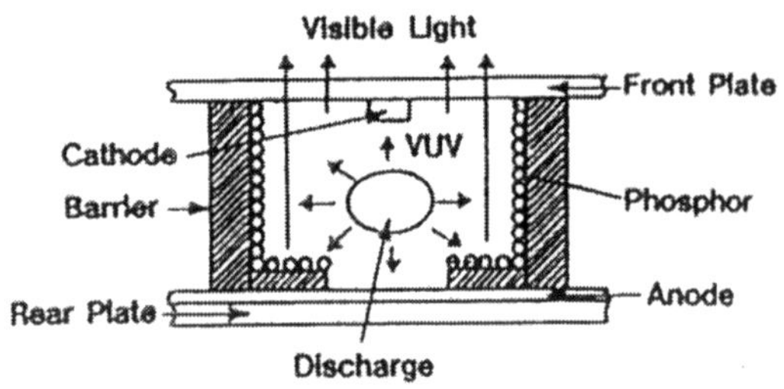

(a)DC PDP

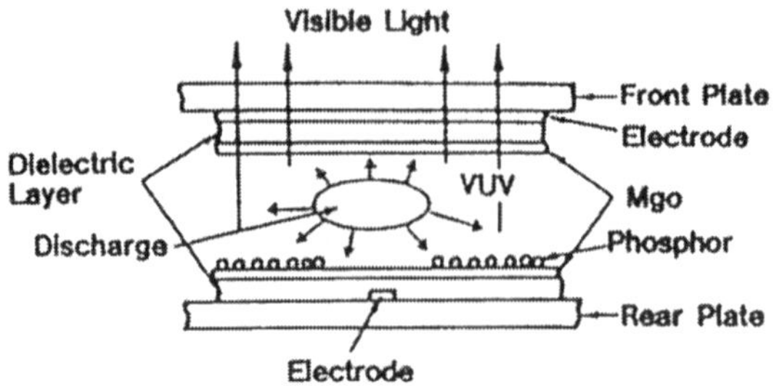

(b)AC PDP

Fig.2 Principle of color PDP

inner surface of the rear plate. The photoluminescence principle converts VUV light into visible light, of which the spectrum depends on the phosphor material.

The AC PDP has a memory function, achieved by controlling the number of charged particles condensed on the MgO layer during short pulse discharge. Although the DC PDP does not have any memory function as it is, a memory function can be introduced by the pulse drive scheme, as described below. The memory function of the PDP ensures high luminance irrespective of the size of the panel. Accordingly, a bright large PDP can be realized.

3 PRESENT STATE OF THE PDP

3.1 Orange Monocolor PDP

Fig.3 is a schematic illustration of an orange luminescent DC PDP for graphic and character displays[4]. The panel consists of two glass plates, front and rear. The front plate has transparent anodes and barriers. The barriers are stacked to the height of 150um by multiple screen-printing with single screen. Only cathodes are screen-printed on the rear plate. Ne-Ar mixture gas fills the space in the panel. The panel is driven with line-at-a-time address for every field time of about 60Hz.

The structure of the orange luminescent AC PDP is the same as that in Fig.2(b) except that it does not have a phosphor layer. As described above, these DC and AC PDPs only have limitd uses at present in specialized equipment. Regarding panel size, a 60-inch AC PDP, the largest monochrome PDP in the world, has been developed for military use[5].

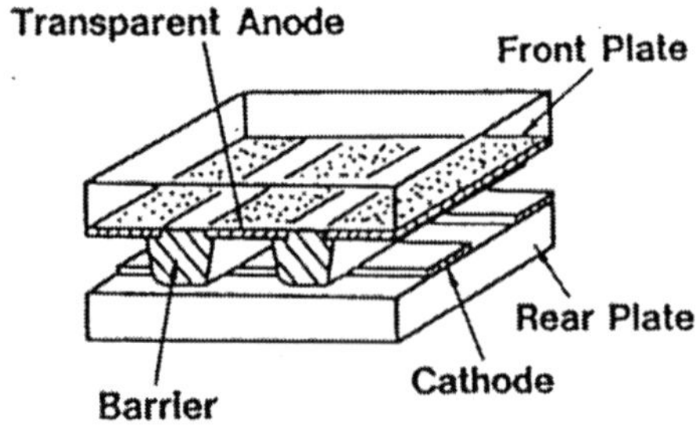

Fig.3 Structure of Orange-yellow DC PDP[4]

3.2 Color DC PDP

Fig.4 shows the structure of a recently commercialized color DC PDP. Phosphors are deposited on the display anodes except in a small area which works as an anode. An auxiliary cell, which is formed by the auxiliary anode, cathode and barrier groove, is introduced into the panel to reproduce good-quality pictures. The specifications and performance of the PDP are shown in Table 1. This color PDP is not used as a TV

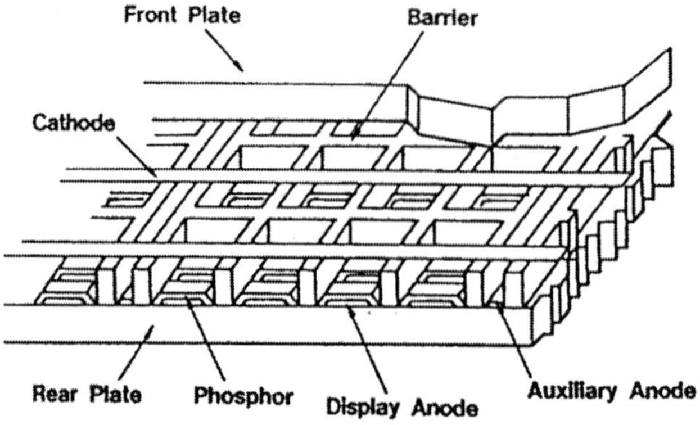

Fig.4 Structure of color DC PDP for information display

Table 1 Specifications and performance of recently developed color PDPs

	Long-Life DC PDP	High-Contrast PDP	Commercialized DC PDP	Commercialized AC PDP	Hi-Vision 40-in. PDP	Commercialized AC PDP
AC/DC	DC Pulse Memory		DC Line Drive	AC Surface Discharge		
Screen Size (mmxmm)	582x333	538x358	528x158	422x319	882x503	407x305
(inch in diagonal)	26	25		21	40	20
Number of Display Cells	896x512	768x512	160x48	640x3x480	840x3x480	640x3x480
Display Cell Pitch (mmxmm)	0.65x0.65	0.7x0.7	3.3x3.3	0.22x0.66	0.35x1.05	0.21x0.63
Number of Gray Levels	256	256	8 Colors	64	256	64*
Peak Luminance (cd/m²)	150	90	100	200	200	200
Displayed Picture	Wide NTSC	Wide NTSC	Information	NTSC	Hi-Vision	NTSC
Announced Year	1993	1991	1994	1992	1994	1994

* 256 with Dither

receiver or PC display, but the display terminals can trans mit news and information to many people.

The details of the color DC PDP for TV display will be described in section 4.

3.3 Color AC PDP

(1)Double Substrate AC PDP

The AC PDP shown in Fig.2(b) is called the "double substrate " AC PDP, because two sets of electrodes are arranged separately on the front and rear plates. Since phosphor is deposited near the electrode in the ordinary double substrate AC PDP, its characteristics tend to degrade with ion bombardment. Recently, this problem of degradation has apparently been solved[6]. The ordinary double substrate AC PDP, however, has no intermediate sheet to separate the cells from each other, as shown in the figure. In such an arrangement, the phosphor in each cell is excited by VUV light from the discharge of the adjacent cell, causing degradation of color purity. To improve color purity, a

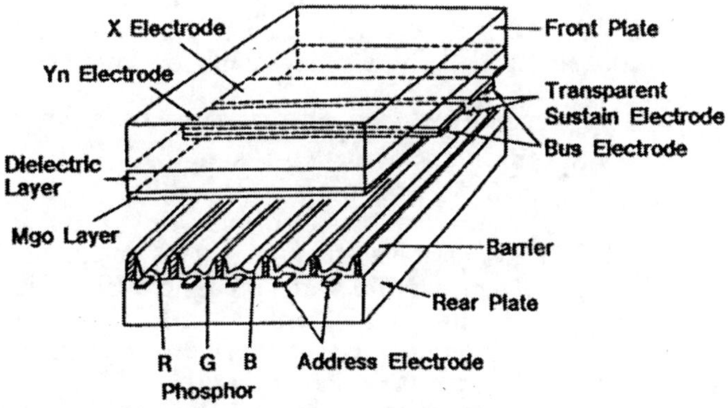

Fig.5 Structure of 21-inch color AC PDP[1]

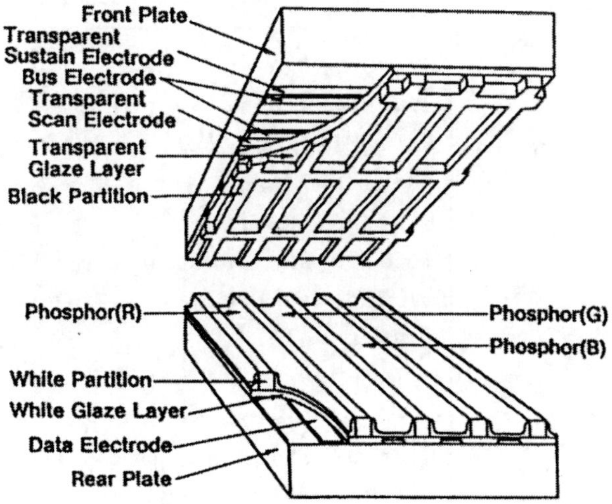

Fig.6 Structure of 40-inch color AC PDP[3]

barrier rib of half of the height of the gap was introduced into the panel between the two glass plates[7]. As a result, a 13-inch double substrate AC PDP with a fine cell pitch of 0.2mm and improved color purity has recently been developed.

(2)Surface Discharge AC PDP

About 10 years ago, a "surface discharge"(or single substrate) AC PDP, in which both the row and column electrodes are placed on the rear plate, was invented to prevent the degradation of phosphor characteristics[8]. Fig.5 shows the structure of the latest 21-inch surface discharge AC PDP, which was commercialized in 1993[1]. The transparent sustain electrodes(X and Yn electrodes) are arranged on the front plate. Barriers and address electrodes are formed on the rear plate. The inner surface of the cell is coated with phosphor on the side wall of the barriers and bottom of the cell. After write access operation with the address and Yn electrodes, sustain discharges are maintained between the sustain electrodes. The specifications and performance of the first commercialized 21-inch color AC PDP are shown in Table 1.

Regarding the surface discharge AC PDP, there are many variations. Fig.6 shows the structure of one of them[3]. The structure of this panel is almost the same as that in Fig.5 except for the addition of the black partition on the front plate and the white graze layer on the rear plate. The specifications are also shown in Table 1. Hi-Vision pictures are displayed on a 40-inch PDP.

Since auxiliary discharge, which is essential for fast write access operation, is always maintained in the cell for the AC PDP, the luminance of the black picture level appears to be comparatively high, resulting in the low contrast of the picture.

4 PROGRESS OF THE DC PDP FOR HI-VISION

4.1 Specifications of the Hi-Vision PDP

Table 2 shows example Hi-Vision flat display specifications based on the CCIR documents[9]. The numbers of display cells and cell pitches in the table derive from both the MUSE transmission system for Hi-Vision and the HDTV studio standard.

Table 3 compares the various possible Hi-Vision flat panel displays. All flat panel displays except for the PDP entail some serious problems for Hi-Vision. In particular, the most serious problem was that of how to realize large displays for Hi-Vision. With regard to feasibility, the table shows that this PDP should be the most suitable for Hi-Vision displays on account of its easy fabrication, especially for large panels, and use of the light emission principle.

Table 2 Specifications of Hi-Vision PDP

Screen size (mmxmm)	55-in.-dia (672x1,195)
Number of Display cells	1,035x(1,496~1,920)x3
Cell Pitch (mmxmm)	0.65x(0.3~0.2)
Peak Luminance (cd/m^2)	200
Contrast Ratio	50:1
Total Power Consumption (W)	300
(Panel & Electrode Drive Circuit	250)
Gray Levels	256
Life (hour)	15,000
Display Thickness (cm)	≤5
Display Weight (Kg)	≤20

4.2 40-inch Hi-Vision PDP

NHK Science and Research Laboratories have being developed DC PDPs for Hi-Vision flat display since the early 1970's. Fig.7 shows the present state of progress of NHK's color DC PDPs. The first breakthrough was to increase luminance using the pulse memory drive scheme. This established a memory function in each cell by means of electronic circuitry, contributing to a dramatic increase of luminance[10]. The next was to establish the large panel fabrication technologies, including a simple flatter configuration and screen printing technologies[11][12]. These breakthroughs opened the way for large full color TV displays and resulted in the development of a 40-inch-diagonal Hi-Vision PDP, the largest in the world[2].

Fig.8 shows the structure of the 40-inch PDP fabricated in 1992. Although the side view of the display cell is different from that in Fig.2(a), the operational principle is the same. Auxiliary cells are arranged between the display cells in order to lower the display cell's breakdown voltage. They cause a priming effect in the display cells through the priming space between the display and auxiliary cells, thus contributing to fast write access operation. The 40-inch panel is driven with the pulse memory drive scheme.

The panel consists simply of two glass plates with no parts that are difficult to make. All components except for the phosphor layers--cathode, anodes, and barriers-- are screen printed as thick films. The phosphors are deposited on the inner surface of the front plate by photo-lithogaghy.

Table 4 shows the specifications of the 40-inch PDP. The panel is 6mm thick, weighs 8kg, and has a Hi-Vision aspect ratio of 16:9. The display cell pitch of 0.65mm is the same as the pixel pitch required for a future 55-inch panel with

Table 3 Comparison of possible Hi-Vision flat displays

	PDP	LCD	Flat CRT	EL	LED
Chromaticity	○	○	○	▲	○
Operating Speed	○	△~○	○	△	○
Dot Pitch	△	○	○	○	○
Screen Size	○	▲	■	▲	▲
Life	○	○	○	○	○
Contrast	○	○	○	▲	○
Luminance	○	○	○	▲	▲
Efficiency	△~○	△	○	▲	△

○Considerable △Some
▲Difficult ■Almost none

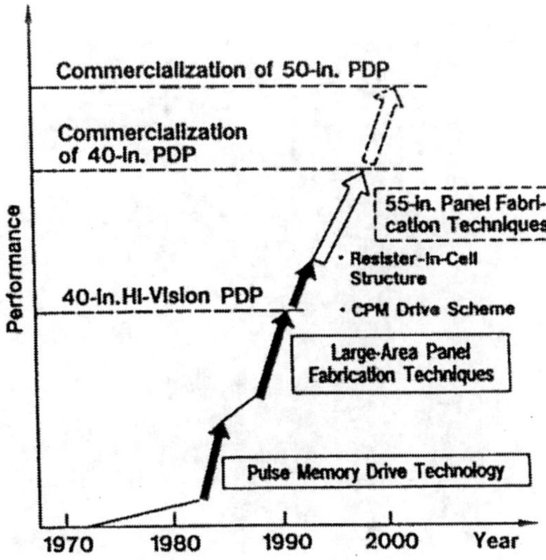

Fig.7 Progress of NHK's color PDPs

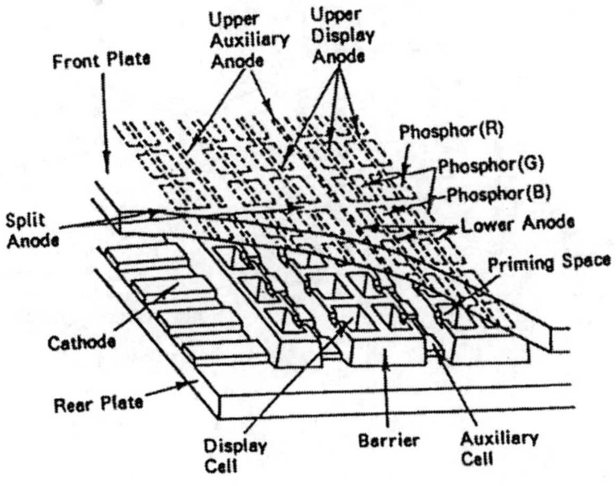

Fig.8 Structure of 40-inch DC PDP[2]

the same as the pixel pitch required for a future 55-inch panel with complete HDTV specifications. In other words, the number of display cells is still insufficient for HDTV demonstrations. We can, however, improve the resolution by displaying the diverse information that corresponds to positions in a picture on the diagonally arranged green cells. The performance of the 40-inch display is also shown in Table 4. As shown in Fig.9, a Hi-Vision picture has been reproduced on a 40-inch wall-hanging PDP.

Table 4 Specifications and performance of 40-inch Hi-Vision DC PDP[2]

Screen Size	(■■x■■)	520x 874
Number of Display Cells		800x1,344(1,075,200)
Average Display Cell Pitch (mm)		0.65
Cell Arrangement		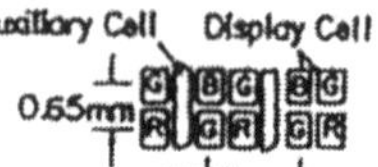 Auxiliary Cell, Display Cell, 0.65mm, 1.3mm
Panel Thickness	(mm)	6
Panel Weight	(kg)	8
Peak White Luminance (cd/m²)		93(27 fL)
Luminous Efficacy(White, lm/W)		0.15
Number of Gray Levels		256
Display Unit Thickness (cm)		8

Fig.9 40-inch Hi-Vision DC PDP

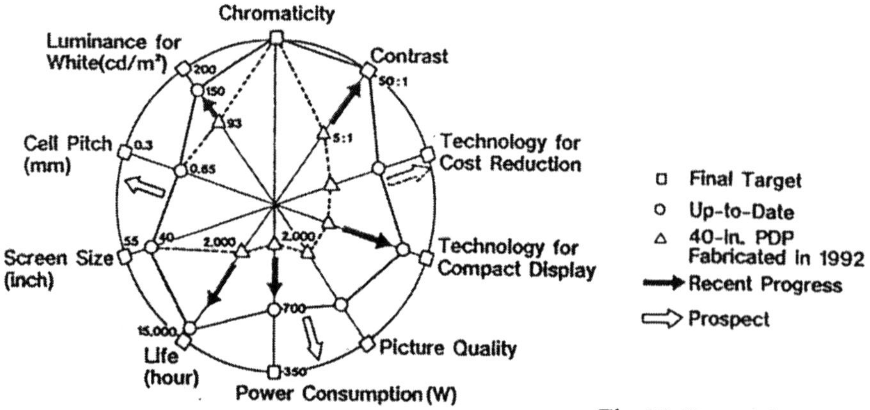

Fig.10 Comparison of present and future DC PDPs

4.3 Present Technologies and Prospects for Pulse Memory DC PDPs

Fig.10 compares the current pulse memory PDPs with both the 40-inch PDP and a final target PDP that we aim to make in the near future. The figure shows that the characteristics

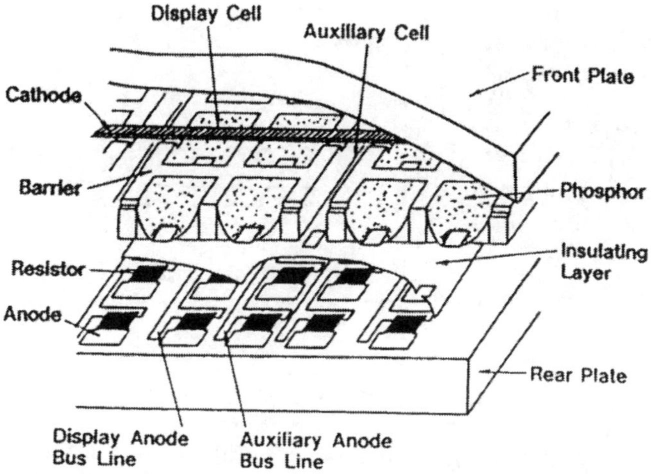

Fig.11 Resister-in-cell structure[14]

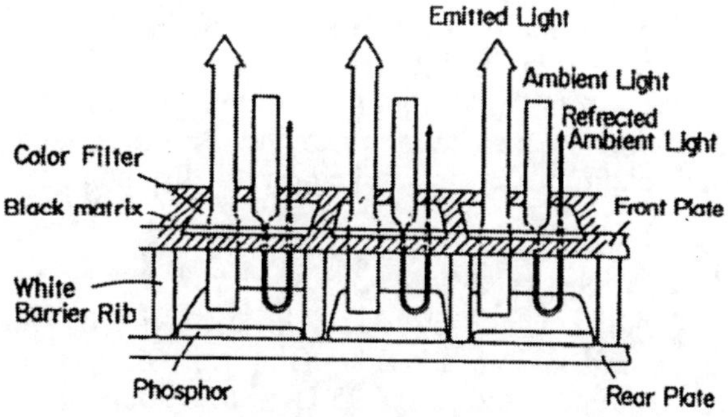

Fig.12 Principle of high-contrast DC PDP[16]

of an up-to-date PDP are considerably better than those of the 40-inch PDP. The main improvements are as follows:

(1)Lifetime and Luminance

In order to improve the life and luminance, a new PDP with a resister-in-cell structure has been developed, as shown in Fig.11[13]. Cathode sputtering, which causes shorter lifetime, can be suppressed by reducing discharge current with the introduction of a resister in series into each cell. To suppress cathode sputtering further, gas pressure in the panel is also increased to 1.5 times the level of the 40-inch panel. As a result, the lifetime exceeds 10,000 hours. A luminance of 150cd/m^2 can be obtained by coating the whole side wall and bottom, except for the cathode in the display cell, with phos.RM 6.20"
phor, as shown in the figure. The development of the phosphor coating technique makes it possible to introduce the new phosphor screen into the panel. A 26-inch PDP has been developed with the features described above and wide-TV pictures can be displayed on the panel[14]. The specifications and performance of the display are shown in Table 1.

(2)Reduction of Power Consumption

The pulse memory PDPs are driven by sustain pulses with a high frequency of 250 kHz. A power which is not useful for discharge is consumed to charge and discharge stray capacity between the panel's electrodes. This power consumption is proportional to the panel's size, and is about 400W for the 40-inch panel. New pulse memory drive techniques(CPM driving) have been developed to reduce this reactive power consumption[15]. Power consumption can be reduced to 80W, one fifth of the level for conventional driving. Furthermore, since the CPM driving circuits are simple, it is comparatively easy to produce drive ICs. Two kinds of drive ICs are now under development.

(3)Contrast Ratio

We have developed color filter technologies which can provide a much larger contrast ratio. The principles are shown in Fig.12. The technologies have been used to fabricate a 25-inch high contrast pulse memory PDP with a contrast ratio of 50:1 under 2001x illumination[16]. The specifications and performance of the display are shown in Table 1.

The cell pitch, which is still 0.65mm, will be reduced to less than 0.5mm by improving the technologies for producing panel parts. The power consumption of the receiver, which is still 700W, will be reduced to 350W through the development of signal processing LSIs. The expected performance is thought to be competitive with that of other large displays, as shown in Fig.13, especially in view of the relatively low cost and high picture quality.

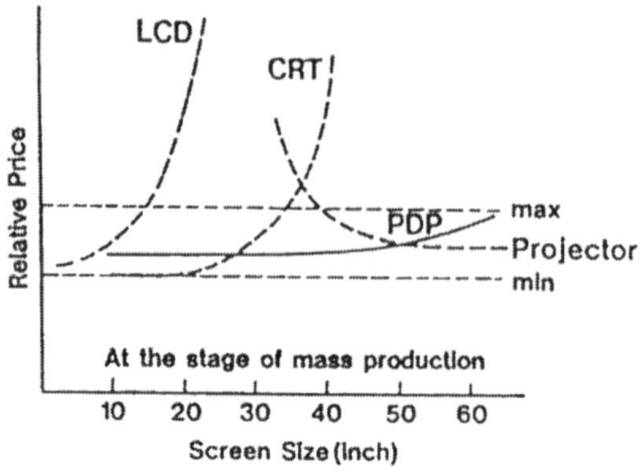

Fig.13 Price prediction of home TV receiver

5 CONCLUSION

The commercialization of the 21-inch AC PDP and success with the Hi-Vision display on two types of 40-inch PDPs suggest that the era of the color PDP will arrive in the near future. To hasten its arrival, better contrast will be needed for the AC PDP, and lower cell pitch for large AC and DC PDPs.

In October 1994, a Hi-Vision PDP Joint Development Consortium was organized to commercialize a 40-inch Hi-Vision DC PDP by 1998, when the Winter Olympic Games will be held in Nagano, and to develop a 50-inch prototype Hi-Vision PDP by 1999 in cooperation with manufacturers having the related technologies. Beautiful, powerful Hi-Vision pictures will be enjoyed on the 40-inch PDP at the next winter Olympic Games.

[REFERENCES]

[1]K.Yoshikawa et al.: Proc. 12th IDRC, pp.605-608 (1992)
[2]T.Yamamoto et al.: SID '93 Dig., pp.165-168 (1993)
[3]T.Nakamura et al.: SID '95 Dig., to be published
[4]Y.Nakagawa : National Technical Report, 33, 1 (1987)
[5]D.K.Wedding,Sr. et at.: SID '87 Dig. ,pp.96-99 (1987)
[6]P.E.Knauer et al.: SID '87 Dig., pp.84-87 (1987)
[7]G.Baret et al.: SID '93 Dig., pp.173-175 (1993)
[8]G.W.Dick et al.: IEEE Trans., ED-23, 4, pp.429-437 (1976)
[9]CCIR Documents 11/1007(Rev.1)-E,1990 and 11/1008-e,1990
[10]H.Murakami,R.Toyonaga : IEEE Tras., ED-29, 6, PP.989-994 (1982)
[11]H.Murakami,T.Katoh : IEEE ED Letters, EDL-6, 3, pp.132-134 (1985)
[12]H.Murakami et al.: SID '91 Dig, pp.713-716 (1991)
[13]T.Sakai et al.: ITEJ Tech.Rep., Vol,15, No.72, pp.7-14 (1991)
[14]H.Kono et al.: SID '95 Dig., to be published
[15]Y.Takano et al.: SID '94 Dig., pp.731-734 (1994)
[16]Y.Takano et al.: SID '92 Dig., pp.720-723 (1992)

RECENT PROGRESS IN CERAMIC MATERIALS FOR LAMP APPLICATION

K. Maekawa/NGK Insulators, Ltd.

Electronic Components Division, Optical Device Department

2-56 Suda-Cho Nagoya 485, JAPAN

1. INTRODUCTION

It is well known to the public that the first lamp applied a ceramic material for an arc-tube was the high-pressure sodium discharge lamp (hereinafter referred to as HPS lamp), which was developed by Louden and Schmidt[1], and introduced in 1965. At present, after 30 years since then, the HPS lamps are used all over the world for outdoor lighting applications, because of their high luminous efficacy and long lives.

In observing the history of lamps since the incandescent lamp developed by Edison, it will be not too much to say that the target of development in the lighting field has been always the improvement in lamp efficacy. Especially, in the development of the high-pressure discharge lamp. It has been eagerly desired to use the translucent or transparent ceramic material, which is excellent in heat resistance and chemical stability, instead of the known glass and/or quartz as an arc-tube material.

The realization of this desire was done by Coble[2] who invented the polycrystalline translucent alumina ceramic (hereinafter referred to as translucent alumina). This invention overthrow the known concept that the polycrystalline ceramic material does not allow to pass the light through it, and, by applying this material instead of the known glass and/or quartz for lamp's arc-tube, the high efficacy and long life HPS lamp appeared in the market.

The luminous efficacy was 105 lm/W in the early HPS lamp, and now it has been improved to 140 lm/W. Although this increased efficacy has been achieved by various improvements in the lamp-manufacturing technology, improvements in characteristics of translucent alumina itself also contributed to this.

In this paper, the various translucent or transparent ceramic materials developed up to now are reviewed and the results of our recent activities on translucent alumina are described.

2. TRANSLUCENT OR TRANSPARENT CERAMIC MATERIALS

With Coble's invention "Transparent alumina and method of preparation" registered in 1962 as US. Patent No. 3026210, numbers of ceramists started to research on the transparency of various polycrystalline ceramic materials, and, since then over the 1980's, a variety of transparent or translucent ceramic materials like yttria[3], magnesia[4], zirconia[5], thoria[6], beryllia[7], spinel[8], mullite[9], yttria-alumina garnet[10] and aluminium nitride[11] have appeared.

However, it is not too much to say that the translucent ceramic material, which is nowadays used practically as arc-tube for the high-pressure discharge lamp, is only alumina. Of course, the application of other ceramic materials like yttria, magnesia, yttria-alumina garnet and aluminum nitride have been investigated. Of those ceramic materials, yttria is more superior in chemical stability than that of alumina and, at 1970's, the realization of the HPS lamp using yttria arc-tube has been attempted[12]. However, it was not practically used for arc-tube of the HPS lamp in place of alumina. Furthermore, the appearance of translucent aluminum nitride is noticed from view points of its good thermal conductivity and lower thermal expansion, and since 1985 some patent applications[13,14] concerning the high-pressure discharge lamp using it as arc-tube material is laid open, but, by now such lamp is not publicized in the market as a general lighting lamp.

3. IMPROVED TRANSLUCENT ALUMINA

It has passed 35 years since the invention of translucent polycrystalline alumina by Coble. During the period of time, as described in the above, many kinds of translucent or transparent ceramic materials have appeared, and it has been proposed to apply some of them for arc-tube material of the high-pressure discharge lamp. However, any high-pressure discharge lamp using other ceramic materials than translucent alumina as arc-tube is not yet realized. Though this reason is not clear, it seems to be due to the characteristics of translucent alumina itself, for example, translucency and corrosion resistance were remarkably improved, and on the other hand, the designing and manufacturing technology of the HPS lamp using alumina arc-tube were also improved, and in the result, even if the arc-tube material of the HPS lamp is replaced with other ceramic materials, such lamp will not show any superiority in lamp characteristics like luminous efficacy and lamp life in comparison with the HPS lamp using alumina arc-tube.

3.1 IMPROVEMENT IN TRANSLUCENCY The improvement in translucency of alumina material lies in making decrease the existence of pores and secondary phase which make scatter and absorb the light in alumina sintered body, and in making larger as possible as the grain-size of sintered body within the range doing not make largely lower its mechanical strength, and many improvement plans have been proposed and executed. Those plans are broadly divided into two categories: one is a review of the additive systems and their adding amount as grain growth inhibitor, and another is the surface reformation of the sintered body.

In translucent alumina of the early stage, for example, the grain growth inhibitor was only MgO and its added amount was 0.2% by weight, comparatively large amount, and it resulted in the alumina sintered body having about 93% in total light transmittance. After that, with the elucidation of mechanism to make ceramic materials translucency, numbers of improvements have been seen in the technical aspects like alumina raw materials themselves[15], manufacturing process conditions from raw material to sintering, and new additive systems other than MgO, and in the results, the present translucent alumina has reached up to more than 95% in total light transmittance.

On the other hand, it is said that, although the improvement of lamps luminous efficacy is sure to be mainly due to the improvement in total light transmittance, the improvement in in-line light transmittance also contribute to it. For example, it is reported[16,17] that, when sapphire, the single crystal of alumina, has been used as arc-tube in place of translucent alumina, such HPS lamp has gained 3 to 5% increase in luminous efficacy. This gain seemed to be due to the high in-line light transmittance of sapphire, because of its total light transmittance being not always higher than that of translucent alumina. In order to make translucent alumina having in-line light transmittance value equal to that of sapphire, some references[18,19] indicated that it was effective to chemically polish the surface of sintered alumina body, and the HPS lamp using chemically polished alumina arc-tube has come into practical use.

However, as cleared in Fig.1a and 1b which show the relations between arc-tube materials, lamp luminous flux, lumen and/or in-line light transmittance. If the translucency of alumina is more than 95% in total transmittance and also more than 2.5% in in-line transmittance, the luminous efficacy of the HPS lamps used such alumina arc-tube reaches to nearly same level as that of the lamps using sapphire or chemically polished alumina arc-tubes. Accordingly, the arc-tube material of the present HPS lamps is al mostly translucent alumina as sintered. In other words, the present translucent alumina has been improved not only in total transmittance but also in-line transmittance by the above-mentioned various improvements.

Today, the additive systems for translucent alumina ceramic materials which are mainly used in the market seems to be two kinds of MgO only and $MgO\text{-}Y_2O_3$. Though it is reported[20] that in $MgO\text{-}Y_2O_3$ system it is possible to obtain the sintered alumina body having a good translucency even in the sintering at lower temperature than that of alumina body added only MgO. However, as shown in Fig.2, the sintered body of $MgO\text{-}Y_2O_3$ system has unhomogeneous grain-structure. Considering from the translucency mechanism of alumina being to control abnormal grain growth, eliminate pores or secondary phases from grain inside and grain boundaries during sintering, and result in a sintered body composed of high-density and homogeneous grain structure, such unhomogeneous grain structure of $MgO\text{-}Y_2O_3$ system's sintered body seems to be not so good; however, even in such grain structure, the translucency of its sintered body is nearly same to that of MgO only system's alumina body, because of the residual pores and the existence of secondary phases in the said body being substantially a very few, and also the body being composed from the large amount of coarse grains, so that the HPS lamp using $MgO\text{-}Y_2O_3$ system's sintered body as arc-tube material results in showing the same luminous efficacy as that of MgO system's alumina arc-tube lamp. Furthermore, as mentioned in the later paragraph, there is a report[21] that this $MgO\text{-}Y_2O_3$ system's translucent alumina is superior to MgO system's translucent alumina in resistivity for sodium corrosion resistance.

3.2 IMPROVEMENT IN SODIUM CORROSION RESISTANCE Transparent yttria, mentioned in paragraph 2, was paid attention as arc-tube material of the HPS lamp because of its excellent resistance against sodium vapor attack. It is well known that the luminous efficacy of the HPS lamp can be improved by raising the wall temperature of arc-tube. However, in case of alumina arc-tube, it reacts with high temperature sodium vapor to form ß-Al_2O3, and Na amount is thus lost to make shorter the HPS lamp life: therefore, it is reported[22] that, to make the HPS lamp of high luminous efficacy and long lives, in case of the standard HPS lamp it is designed to be less than 20 W/cm^2 (1150°C) in the wall load (temperature) of alumina arc-tube to control sodium loss.

On the contrary, transparent yttria is stable for the high-temperature sodium vapor, and it is reported[23] that the HPS lamp using yttria as arc-tube material has reached up to 165 to 170 lm/W in the luminous efficacy because of yttria arc-tube being possible to operate at its wall-load (temperature) of 40 W/cm^2 (1450°C) without any sodium loss. It is presumed that the reason why this excellent yttria has been not yet applied for arc-tube of the HPS lamp is not only because of its high price, but the primary reason may be in that the efficacy and life of the HPS lamp using alumina arc-tube has been remarkably increased by the improvements[24] in technical aspects like arc-tube design, kinds or pressure of the starting gas, electrode design etc. and also in sodium corrosion resistance of translucent alumina itself.

According to the published papers[25,26] concerning the reaction between sodium and translucent alumina, one paper indicated that this reaction can be observed even in sapphire whose reaction differs obviously on crystal faces, and therefore alumina is essentially unstable from a view point of thermodynamics; however, in case of translucent alumina, the degree of sodium corrosion largely differs depending on additive systems, additive contents and/or sintering atmosphere, and, for example, it is concluded to be more better in case of translucent alumina with $MgO-Y_2O_3$ system than MgO system, or with MgO system whose content being as possible as small, free from CaO and sintered in vacuum not hydrogen gas atmosphere.

Of patents registered or applied up to now, patents whose object is the improvement in sodium corrosion resistance of translucent alumina are only two of NL-8502457A and EP-PA3119959; however, US. Patent No.4567396 concerning HPS lamp describes to utilize good Na corrosion resistance of translucent alumina registered as US. Patent No.4285732. From those patents and papers, the favorable additive systems seem to be 1) $MgO-Y_2O_3$, 2) $MgO-ZrO_2$ or HfO_2 or (ZrO_2+HfO_2), 3) $MgO-Er_2O_3$, and $MgO-ZrO_2-Y_2O_3$. The followings are our evaluation test results about sodium corrosion resistance on translucent alumina of various additive systems described in the above.

At first, the arc-tube's inner surface of the HPS operated at the wall temperature of about 1300°C for 6000 hours, where, this alumina arc-tube contain about 0.06% by weight of MgO. As shown in Fig.3, abnormal crystals have been formed in the portion which sodium penetrated into the tube-wall, and it has been confirmed that sodium elements were detected from those crystals, and those are ß-alumina by X-ray diffraction analysis. This result is the similar to the known experimental result.

Fig.4 shows the change of luminous maintenance in case of HPS lamps using various alumina arc-tube designed for 150W use being overload-operated as 250W HPS lamp, and Fig.5 shows a typical example of blackened lamp. In the above-mentioned standard MgO system's alumina arc-tube it shows the most rapid deterioration to drop its efficacy by about 40%, but, in case of another alumina of whose MgO content is 0.04% by weight and sintering atmosphere is vacuum, its dropping rate is 9%. Although the addition of Er_2O_3 with MgO has a little effect to maintain its efficacy. It is not so good in comparison with other alumina except for the standard one. The most effective systems are $MgO-Y_2O_3$ and $MgO-ZrO_2$, and their drop efficacy were not almost observed. However, even though an alumina material belong to $MgO-Y_2O_3$ system, if it is sintered in hydrogen atmosphere and also MgO content is more than 0.03% by weight, the effect of added Y_2O_3 is not so much, and it was confirmed that the vacuum-sintered $MgO-Y_2O_3$ system's alumina as shown in Fig.4 is excellent.

Fig.6 shows the micro-structures of these alumina arc-tubes before operation. It is clear that the microstructures of alumina sintered in hydrogen are composed of distinct uneven grains and wider grain boundaries, and in case of vacuum sintering, it becomes to a comparatively flat and dense structure. The reason why the vacuum sintered $MgO-Y_2O_3$ system alumina showed a good result is considered as follows: as reported by G.D. With et al., the vacuum sintered alumina body having the above mentioned micro-structure is the grain boundaries in a less reactive state than that of the hydrogen sintered body, and more over, as shown in Fig.7, Mg elements in the vacuum sintered alumina tube distribute significantly in the central portion of tube's wall but a very few at the vicinity of inner and outer surfaces of the tube. As result of those states in microstructure of $MgO-Y_2O_3$ system alumina, sodium penetration along grain-boundaries is significantly suppressed.

However, the sintering in hydrogen is best suited to obtain commercially translucent alumina in comparison with vacuum sintering, and, therefore, it is more interested in MgO-ZrO_2 system alumina which is possible to sinter in hydrogen and has homogeneous micro-structure. And then, in order to clear the effect of ZrO_2 additives, the further evaluation was executed about translucent alumina of MgO-ZrO_2 system, and in addition to these, of MgO-ZrO_2-Y_2O_3 system whose additional Y_2O_3 may have more improved sodium corrosion resistance.

Fig.8 shows the change of luminous maintenance about HPS lamps using various alumina arc-tubes of MgO-ZrO_2 and MgO-ZrO_2-Y_2O_3 systems operated at the same conditions as described in the above. In case of MgO-ZrO_2 system, the higher the ZrO_2 content is, the smaller the drop of luminous efficacy becomes, but even the best one it is in the same level as MgO-Y_2O_3 system's alumina or sapphire as a reference sample. In contrast to this, MgO-ZrO_2-Y_2O_3 system alumina is much less drop in luminous efficacy, and this means that Y_2O_3 addition of very small amount is very effective to suppress sodium penetration into arc-tube wall.

Fig.9 shows the micro-structures of these alumina arc-tubes before operation. In MgO-ZrO_2 system, with increase of ZrO_2 content its grain size of the sintered body becomes significantly huge to make lower its mechanical strength, but, in limiting Y_2O_3 content less than 0.01% by weight, such abnormal grain growth which is apt to occur in MgO-Y_2O_3 system alumina has not observed in MgO-ZrO_2-Y_2O_3 system alumina, and its sintered body is comparatively uniform in the micro-structure and also good in the translucency and mechanical strength. Although the existence of ZrO_2 or Y_2O_3 or ZrO_2+Y_2O_3 in translucent alumina was effective to improve sodium corrosion resistance. It was impossible to make translucent the alumina added only one or them without MgO, because of such materials having caused the extremely abnormal grain growth as remained many pores. Therefore, the addition of minimum amount of MgO is indispensable to obtain translucent alumina for arc-tube of HPS lamp.

The reason why sodium corrosion resistance is improved by adding ZrO_2 or Y_2O_3 with MgO is not clear yet, but it is supposed to be as follows: The phenomenon that the blackening of the HPS lamp's outer envelope which occurs simultaneously with the drop of lamp luminous efficacy is considered that, as shown in the following formula (1), sodium component filled in arc-tube reacts with alumina to form ß-alumina, and simultaneously the generated Al metal diffused into the wall of alumina arc-tube to pass through the wall and then evaporate from the outer surface of arc-tube to deposit on the inner surface of lamp outer envelope;

$$2Na + (1/3 + 11)Al_2O_3 \rightarrow Na_2O \cdot 11Al_2O_3 + 2/3Al \quad \text{.........................} \quad (1)$$

Fig.10 shows the deposit material on the inner surface of lamp outer envelope, and it has been confirmed that those material is Al metal by a thin-film X-ray diffraction analysis.

On contrary to this, it is supposed that the reaction of Formula (1) is suppressed by the existence of ZrO_2 or Y_2O_3 in alumina arc-tube. As shown in Fig.11, on the both inner surfaces of MgO-Y_2O_3 and MgO-ZrO_2-Y_2O_3 systems alumina arc-tubes, it is not able to find any formation of ß-alumina crystal, but some of whitish fine particle crystallize in their grain-boundaries. From EDS analysis, Y and/ or Zr+Y elements were detected from these particles. Although those particles are not observed in alumina arc-tubes before operation, they exist already in the tubes after 1300 hours operation and remain without any change even after 6000 hours operation. Supposing from those results, it is considered that those particles containing Zr and/or Y crystallized in grain-boundaries obstruct Al metal diffusing along grain boundaries, and, by this obstruction, Al metal can not diffuse and pass-through the wall of alumina arc-tube to suppress the progress of Formula (1) reaction. Therefore, the formation of ß-alumina crystal also is suppressed and the lamps are maintained so as to make not lower the luminous efficacy without causing the blackening of lamps.

The formation mechanism of the above particles is not clear yet, but, for example, it is presumed to form such particle due to the following reaction;

$$2Na + 1/2ZrO_2 + 11Al_2O_3 \rightarrow Na_2O \cdot 11Al_2O_3 + 1/2Zr \quad \text{......................(2)}$$

The formed particle is still unknown which is a metal or oxide, so that the adaptation of Formula (2) is not also proved. However, in case of MgO content being too much in alumina arc-tube, it is considered that the reaction due to Formula (1) proceeds preferentially to obstruct the formation of those particles in grain-boundaries, so that it is not able to sufficiently prevent sodium attack against the grain-boundary and/or grain itself of alumina arc-tube.

4. CONCLUSION

The description is given in the above about translucent alumina arc-tube materials which have the most important role for the realization of high efficacy and long life HPS lamp, a target of lamp engineers, especially about recent activities on the improvements of translucency and sodium corrosion resistance of translucent alumina. In recent year, passed 35 years already since the invention of translucent alumina by Coble, it is hardly to find the patents or papers describing about improvements in its translucency and sodium corrosion resistance. However, as described in detail in paragraph 3.2, $MgO\text{-}ZrO_2\text{-}Y_2O_3$ system's alumina arc-tube material has characteristics which exceed those of prior inventions of sapphire, so that the appearance of HPS lamp using this alumina is expected.

On the other hand, in the high-pressure discharge lamps, the light source in the future, which is most expectable for the widest range of illumination uses, is said to be metal-halide lamps. This lamp is inferior to HPS lamp in luminous efficacy and life because of the arc-tube material being quartz, but superior to HPS lamp in color rendition. From last autumn, Philips has already introduced to the market the metal halide lamps so-called "Master Color Lamp" in which translucent alumina arc-tube is used in place of quartz one, so that it is predicted that other manufacturers will also place the ceramic metal-halide lamps on the market and after this year following Philips lighting Co. However, it is said that the metal-halide components have stronger corrosion behavior than Na, and, in order to realize the metal-halide lamps being more superior to the HPS lamps in efficacy and life, from now it is considered to be an important thing to develop chemically more stable translucent or transparent ceramic materials such as yttria, yttria-alumina garnet and the others.

REFERENCES

1) W.C. Louden and K. Schmidt; Illum. Eng., 55, 1962, p.696

2) R.L. Coble; US. Patent 3026210, 1962

3) P.J. Jorgensen and R.C. Anderson; J. Am. Ceram. Soc., 50, 1967, p.553

4) G.D. Miles et al.; Trans. British Ceram. Soc., 80, 1967, p.319

5) K.S. Mazdiyasni; J. Am. Ceram. Soc., 50, 1967, p.532

6) P.J. Jorgensen and W.G. Schmidt; J. Am. Ceram. Soc., 53, 1970, p.24

7) K. Kodaira and M. Koizumi; Mat. Res. Bull., 6, 1971, p.261

8) R.J. Bratton; J. Am. Ceram. Soc., 57, 1974, p.283

9) S. Prochazka and F.J. Klug; US. Patent 4418025, 1983

10) G.DE. With and H.J.A. Vandick; GB. Patent 2138802, 1984

11) N. Kuramoto and H. Taniguchi; J. Mat. Sci. Letters 3, 1984, 471

12) W.H. Rhodes; 2nd int. symp. on incoherent light sources, Enschede, Netherlands, 1979, p.103-104

13) Brunner Dieter et al.; DE. Patent Appl. 3840577.6, 1989

14) Verweij Hendrik; NL. Patent Appl. 9000687, 1990

15) T. Hashimoto and M. Hama; J. Function Mat. Jpn., 5, 1982, p.23

16) W.R. Campbell; J. Illum. Eng. Soc., 1, 1972, p.281

17) J.D. Rickman; J. Appl. Phys., 48-9, 1977, p.3733

18) Garland E. Scott, Jr. and Michael K. Levenson; US. Patent 3935498, 1976

19) M. Kaneno and I. Oda; Mat. Sci. Monograph, 6,1980, p.1114

20) G. Toda et al.; Powder & Powder Metallagy, 21-3, 1974, p.76

21) R.J. Tiernan and D.B. Shinn; GTE Lighting Products, Sylvania Lighting Center Report, 1985

22) N. Saito et al.; J. Illum. Eng. Inst. Jpn.; 63-5, 1979, p.272

23) J. F. Waymouth and E. F. Wyner; J. Illum. Eng. Soc., 10, 1981, p.237

24) K. Otani et al.; J. Illum. Eng. Inst. Jpn.; 64-10, 1980, p.577

25) G.DE With et al.; J. Mat. Sci., 20, 1985, p.1215

26) V. Rohe and C.M. Shoene; 8th Ann. Conf. on composites and Advanced Ceram. Mat., 1984

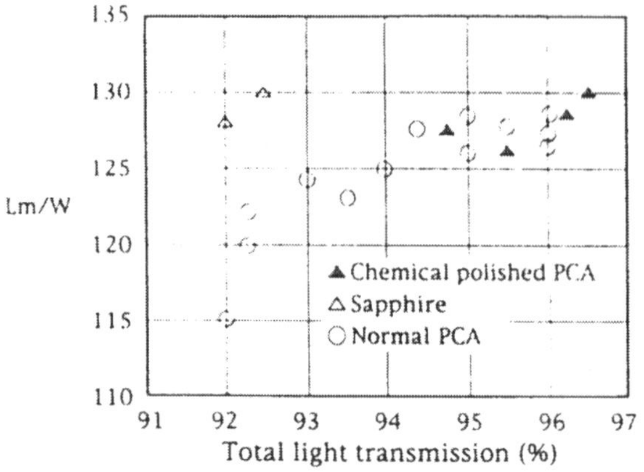

Fig.1a RELATION OF TOTAL TRANSMISSION AND LAMP' s EFFICACY

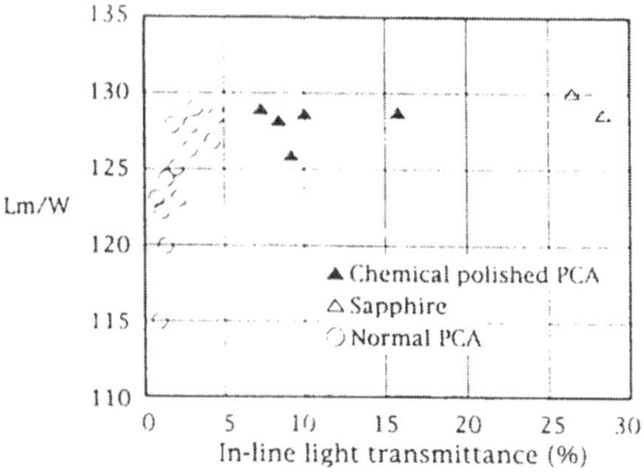

Fig.1b RELATION OF IN- LINE TRANSMISSION AND LAMP's EFFICACY

MgO 0.05wt%

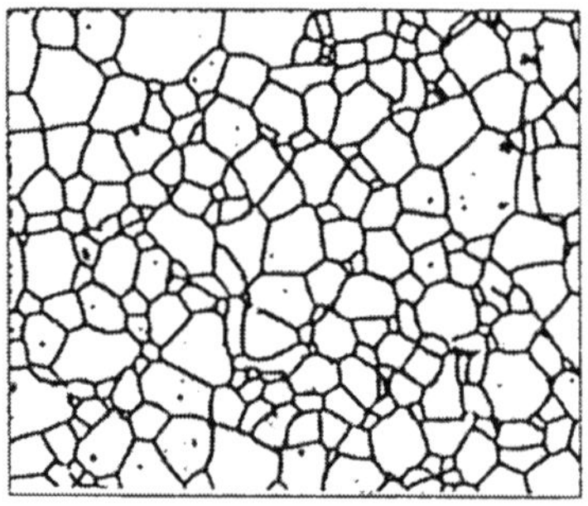

(Light transmission)
Total ; 96.5%
In-line ; 3.5%

MgO 0.03wt%+Y_2O_3 0.01wt%

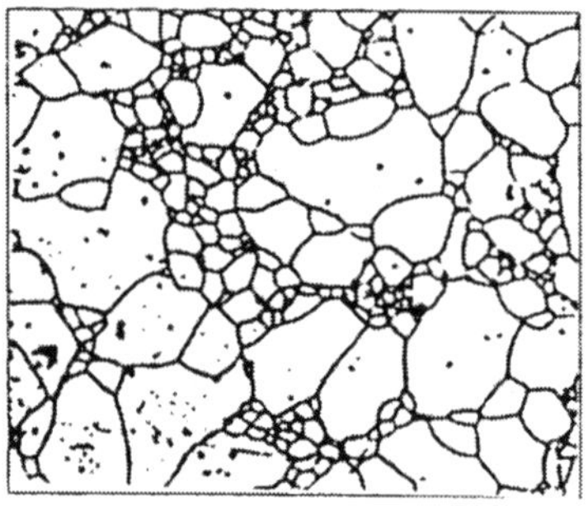

(Light transmission)
Total ; 95.8%
In-line ; 2.5%

Fig.2 TYPICAL MICROSTRUCTURES OF ALUMINA CERAMICS

Crystal structure

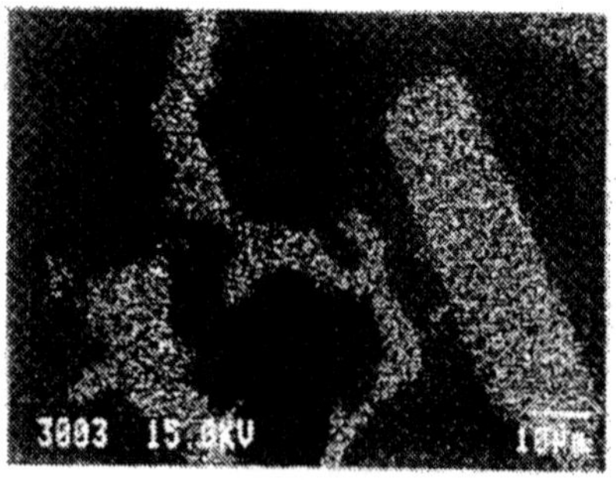

Na distribution

Fig.3 CRYSTAL STRUCTURE OF Na-PENETRATED ARC-TUBE

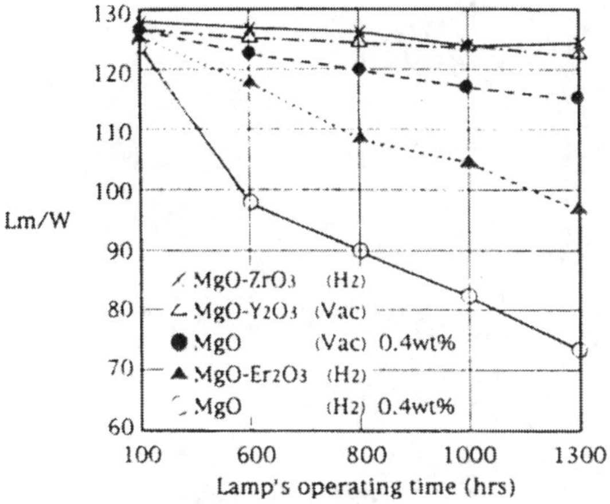

Fig.4 RELATION OF LAMP's OPERATING TIME AND EFFICACY CHANGE

Fig.5 TYPICAL EXAMPLE OF BLACKEND HPS LAMP

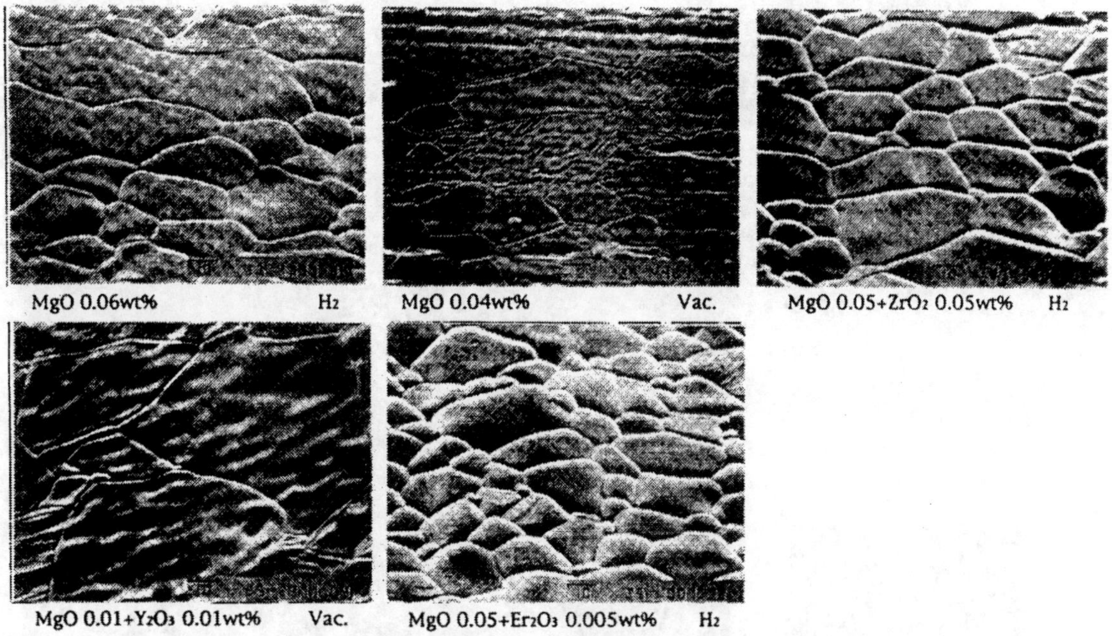

Fig.6 MICROSTRUCTURES OF PCA TUBES COMPRISING FROM VARIOUS ADDITIVE SYSTEMS

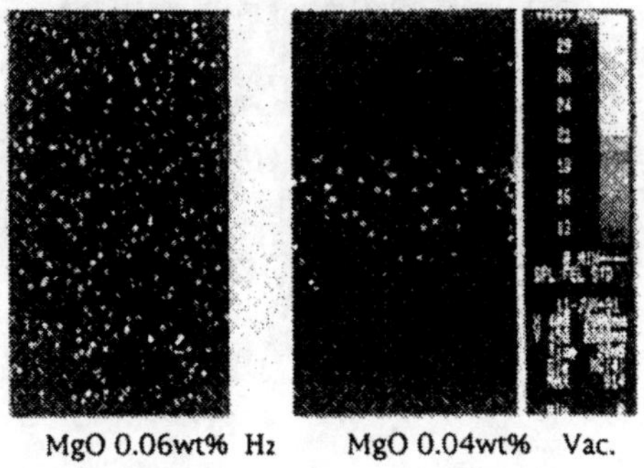

Fig.7 Mg DISTRIBUTION OF PCA TUBES

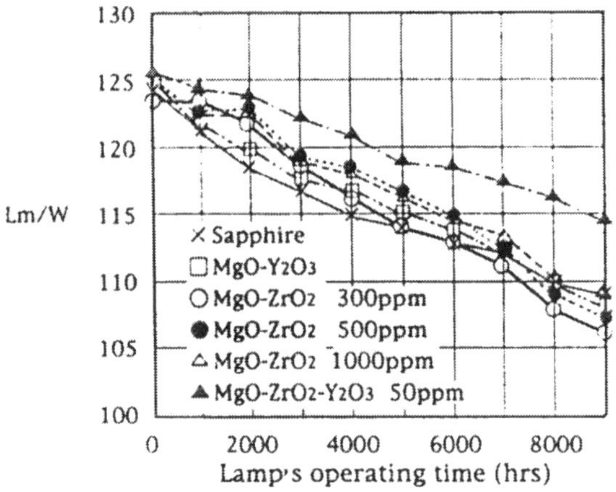

Fig.8 RELATION OF LAMP's OPERATING TIME AND EFFICACY CHANGE

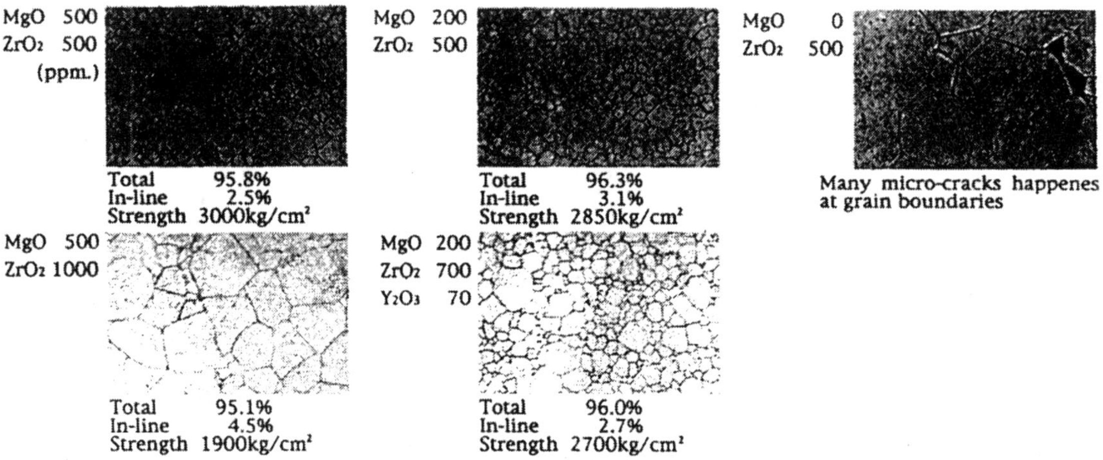

Fig.9 EFFECT OF MgO, Y_2O_3 And ZrO_2 ADDITIVES ON PCA TUBE's MICROSTRUCTURE AND PROPERTIES

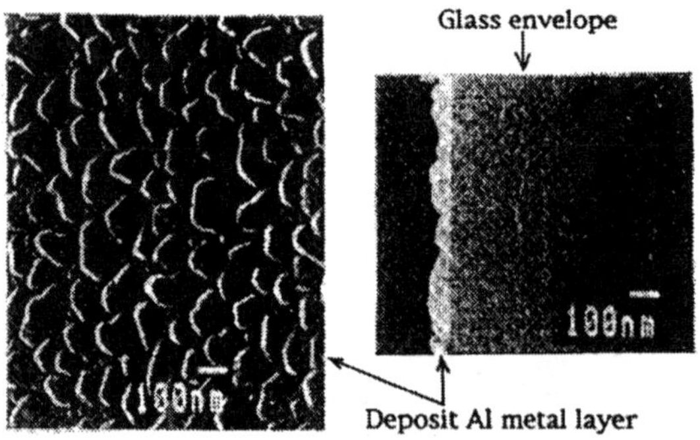

Fig.10 Al METAL DEPOSITED ON THE LAMP's OUTER-ENVELOPE

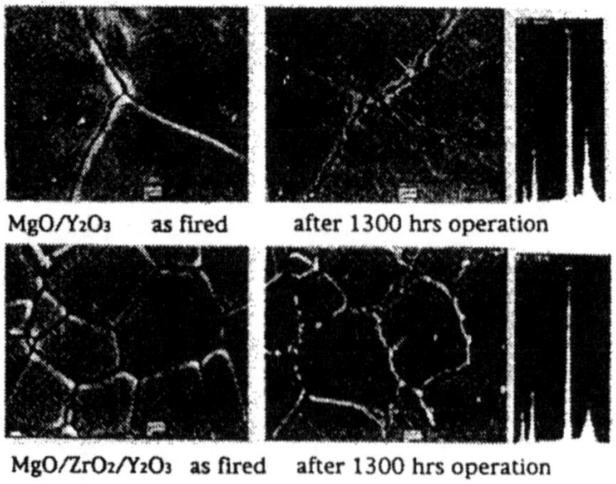

Fig.11 SEM AND EDS ANALYSIS ON THE INNER-SURFACE OF ALUMINA ARC TUBES BEFORE AND AFTER TESTING

THE CHARACTERISTIC IMPROVEMENT OF THE CERAMIC METAL HALIDE LAMP

Y.Takeji, S.Taniguchi, S.Mori, Y.Hida, K.Nakano, J.Honda, H.Takasu, H.Nagai' and Dr.K.Hayashi'
Japan Storage Battery Co., Ltd., Nishinosho, Kisshoin, Minami-ku, Kyoto-city, 601, Japan
'TOTO Ltd., 2-1-1, Nakashima, Kokurakita-ku, Kitakyushu-city, 802, Japan

Metal halide lamps with translucent alumina arc tubes are superior than lamps with conventional quartz arc tubes in that with alumina arc tubes, there is no loss of sodium and the lamps can be operated at higher temperature etc. As a result, significant improvements on the lamp design and efficacy can be expected with translucent alumina arc tubes. However, the most difficult technical question in dealing with alumina arc tubes is the sealing of these tubes. Although there have been several methods including the utilization of cermet proposed to produce a seal sufficiently resistant to the chemical attack from the halide filling and still maintains a reliable seal with the alumina tube [1,2], a technique with satisfactory result has not been found until now. Viewing the needs involved, a sealing method utilizing electricity conductive platinum films has been developed in the course of this project to improve ceramic metal halide lamps with Sc-Na-I filling.

Presently, the newly developed sealing method has been tested on ceramic metal halide lamps and electrically conductive platinum film has been proved to work in these arc tubes with no need for any outer bulbs. The apparent advantages of working without the extra parts such as outer bulb and lead wires are that the loss of the intensity of the radiation from the arc tubes is significantly decreased and that the volume of the whole lamp is largely decreased. Therefore, a far more compact lamp with significantly increased efficacy can be obtained through the alumina arc tube and the newly developed sealing technique. During the life test of these ceramic metal halide lamps, white spots on the inner surface of the upper wall of the lamps' central section have been found with lamps positioned horizontally. Analyses of these spots showed them to be depositions of metallic Sc. It has been known that ScI_3 in the Sc-Na-I filling of quartz arc tubes reacts with the SiO_2 of the tube wall to form $Sc_2Si_2O_7$, and depletion of Sc during lamp operation results due to the reaction [3]. With alumina arc tubes, similar depletion of Sc has also been observed in experiments due certain reaction between the ScI_3 in the filling and the Al_2O_3 of the tube wall. In both cases, it is thought to be important to prevent a bowing arc cased by thermal convection within the arc tube. To prevent the arc from bowing, it is effective to a certain degree to optimize the amounts of the filling material and the design of the arc tubes.

The latest development in this project is a 7% increase in efficacy achieved with 250W Sc-Na-I ceramic metal halide lamp operated with no outer bulb as compared to the equivalent version of the conventional outer bulbed metal halide lamps with quartz arc tubes.

References

[1]Brown,K.E., Chalmers,A.G., Wharmby,D.O. :Tin Sodium Halide Lamps in Ceramic Envelopes [Journal of CIBS 6 pp.1-28 (1981)]

[2]Dr.Jungst,S., Maekawa,K., Asano,O., Dr.Heider,J., European Patent 60,9477(1994)

[3]Russell T. D. :Silica and Tungsten Corrosion in Discharge Lamps [LS6 PP.157-159]

Multiplet Structure Calculations for Rare Earth Ions

Shinichi Itoh and Naoyoshi Nameda

Department of Information and Computer Science, University of Kagoshima,

1-21-40 Korimoto, Kagoshima-shi, 890 Kagoshima

As fluorescent materials, a series of rare earth (RE) ions has been investigated about the luminescence spectra and applied to a light source of instruments by researchers. The sharp luminescence spectra are observed due to the intra-transitions of $4f$ electrons in the $4f$ orbitals because the $4f$ orbitals are shielded by outer $5d$ and $6s$ orbitals and maintain the narrow shape even in an applied electric potential field. RE doped fluorescent materials are utilized in lamps for the suitable feature of presenting naturally coloured spectra and of that the luminescence spectra of RE does not much depend on the host materials and temperature. Multiplet energy structures for REs are constructed by the electronic configurations of $4f^n$ electrons of REs (n=1 $\sim$ n=13) and show the complex energy levels in the region of energy in far-infrared and up to near-ultraviolet. Because Coulomb and spin-orbit interactions are working on the $4f$ electrons and the complex multiplet energy structures for a RE ion are generated by these interactions. Thus, it is important to take into account these interactions for obtaining the multiplet energy levels of REs.

In this report, we obtain the multiplet energies for several RE ions by *ab initio* molecular orbital calculations including the two interactions mentioned above. Several approaches have been done for calculations of multiplet terms for REs. Among them *ab initio* calculation for REs which doesn't include any parametrization in the calculations is performed by several researchers in these days and considered to be a powerful approach to the problems for luminescence. Because of a limitation of the ability of computations, the lower lying multiplet energies for Ce^{3+}, Nd^{3+}, Gd^{3+}, Eu^{3+} and Er^{3+} ions are calculated. The aims of this calculation are, first, to obtain electronic structures of REs in which relatively strong luminescence spectra can be observed, and second, to verify the precision in the present molecular orbital calculation for applying this calculation to further realistic RE doped fluorescent materials. The program package COLMBS at Insutitute of Molecular Science, Okazaki, Japan is used in this calculation.

The strong and sharp luminescence is caused by the intra-excitation $4f^n \rightarrow 4f^n$ in the region of energy to infrared and visible for Nd^{3+}, Eu^{3+} and Er^{3+} ions and to ultraviolet for Gd^{3+}. As for Ce^{3+}, a strong and broad luminescence is observed in ultraviolet region by the inter-excitation $4f^1 \rightarrow 4f^0 5d^1$. Ce^{3+} has been applied to fluorescent compounds in lamps for the sensitivity of an absorption for ultraviolet rays by arc emission and efficiently transmit the energy of ray to the center of fluorescent light source. For the two excitations, spin multiplicity for $4f^n$ electrons is not unchangeable (not always $\Delta S = 0$) both in initial and final states by selection rule under spin-orbit interaction. We estimate the possibility of transitions for the multiplet terms for RE ions with use of the obtained electronic spin states in the present case.

The calculated multiplet energies are in good agreement with another calculations in the book [S. Fraga *et.al.*, *Handbook of Atomic Data*, Elsevier, 1976] and experiments. This shows that spin-orbit interaction is effective to the multipets of REs. Especially for Gd^{3+}, the lowest luminescence exists in the high energy region of near-ultraviolet and the calculated multiplet energies are $\sim$ 42,000 cm^{-1}. This is caused by the characteristic case of half filling $4f^7$ configurations of Gd^{3+} and the ground state is much more stable than the first excited state due to the lower Coulomb energy in the ground state. Different spin multiplicities from that of the ground state are largely mixed in the first excited state and $\Delta S = 0$ is not satisfied for Gd^{3+}. Although the spin multiplicities are not restricted to $\Delta S = 0$ according to the selection rule under spin-orbit interactions, it is reported that the intensity of luminescence spectra of Gb^{3+} is not so strong comparing with those of another ions by another group. This shows that selection rule for spin multiplicity is still effective for RE ions.

Some of the calculated multiplet energies for ions are slightly overestimated comparing with luminescence spectra of RE doped fluorescent materials. The discrepancy between the present results and experimental results is occurred by neglect of the crystal field effect which interacts with $4f$ orbitals. The $4f$ orbitals expand to more outside by the interactions with crystal field and this will present lower multiplet energy of $4f$ electrons. For Er^{3+} the first excited energy is calculated as 7127 cm^{-1} which is a little larger than 6482 cm^{-1} which is that of Er^{3+} in $LaCl_3$. The crystal field effect is considered to be about 10 $\sim$ 20 percents of the multiplets for ions.

DESIGNING AND SYNTHESIS OF BLUE PHOSPHORS WHOSE SPECTRAL PROPERTIES ARE CHANGING CONTINUALLY AND ITS INFLUENCE ON CRI OF THREE BAND PHOSPHORS BLEND

Jinggen Huang[a], Xinghai Yu[a], Maofu Tong[b], Jianping Jiang[b] and Zuquan Cai[c]
a. Institute of Materials Science. Fudan University. Shanghai,200433.China.
b. Yue Long Non-Ferrous Metals Ltd.Co.,Shanghai,200949,China.
c. Institute for Electric Light Sources. Fudan University,Shanghai,200433.China

In this present study, two series of blue band phosphors are synthesized, based on the blue emitting phosphor $BaMgAl_{10}O_{17}$:Eu, by forming solid solutions of two compounds with the same structure or non-stoichiometric compounds by changing the content of cation. the peak position and the profile of emission spectrum are separately changeable continually in each series and the half width remains the same. With these blue band phosphors, we have studied the relation between the peak wavelength, the shape (green tail at longer wavelength) and the color rendering index (CRI) of three band phosphors blend, by means of computer simulation.

1. Blue band phosphors

Two series of blue band phosphors were prepared. They are as follows:

(a). $BaMgAl_{10}O_{17}$ - $SrMgAl_{10}O_{17}$:Eu solid solution series (below called as series A). In this series, the phosphors exhibit similar emission properties such as the spectral shape and the half width. The emission spectra display difference only in the position of the emission peak that goes up from 450 nm to 463 nm as increasing $SrMgAl_{10}O_{17}$ content (see table 1).

(b). $BaMg_xAl_{10}O_{16+x}$:Eu non-stoichiometric series (below called as series B). In this series, the Eu^{2+} emission spectra have the same peak position and half width and differ only in the intensity of the tail at longer wavelength. Thus the difference is mainly demonstrated in the value of y chromatic coordinate (see table 2).

2. Computer Simulation

In the computer simulation, the blue phosphors used were the A and B series of phosphors, the green phosphor was $(Ce,Tb)MgAl_{11}O_{17}$ and the red phosphor was Y_2O_3:Eu. All the spectra were normalized in order to singly compare the influence of emission of the blue phosphor on characteristics of the three band phosphors blend.

The results of the computer simulation are given in tables 3 and 4. Tables 3 and 4 show that not only the emission peak wavelength but also the y chromatic coordinate of the blue band phosphor exerts great influence on Ra of the three band phosphors blend. Ra increases with y value. Thus, in studying the effect of the blue phosphor on the three band phosphors blend, only considering emission wavelength is insufficient, the influence of the y chromatic should be considered.

Table1. $[(1-x)BaMgAl_{10}O_{17} - x(SrMgAl_{10}O_{17})]:Eu^{2+}$

No.	x	Blue phosphors	Peak λmax (nm)	Half width (nm)	color coordinate x	color coordinate y
A1	0	$BaMgAl_{10}O_{17}:Eu^{2+}$	454	53	0.1467	0.0623
A2	0.2	$Ba_{0.8}Sr_{0.2}Mg_{10}O_{17}:Eu^{2+}$	456	53	0.1438	0.0671
A3	0.4	$Ba_{0.6}Sr_{0.4}Al_{10}O_{17}:Eu^{2+}$	457	53	0.1429	0.0774
A4	0.6	$Ba_{0.4}Sr_{0.6}Al_{10}O_{17}:Eu^{2+}$	461	54	0.1405	0.1322
A5	0.8	$Ba_{0.2}Sr_{0.8}Al_{10}O_{17}:Eu^{2+}$	463	56	0.1412	0.1322

Table2: $BaMgxAl_{10}O_{16+x}:Eu^{2+}$

No.	x	Blue phosphors	Peak λmax (nm)	Half width (nm)	color coordinate x	color coordinate y
B1	2.0	$BaMg_2Al_{10}O_{18}:Eu^{2+}$	452	54	0.1484	0.0559
B2	1.5	$BaMg_{1.5}Al_{10}O_{17.5}:Eu^{2+}$	453	54	0.1488	0.0632
B3	1.2	$BaMg_{1.2}Al_{10}O_{17.2}:Eu^{2+}$	453	54	0.1488	0.0777
B4	1.0	$BaMg_{1.0}Al_{10}O_{17.0}:Eu^{2+}$	453	54	0.1498	0.0800
B5	0.8	$BaMg_{0.8}Al_{10}O_{16.8}:Eu^{2+}$	453	54	0.1521	0.0832
B6	0.5	$BaMg_{0.5}Al_{10}O_{16.5}:Eu^{2+}$	453	60	0.1555	0.0907

Table3: Influence of the emission peak wavelengthof blue phosphors on the Ra ofthe three band phosphors blend

No.		A1	A2	A3	A4	A5
Blue phospor color	x=	0.1467	0.1438	0.1429	0.1405	0.1412
coordinates	y=	0.0623	0.0671	0.0774	0.1071	0.1322
Peak wavelength(nm)		454	456	457	461	463
Blend powder color	x=	0.3686	0.3685	0.3686	0.3686	0.3686
coordinates	y=	0.4051	0.4051	0.4051	0.4051	0.4051
Color temp.Tc(K)		4490	4490	4490	4490	4489
C R I (Ra)		80	80	81	83	85
Three band	R%=	38.00	38.31	38.62	39.49	40.12
phosphors	G%=	47.00	46.55	45.85	43.66	41.87
composion	B%=	15.00	15.14	15.53	16.85	18.01

Table4. Influence of the longer wavelength tail of $BaMg_xAl_{10}O_{16+x}:Eu^{2+}$ on Ra of the three band phosphors

No.		B1	B2	B3	B4	B5	B6
X=		2.0	1.5	1.2	1.0	0.8	0.5
Blue phospor color	x=	0.1484	0.1486	0.1488	0.1498	0.1521	0.1555
coordinates	y=	0.0589	0.0631	0.0777	0.0800	0.0832	0.0907
Peak wavelength(nm)		452	453	453	453	453	453
Blend powder color	x=	0.3684	0.3684	0.3684	0.3684	0.3684	0.3684
coordinates	y=	0.4030	0.4030	0.4030	0.4030	0.4030	0.4030
Color temp.Tc(K)		4485	4483	4483	4484	4484	4484
C R I (Ra)		78	79	79	80	80	80
Three band	R%=	38.00	38.15	38.38	38.44	38.32	38.08
phosphors	G%=	47.00	46.56	45.98	45.63	45.47	44.93
composion	B%=	15.00	15.29	15.64	15.93	16.21	16.99

THE RISE OF Ra BY IMPROVEMENT OF BLUE ALKALINE EARTH ALMINATE PHOSPHOR

T. Hisamune[a] N.Kijima[b] S.Fujino[a] Y.Oguri[a]
[a]Kasei Optonix,Ltd., 1060 Naruta, Odawara, Japan
[b]Yokohama Research Center, Mitsubishi Chemical Corp., 1000 Kamoshida-cho, Aoba-ku, Yokohama, Japan

About 20 years ago, Verstengen surveyed the phosphors of which host lattice is hexagonal alminate, and found the efficient blue phosphor $BaMg_2Al_{16}O_{27}:Eu^{2+}$ (BAM) [1]. Verstengen et al. made the fluorescent lamp that has high luminous efficacy and high color rendering indices by mixing this blue BAM, red Y_2O_3:Eu, and green (Ce,Tb)$MgAl_{11}O_{19}$ and reported that a small increase of average color rendering index Ra is achieved when Mn^{2+} is incorporated in BAM:Eu^{2+}[2]. Since then BAM has been used as a blue phosphor for tricolor fluorescent lamp and various kinds of improvement have been tried [3]. Concerned about the composition of BAM, it had been thought for long time that the ratio of (Ba,Eu):Mg:Al could be changed by large range [4]. But Smets et al. reported that when $BaMg_2Al_{16}O_{27}$ is fired, $MgAl_2O_4$ and Al_2O_3 are formed as secondary phases, and a pure compound is found to be formed for the composition $BaMgAl_{10}O_{17}$ [5]. So it can be said that past many compositions tried to improve BAM, might contain secondary phases, and no investigations have been carried out to get the optimum condition of Ba^{2+}, Sr^{2+}, Eu^{2+} and Mn^{2+} ratio to improve tricolor fluorescent lamp. We adopted the formula $Ba_{1-x-y}Sr_xEu_y\,Mg_{1-z}Mn_zAl_{10}O_{17}$ for prevention of forming secondary phases. We studied the emission properties and simulated the luminous efficacy and Ra of tricolor fluorescent lamps containing this blue phosphor. We found that it is possible to improve Ra of tri-color fluorescent lamp with little lowering luminous efficacy.

The samples were prepared by solid state reaction. The appropriate carbonates and oxides were thoroughly mixed and fired at 1450-1600℃ in a reduced atmosphere. For prevention of forming $EuAlO_3$ as a secondary phase, carbon was used to confirm the reduced atmosphere. AlF_3 was used as a flux for crystal growth. The phases of samples were checked by X-ray diffraction to confirm that secondary phases were not formed. For estimating luminous efficacy and Ra, we mixed spectra of BAM, Y2O3:Eu, LaPO4:Ce,Tb, in appropriate ratio and fixed strength of Hg lines to simulate objective color temperatures.

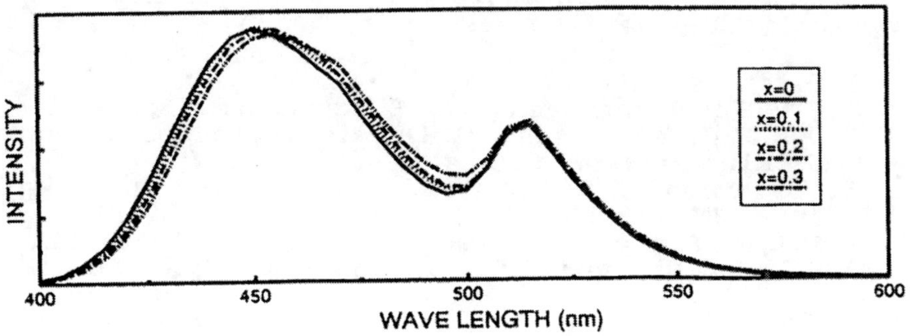

Fig.1 Spectra of $Ba_{0.9-x}Sr_xEu_{0.1}Mg_{0.98}Mn_{0.02}Al_{10}O_{17}$ in various x.

The spectra of BAM:Eu^{2+},Mn^{2+} in various Sr^{2+} contents are shown in Fig.1. As Sr^{2+} increases, the peak of Eu^{2+} emission shifts to longer wavelengths. This shift is similar to the one when Eu^{2+} content increases [6]. On the other hand, the peak of Mn^{2+} emission does not shift.

These properties are explained by configuration coordinates as shown in Fig.2. The more Eu^{2+} or Sr^{2+} is substituted, the smaller the distance of Eu^{2+}-O^{2-} becomes, because the ionic radii of them are smaller than that of Ba^{2+}. This causes the stokes-shift bigger (state b in Fig.2) than that of BAM that contains less Eu^{2+} or Sr^{2+} (state a), because negative oxygen ions are closer to Eu^{2+}. This change occurs inter the spinel blocks and does not affect Mn^{2+} in the blocks, so the peak of Mn^{2+} emission does not shift. It is thought from this model that the probability of nonradiative transition from state b at high temperature is bigger than that from state a. In practice we found that the brightness declines as temperature increases, and the more Eu^{2+} or Sr^{2+} content is, the bigger the decline ratio of brightness is slightly, though it is small enough not to be problems for ordinary lamps.

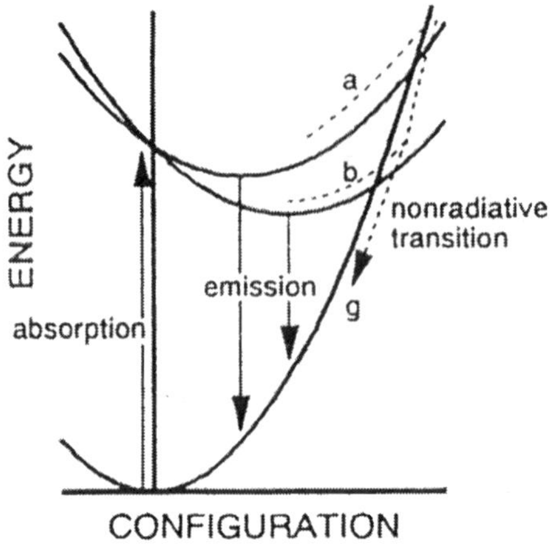

Fig.2 Configuration coordinate model.

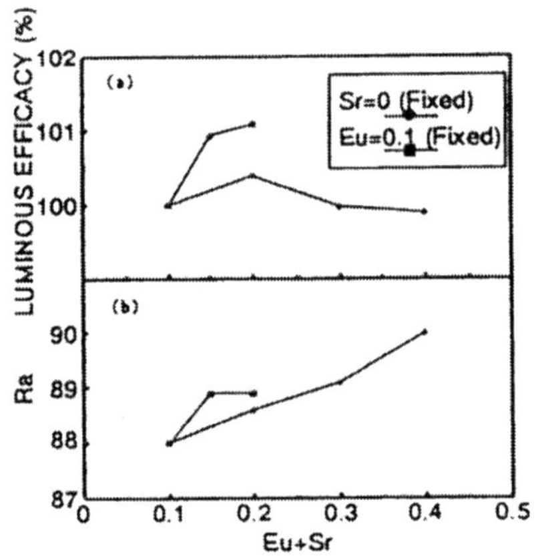

Fig.3 Luminous efficacy(a) and Ra(b).

The properties of tricolor fluorescent lamps owe to the shape of the spectrum of a blue phosphor remarkably. Fig.3 shows the luminous efficacy and Ra of simulated 6500K tricolor fluorescent lamps containing BAM in various Eu^{2+} and Sr^{2+} content. Luminous efficacy is a relative value to the conventional $Ba_{0.9}Eu_{0.1}Mg_{0.98}Mn_{0.02}Al_{10}O_{17}$. We found it is possible to raise Ra with little lowering luminous efficacy by co-activating Mn^{2+} and increasing Sr^{2+} or Eu^{2+} content. Particularly by Sr^{2+} substitution, one would be able to achieve Ra=90 with little cost increase.

References

1. J.M.P.J.Verstengen, J. Electrochem. Soc., **121** (1974) 1623
2. J.M.P.J.Verstengen, D.Radielovic, and L.E.Vrenken, J. Electrochem. Soc., **121** (1974) 1627
3. *For example;* K.Iwama and T.Yamamoto, Jpn. Kokai Tokkyo Koho, JP 56/86892 (1981)
4. A.L.N. Stevels, J. Luminescence, **17** (1978) 121
5. B.M.J Smets and J.G. Verlijsdonk, Mat. Res. Bull., **21** (1986) 1305
6. G.Kunzler, W.Tews, G.Herzog, and W.Blumchen, Z. Phys. Chem., **127** (1990) 1181

RED EMITTING PHOSPHORS FOR THREE-BAND FLUORESCENT LAMPS

Y. Sakakibara*, A. Taya*, N. Matsuda**, H. Takemura**
* Toshiba Lighting & Technology Corporation
1-201-1, Funakoshi-cho, Yokosuka-shi, Kanagawa 237 JAPAN
**Toshiba Corporation
72, Horikawa-cho, Saiwai-ku, Kawasaki-shi, Kanagawa 210 JAPAN

Three-band fluorescent lamps with high efficacy and good color rendering characteristics are used widely.(1) The General color rendering index(Ra) of these lamps are more than 80, but some of the special color rendering indices are poor and need improvement. These include R_9(red) which is about 30(for a 5000K lamps). In general, europium activated yttrium oxide (Y_2O_3:Eu) is used as the red component. The emission spectrum of Y_2O_3:Eu suited to achieve high efficacy when it is mixed with blue and green emitting phosphors, but cannot yield higher R_9. We therefore examined several other red emitting phosphors in order to improve R_9.

It is known that the R_9 value can be improved by mixing manganese activated magnesium fluorogermanate, but the lamp efficacy is much decreased. This is because the peak wavelength is too long(658nm). Hence we investigated alternative red emitting phosphors. The phosphors selected for study are those which have longer peak wavelength than Y_2O_3:Eu similar chemical compositions to practical phosphors(Table 1).

Table 1 Studied red emitting phosphors

composition	peak wavelength
Y_2O_2S:Eu	626nm
La_2O_3:Eu	626nm
monoclinic Gd_2O_3:Eu	623nm
Y_2O_3:Pr	630nm

Cubic Gd_2O_3:Eu shows the same emission spectrum as Y_2O_3:Eu. On the other hand, monoclinic Gd_2O_3:Eu (m-Gd_2O_3:Eu; high temperature phase) has a different emission spectrum.(2)(3)

By powder measurements, we found that the efficiencies of

Y_2O_3:Pr and La_2O_3:Eu are too poor. In addition, La_2O_3:Eu is highly hygroscopic. Y_2O_2S:Eu is widely used for CRT, and is excited efficiently by 254nm UV. When applied to lamp, however, a rapid blackening of lamp is observed, (supposedly by interaction of free sulphur and mercury). The m-Gd_2O_3:Eu is stable and causes no problem in application to fluorescent lamps.

Figures 1 and 2 show the emission spectra of lamps coated with m-Gd_2O_3:Eu and with a mixture of three phosphors respectively. Table 2 shows the characterictics of lamps coated with blends including m-Gd_2O_3:Eu. By Replacing Y_2O_3:Eu with m-Gd_2O_3:Eu, R_9 is improved. At the same time the efficacy decrease, but to a smaller extent than by addition of Mn activated magnesium fluorogermanate. We expect that the improved lamp is suitable for illuminating foods, plants, fishes and so on.

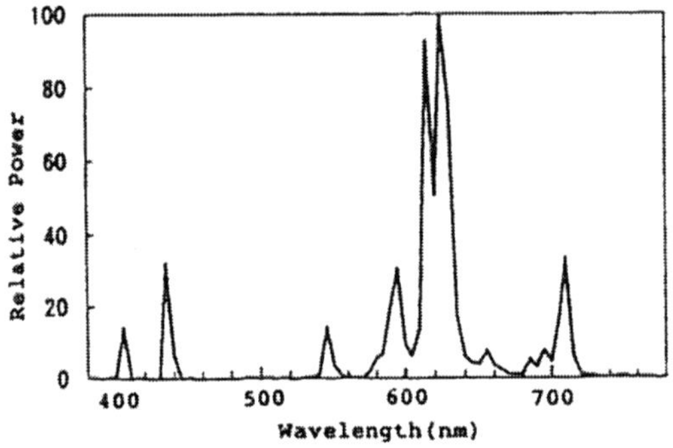

Fig.1 Emission Spectrum of m-Gd_2O_3:Eu coated lamp

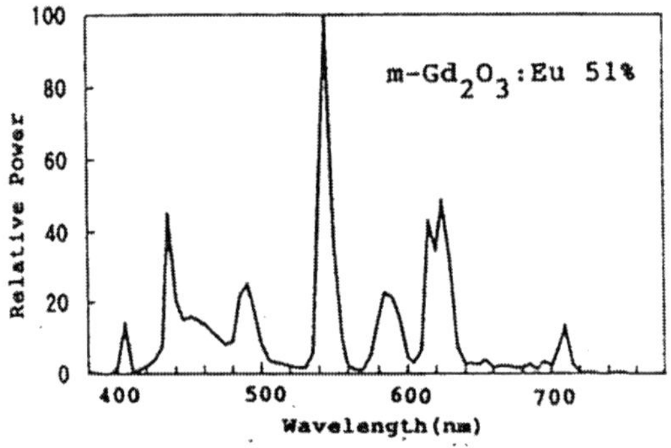

Fig.2 Emission Spectrum of blend-coated lamp

Table 2 Characteristics of blend-coated lamps(18W T9,5000K)

Phosphor ratio(%)				Relative	R_a	R_9
SCA	LAP	YO	GDO	Efficacy		
28	32	40	0	100	84	27
25	29	23	23	90	85	63
23	26	0	51	79	84	92

SCA: $(Sr,Ca,Ba)_5(PO_4)_3Cl$:Eu
LAP: $(La,Ce,Tb)PO_4$
YO : Y_2O_3:Eu
GDO: monoclinic Gd_2O_3:Eu

References

(1) W.A.Thornton: J.Illum.Engng.Soc., 2, 29 (1972)
(2) R.S.Roth and S.J.Schneider: J.Research NBS, 64A, 309 (1960)
(3) L.Ozawa: "Cathodoluminescence", p.172, VCH Verlagsgesellschaft (1990)

LUMINANCE DEGRADATION BY FORMING Eu^{3+} ION IN Eu^{2+} DOPED BARIUM MAGNESIUM ALUMINATE PHOSPHORS

Shozo OSHIO and Tomizo MATSUOKA
Lighting Research Laboratory, Matsushita Electric Industrial Co., Ltd.
3-4, Hikari-Dai, Seika-Cho, Soraku-Gun, Kyoto 619-02 JAPAN

If divalent europium ions are doped in $BaMgAl_{10}O_{17}$ compound, it will become an efficient blue phosphor, i.e. $BaMgAl_{10}O_{17}:Eu^{2+}$. This emission, peaking around 450 nm, is due to parity allowed $4f^6(^7F)5d-4f^7(^8S_{7/2})$ transition for Eu^{2+} ions. $BaMgAl_{10}O_{17}:Eu^{2+}$ phosphor has been applied to fluorescent lamps or plasma display panels as the blue component. However, blue emission from $BaMgAl_{10}O_{17}:Eu^{2+}$ tended to decrease with time when the phosphor was excited by ultraviolet light. The emission also decreased following anneal in an oxidizable atmosphere, such as air. To alleviate these decreases, we have studied the causes in detail. In this report, luminance degradation of $BaMgAl_{10}O_{17}:Eu^{2+}$ phosphor following anneal will be presented.

Figure 1 exhibits the rate of maintaining the luminance of Eu^{2+} doped phosphors as a function of the annealing temperature with reference to phosphors doped with other ions. The phosphors were excited by 253.7 nm following anneal at each temperature shown in Fig. 1 for one hour in air. It is clear that the degradation is caused by the Eu^{2+} ion. This is because not only $BaMgAl_{10}O_{17}:Eu^{2+}$ but also the other phosphors which are activated by Eu^{2+} decrease in rate following anneal, while the phosphors activated by Tb^{3+}, Mn^{2+} and Eu^{3+} keep the initial intensity even if they are heated up to 1000℃, as shown in Fig. 1.

Figure 2 shows a comparison of the photoluminescence spectra of $BaMgAl_{10}O_{17}:Eu^{2+}$ with and without anneal at 1000℃. A Eu^{3+} emission of 610 nm appeared following anneal over 800℃. At the same time the red emission intensity, separated from the total

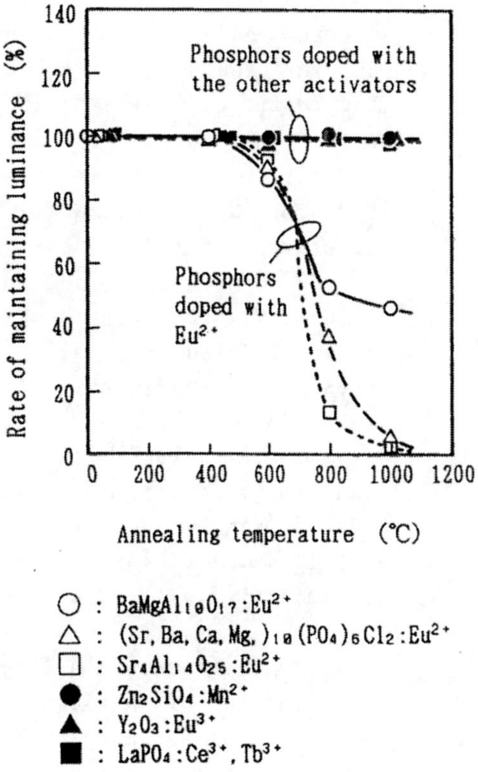

Figure 1 Luminance degradation in Eu^{2+} doped phosphors.

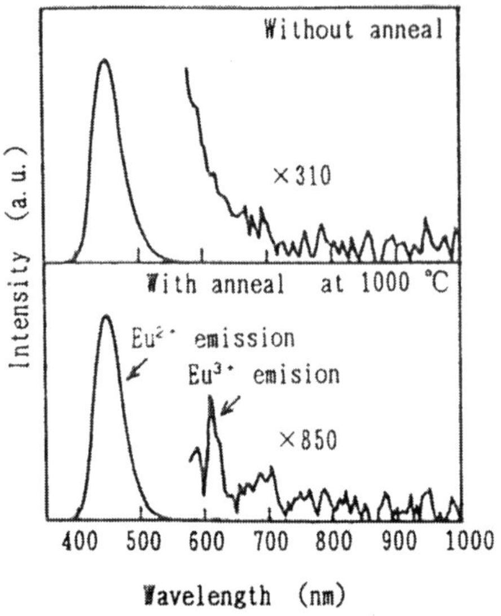

Figure 2 Appearance of Eu^{3+} emission with anneal.

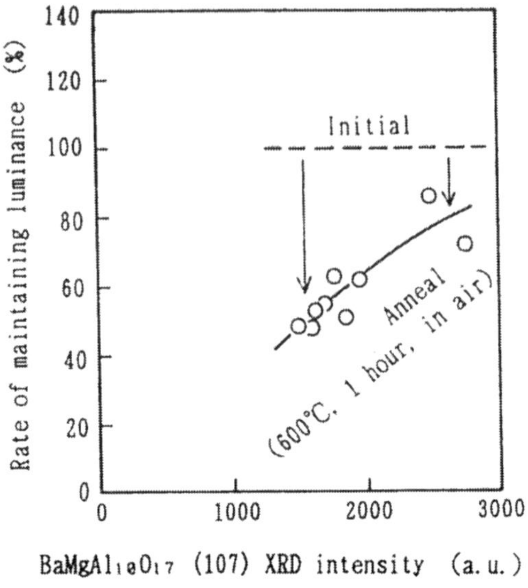

Figure 3 Alleviating luminance degradation with XRD intensity.

by a filter, increases following the anneal. Similar behavior was also observed in other Eu^{2+} doped phosphors described above. Furthermore, when $BaMgAl_{10}O_{17}:Eu^{2+}$ phosphor is annealed in a nitrogen atmosphere, the degradation can be alleviated. From these results, we estimated that Eu^{3+} ions would appear in this phosphor during anneal by the oxidation, and result in luminance degradation following anneal. These Eu^{3+} ions would be included in another formed compound which appears to be $EuMgAl_{11}O_{19}$. Further analyses by XRD (X-Ray Diffraction), ESR (Electron Spin Resonance) and XANES (X-ray Absorption Near Edge Structure) offered evidence to support these hypotheses.

In order to find an excellent $BaMgAl_{10}O_{17}:Eu^{2+}$ which resists oxidation, the relationship between the rate of maintaining the luminance and crystallinity has been investigated concerning similar phosphors prepared by various methods. As shown in Fig. 3, the rate of maintaining luminance following anneal at 600℃ for one hour in air increased with the intensity of the $BaMgAl_{10}O_{17}$ (107) XRD. This suggests that the phosphors with excellent crystallinity will resist oxidation.

In conclusion, the Eu^{3+} ion formed in $BaMgAl_{10}O_{17}:Eu^{2+}$ following anneal in air results in luminance degradation. Creating excellent crystallinity will alleviate the degradation.

The authors would like to express their thanks to Drs. H. Shibata and K. Iwama of Matsushita Electronic Corporation for their support of this study, and Dr. Y. Yoshioka and Mr. Y. Umetani of Matsushita Technoresearch, Inc., for ESR and XANES analyses and for their helpful discussions.

FEASIBILITY OF THE TWO PHOTON EMISSION ATOMIC PHOSPHOR

Makoto Toho and Hideyoshi Kimura
<Matsushita Electric Works Ltd. Lighting Business Group>
1048,Kadoma, Osaka 571, Japan

1. Fluorescent lamp and Phosphor

The efficiency of fluorescent lamps is about 24%. The largest loss of the lamp is the energy loss of the phosphor, because the present phosphor is one photon emission type and the efficiency goes down to a half through the transfer from UV to visible light. If a high efficient "two photon emission phosphor" was realized, the efficiency of the lamp would become 2 times higher.

2. Concept of "the Two Photon Phosphor"

There exist some metal atoms which have possibility of two photon visible emission by about the same 5 eV excitation as Hg UV 254 nm energy. For example the candidate metals are Na, Fe, Mn, Al, V, In, Sr, Li, etc. However these metal atoms exist as atom only in high temperature.

So in order to realize the two photon phosphor, we had the following idea concepts.

1. To fix an atom with two photon emissivity in atomic state in a micro-pore of the solid matrix material at the room temperature.
2. To shift and broaden the excited energy state only a little to increase the possibilities of the excitation and non-radiative transition to a desired energy state.
3. A desirable atom should have the possibility of two photon emission with red-like and blue-like light to emit white light. It will be also possible to blend a few atoms.

3. Rough Analysis of the Two Photon Emission Probability

We roughly analyzed the probability of two photon visible emission of Na atom as one of candidate atoms, using a very simplified only energy state diagram such as Fig.1.

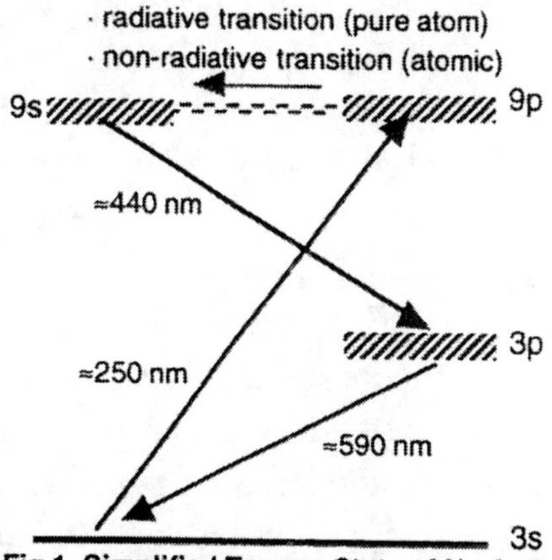

Fig.1. Simplified Energy State of Na Atom

Atomic Na is fixed in a pore of the matrix. Considering such atomic state, the energy states of Na atom are broaden and shifted slightly. So, the first excited state (9p) is overlapped with the second excited state (9s) and non-radiative transition probability from 1st(9p) to 2nd(9s) becomes vary large. If we assume that more than a half of re-emission of the excited UV would be resulted in two photon transition process through re-absorption to the first excited state (9p), the two visible photon transition probability of the cascade emission of (9s →3p) and (3p→3s) will be estimated to about 70~80 %.

4. Fabrication of "the Atomic Phosphor" and Its Evaluation

We selected a gel-type silica (or alumina) as the matrix material. Because it has a high transparency from UV to Visible light, and a possibility to fabricate spherical micro-pores for the atomic fixation of atom. We can get the pored silica which has very sharp pore distribution by control of the fabrication condition.

We tried a fabrication of the "atomic phosphor" with Fe/silica as first step. Na atom is one of ideal atoms, but we considered it as the next target because of its strong chemical activity.

The fabrication process is as follows. Mix Fe nitrate solution and tetraethoxysilan. By starring silica-sol which contains Fe nitrate is formed. Coat the silica-sol on a substrate. Heat the substrate to decompound the Fe nitrate. Heat more in the deoxidized atmosphere to reduce Fe oxide and sintering.

We made some samples with several Fe densities (Si:Fe=10:1 ~ 1000:1). Low Fe density sample of "100:1" to "1000:1" showed a fluorescent emission. In a sample of "1000:1", existence of atomic state of Fe was observed by absorbed spectrum of three sharp lines peculiar to Fe atom as shown in Fig.2.

However, the fluorescence was a broad emission, which is different from the two photon emission. This excitation/emission feature was kept a few days under the room air condition. But it disappeared after that, maybe due to an oxidation reaction.

5. Results

We could expect a high efficient two photon emission of a proper atom with an atomic state as a result of a rough analysis of Na atom, which was selected from some possible metal atoms having possibility of two photon emission under the Hg 254 nm UV excitation.

We tried to fabricate the atomic phosphor of Fe/Silica using the sol-gel thin film technology and we confirmed that Fe atom was fixed in atomic state in the case of the ratio of "Si:Fe = 1000:1".

Furthermore, we observed a broad fluorescence, but did not observe any two photon emission fluorescence.

At the present time, we have several issues on this theme. However, it will be very valuable for the future development of a high efficient two photon phosphor that the atomic state fixation of Fe atom under the room temperature was realized.

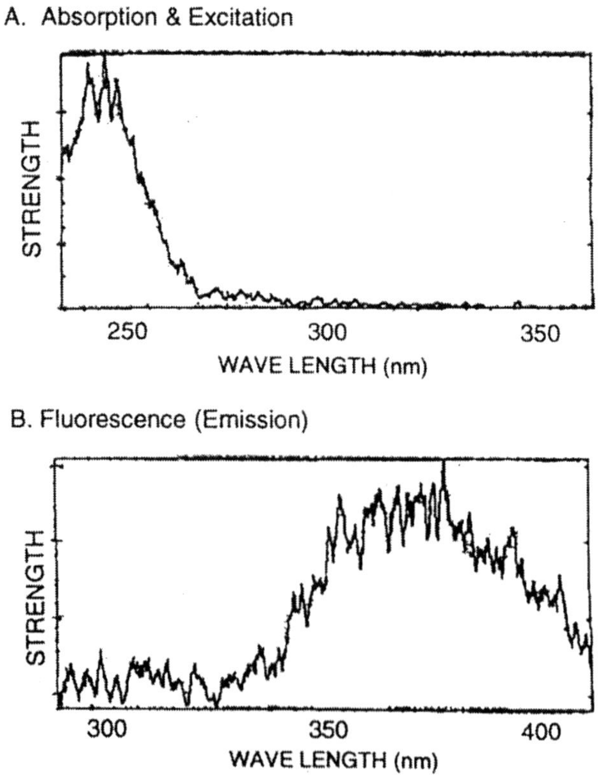

Fig.2 Phosphor Characteristic of the Experimental Fe/Silica Sample

OPTICAL FILTERS ON LINEAR HALOGENLAMPS PREPARED BY DIP COATING

Dr. G. Hebbinghaus, G. Frank and Dr. C.J.M. Denissen
Philips Forschungslaboratorium Aachen, Weißhausstraße,
52066 Aachen, Germany

Optical coatings are of increasing importance in the lamp industry. Application of selectively reflecting or absorbing multilayer interference filters on the bulb of incandescent halogen lamps modifies the lamp emission spectum according to customer's needs or results in considerable energy savings. Interference filters are layer stacks of materials with different refractive indices. In the simplest case, alternating layers of two materials with high and low refractive index are used. Layers of medium refractive index or layers with spectrally selective absorption allow for improved or simplified filter designs in special applications. High temperature stability, low infrared absorption and low mechanical stress are important characteristics for coatings applied on the bulb of halogen lamps.
Several deposition processes like e.g. PVD, LPCVD, sputtering and dip coating are used in lamp industry. For economical reasons non vacuum technologies like dip coating are especially attractive for the coating of bulky linear halogen lamps. In the dip coating process homogeneous thin films are deposited on a substrate by pulling it out of an sol-gel solution of metal organic or inorganic salts. Mixed materials having improved properties are achieved by mixing the different salts with the desired metal stoichiometry. The layer thickness is controlled by the concentration, viscosity and pulling rate. Dense oxide films are obtained by annealing the dried gel films at high temperature for some minutes. After the substrate has cooled down, the next layer is added by dipping.
Dip coated floodlight lamps with up to 42% energy saving have been realized (ref. 1). Here the mixing of two materials in the dipping bath allowed to make a material of high refractive index with good mechanical stress properties. This permitted us to realize filters of up to few um total thickness.
The design of infrared reflecting optical filters with a wide transmission range for visible light can be realised by introducing thin layer pairs of high and low refractive index in the layer stack or by substituting such a pair by just one layer with the correct medium refractive index. Here the dip coating process allows to make almost any needed medium refractive index material by mixing materials with high, medium and low refractive indeces.
Several layers were combined in a filter to block the visible light.

The picture shows the filter transmittance, absorptance, and the reflection seen from inside and outside of the lamp. The advantage of combining selective absorption and reflection in a multilayer filter are: Less layers are needed compared to using only reflecting materials; the absorption is mainly concentrated in the low wave length region, where the halogen lamp has not too much energy to heat up the lamp bulb; and the transmission in the infrared is very high. The light reduction, the colour and the colour rendering of the burning lamp can be adjusted by the filter design, and the reflection of the lamp seen from outside can be designed to give an attractice appearence (e.g. golden colour).
Summarizing: coatings with a large variety to tailor lamp properties can be realized by dipping.

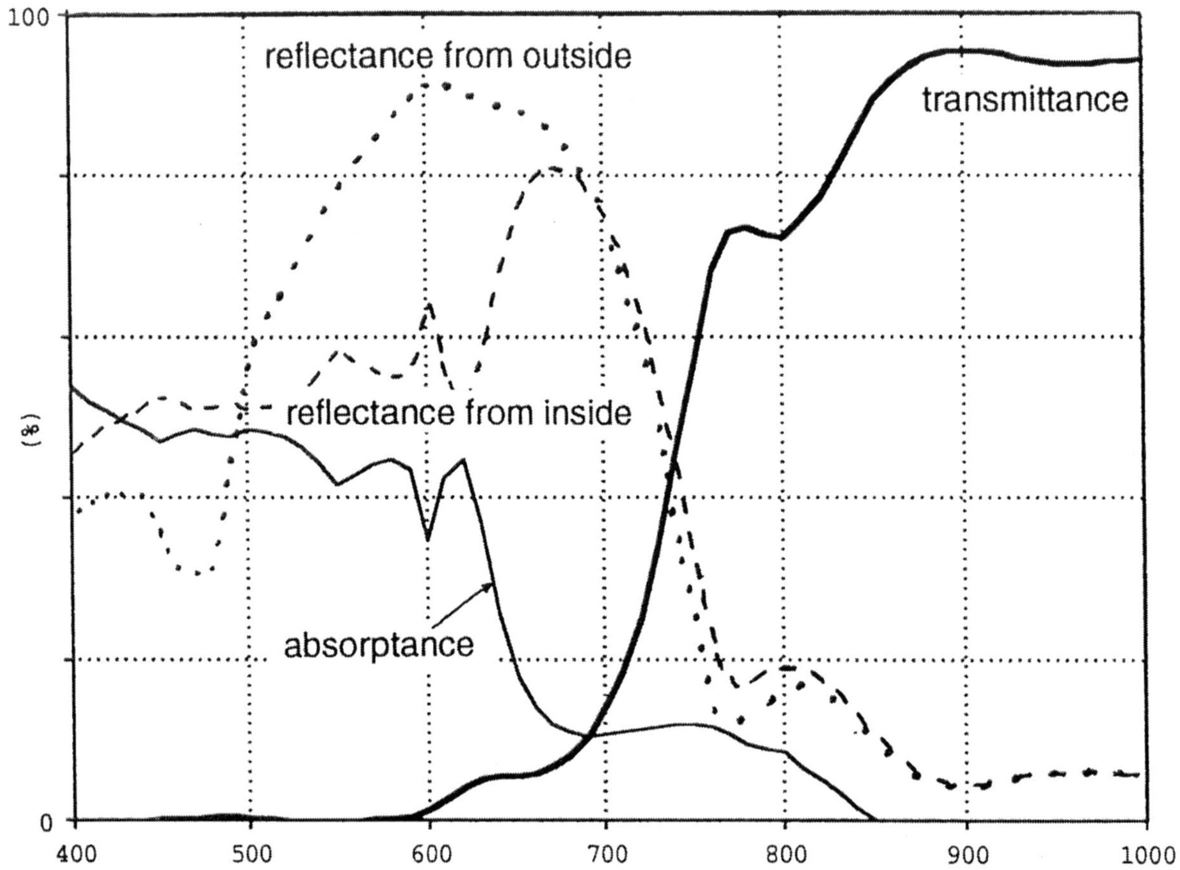

references:
1) G. Hebbinghaus, G. Frank , Contrib. light source 6 symposium Budapest 92

PREPARATION OF CERIUM OXIDE THIN FILMS AND THEIR OPTICAL PROPERTIES

Hiroshi Maiwa and Noboru Ichinose*
Shonan Institute of Technology, 1-1-25 Tsujido-Nishikaigan, Fujisawa, Kanagawa, 251
*Waseda University, 3-4-1 Ohkubo Shinjuku-ku, Tokyo, 169

1. Introduction

Cerium oxide (CeO_2) with a fluorite structure exhibits inertness toward moisture, compared to the sequioxides of the lanthanide auch as lanthanum oxide (La_2O_3). This material absorbs ultraviolet (UV) light. As a result, CeO_2 thin films are considered favorably for coating for shielding UV light. In this paper, we report the preparation of CeO_2 thin films by RF magnetron sputtering and sol-gel spin-on methods. The crystal structure and optical properties of the films have been studied. The relation between substrate temperature, film thickness and visible or ultraviolet spectrum is discussed.

2. Experimental

Cerium oxide thin films were prepared on fused quartz substrate by RF magnetron sputtering and sol-gel spin on methods. The sputtering target was a 3-inch disk of CeO_2 ceramic. In case of sol-gel spin-on methods, three weight percent solution was used for coating. Thin film crystallography was determined by X-ray diffraction (monochromated CuK_α radiation). Surface morphology was observed by a field-emission scanning electron microscope (FE-SEM). The optical transmittance of the films was measured at room temperature in the wavelength range from 190nm to 910nm with a dual beam spectrophotometer.

3. Results and Discussion

3.1 Sputter-deposited films

X-ray diffraction patterns of the suptter-deposited CeO_2 thin films at various substrate temperature are shown in Fig.1. Crystallized CeO_2 could be obtained without substrate heating. By heating the substrate , the diffraction intensity from the (111) peak increased relative to the other peaks. Lattice constant of the CeO_2 thin films deposited at room temperature, 420℃ and 580℃ was 0.556nm, 0.551nm and 0.547nm, respectively. Compared to the bulk value of 0.541nm, their lattice constant is larger, suggesting that the prepared film possesses defects. The film deposited at room temperature consists of fine grains of about 40~50nm.

Figure 2 shows the optical transmittance of 630nm-thick CeO_2 thin films deposited at various temperature. The film deposited at room temperature absorbed light in the wavelength close to the absorption edge, compared to the other films. This is probably due to the poorer crystalline of the film deposited at room temperature. The optical transmittance of various thick CeO_2 thin films deposited at room temperature is shown in Fig.3. Even 20nm-thick films absorb around

70% UV light. With increasing film thickness, the absorption of the UV light increased, which suggests that the absorption of the UV light occurred in the film itself. On the other hand, considering oscillation, absorption in the visible region is not dependent on the film thickness. This suggests that the absorption in the visible region did not occur at the film itself. It is estimated that absorption occurred at surface due to the roughness. Oscillation is considered to be not desirable for practical application. It is considered to be able to delete the oscillation by the formation of multi-layer thin films.

3.2 Sol-gel deposited films

X-ray diffraction revealed that sol-gel deposited CeO_2 films had lower crystallinity, compared to the sputter-deposited films. Transparent film obtained in the films annealed higher than 450℃, as shown in Fig.4. The films annealed at 300℃ were yellowish. Thickness dependence was almost same tendency with the sputtered films.

4. Conclusion

Crystallized and transparent CeO_2 films could be obtained on sputter-deposited films without substrate heating. The prepared films absorbed UV light. Considering application to the shielding UV light, sputtered-deposited CeO_2 is suitable because substrate temperature during deposition could be maintained to be lower.

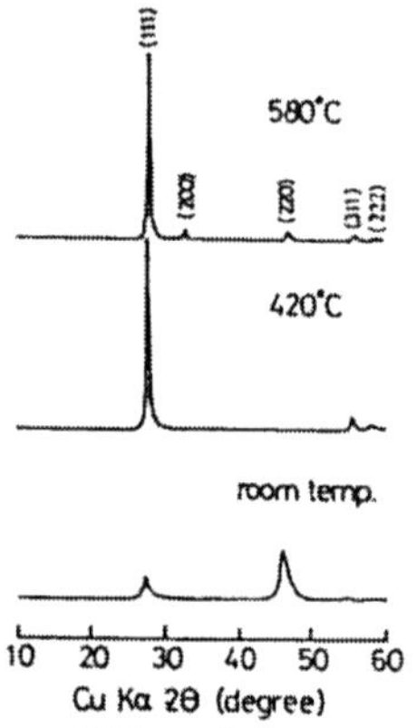

Fig.1 X-ray diffraction patterns of CeO_2 thin films.

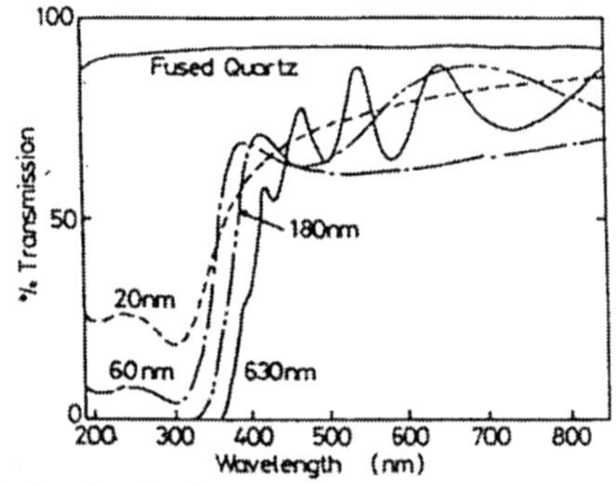

Fig.3 Optical transmittance of sputter-deposited films on fused quartz substrate.

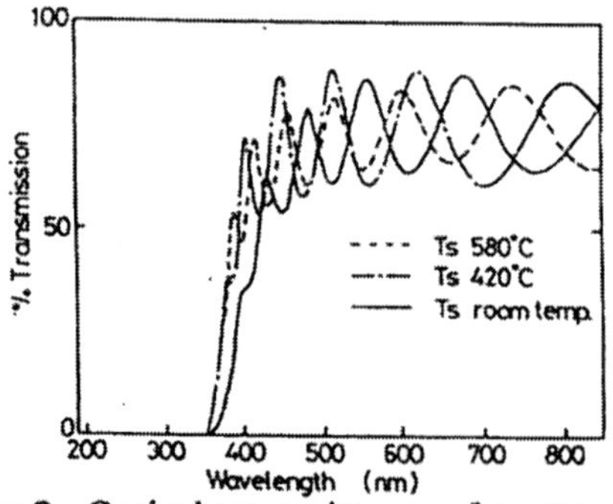

Fig.2 Optical transmittance of sputter-deposited 630nm-thick films on fused quartz substrate.

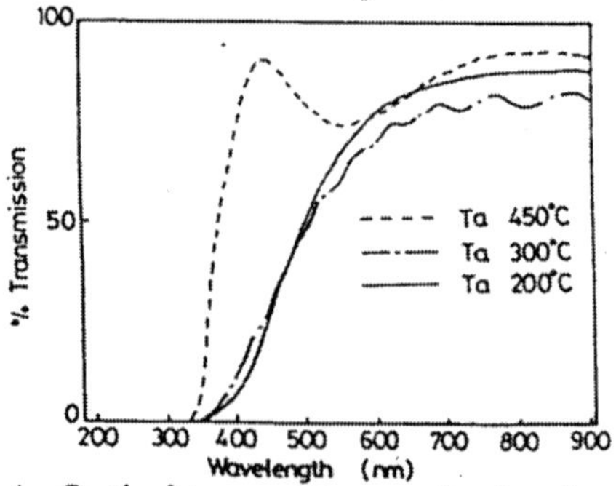

Fig.4 Optical transmittance of sol-gel deposited 200nm-thick films on fused quartz substrate.

IMPROVED MERCURY SOURCE AND GETTER SUITABLE FOR FLUORESCENT LAMPS MANUFACTURING

S. P. GIORGI, A. SCHIABEL, C. BOFFITO
SAES GETTERS S.p.A., Via Gallarate 215, I-20151 Milan, ITALY

INTRODUCTION

The combination of the Ti_3Hg [1] intermetallic compound with the $ZrAl_3$ [2] getter alloy, commercially available under the trade name of GEMEDIS [3,4], has been extensively used for over fifteen years as a safe source of mercury and as a getter in the fluorescent lamp industry. However, in some cases, for example circular fluorescent lamps, some steps of the manufacturing process are very severe (high temperatures in an oxidising atmosphere); this introduces some difficulties to fully exploit the GEMEDIS characteristics. These process steps lead to an important reduction of the actual yield of mercury which cannot be fully compensated by more drastic activation conditions. Such conditions are in any case not fully compatible with practical production requirements. Some reduction in mercury yield is also observed in other types of lamps where a certain oxidation of GEMEDIS can occur even if the conditions are less severe than in circular lamps. Moreover, since the recent trends in regulations concerning environmental protection address the problem of minimising the actual quantity of mercury present in a lamp, a new investigation to overcome these limitations becomes important. Studies have been carried out to tackle the above mentioned problems and to develop a mercury dispenser having an improved yield combined with good gettering activity. This gettering activity is required to take care of the residual impurities (particularly H_2) which can affect the correct ignition conditions and operation of a lamp. The improved mercury yield reduces excess mercury in the lamp and, therefore, makes this type of mercury dispenser more friendly towards the environment.

RESULTS

The investigations were first focused on understanding of the effect of the oxidising conditions on GEMEDIS and then on various materials which could be added to the Ti_3Hg intermetallic compound as promoters of the mercury releasing reaction. The yield characteristics were investigated for a common configuration of GEMEDIS, e.g. the strip configuration [4], where the getter material is deposited on one side and the mercury containing compound with the promoter is deposited on the other side. The yield characteristics have been studied both on untreated strips and on strips submitted to treatments simulating drastic manufacturing process conditions. In particular the samples were heated in oxidising conditions (air) at from 400 °C up to 600 °C. The residual gettering activity (focused on H_2) was also measured after the various treatments. The characteristics of the standard GEMEDIS were also studied in the same conditions as reference. The results concerning yield are shown in Figures 1 and 2. Figure 1 shows the mercury yield for the newly developed High Yield GEMEDIS (HYG) and for the standard GEMEDIS as a function of time at the activation temperature of 900 °C; the curves refer to samples submitted to heating in air at 400°C for 45 seconds before activation and to untreated samples. Figure 2 shows the mercury yield of the HYG and standard GEMEDIS as a function of activation temperatures, for an activation time of 30 seconds, after heating in air at 400 °C for 45 seconds. As can be seen the HYG exhibits a high yield, over 90%, also after the heat treatment in air and even at short activation times, whereas the standard GEMEDIS not only exhibits a lower yield without heat treatment but it is heavily affected by this treatment. It is also possible to see that the yield for HYG is less affected by the activation temperature compared to the standard GEMEDIS. A yield curve very close to the upper curve of Fig 1. was also obtained, with suitable promoters, after heat treatment in air at 600°C for about 30 seconds. The results obtained concerning the residual gettering activity show that it is still good in the test conditions with activation temperatures above 750 °C for more than 10 seconds.

Figure 1 % Hg yield as a function of Time of heating at 900°C

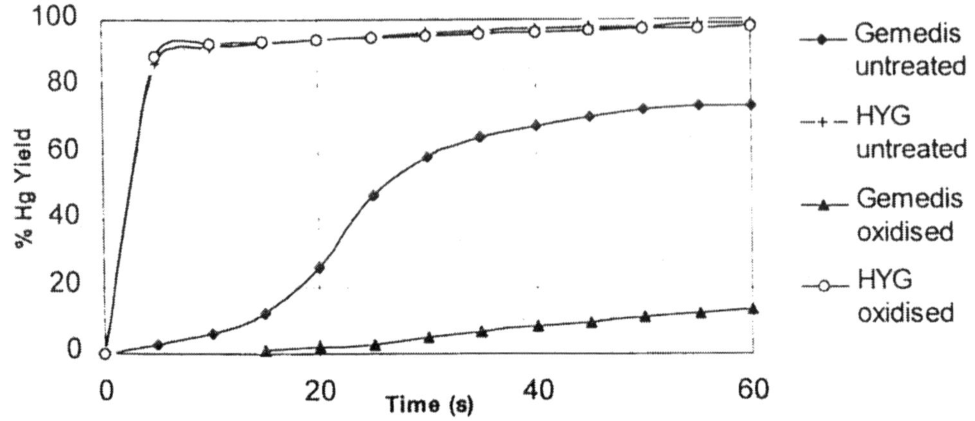

Figure 2 % Hg yield as a function of Temperature for 30 seconds heating time

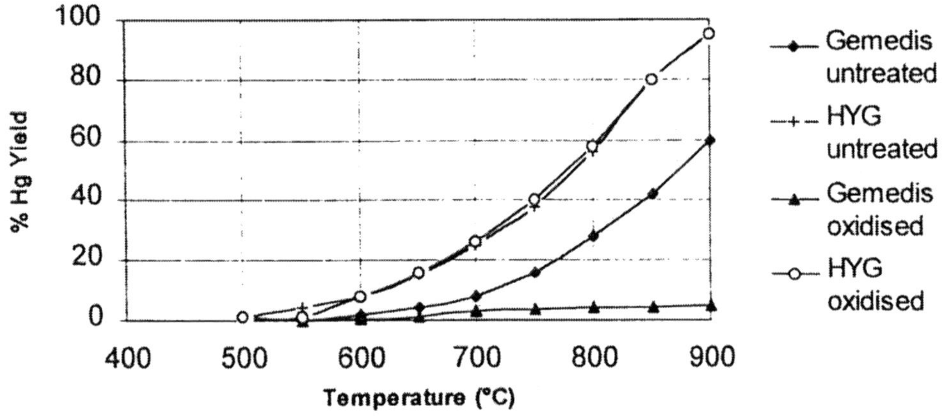

CONCLUSIONS

The effect of heavy oxidation on GEMEDIS has been studied and the related problem of Hg yield reduction has been overcome by developing suitable promoters to be combined with the Ti_3Hg intermetallic compound. The improved GEMEDIS (HYG) with higher mercury yield can meet the new trends in environmental regulations towards the reduction of the mercury content in lamps.

REFERENCES

(1) A. Barosi, E. Rabusin, M. Rebaudo: Pro. of the 1970 IEEE Conf. onElectron Device Techniques, p. 212-214

(2) A. Barosi, E. Rabusin: Japan J. Appl. Phys. Suppl. 2, Pt.1 (1974) p. 49-52

(3) P. Della Porta, E. Rabusin: Residual gases in electron tubes, Florence 1971, Academic Press (1972) p. 123-133

(4) GEMEDIS is a Registered Trade Mark of SAES GETTERS

A STUDY OF MERCURY CONSUMPTION IN FLUORESCENT LAMPS.

H.Tomioka, T.Higashi and K.Iwama
Light R&D Center, Matsushita Electronics Corp.
1-1, Saiwai-Cho, Takatsuki, Osaka 569, Japan

Recently, global environment pollutions have been regarded as a serious human problem in every field. Concerning light sources, reduction of mercury contained in fluorescent lamps is raised in the risk reduction plan by the chemicals group of OECD(Organization of Economic Co-operation and Development). Also in Japan, JELMA(Japan Electric Lamp Manufacturer's Association) has been working to reduce the quantity of mercury-dose in fluorescent lamps, aiming at achieving mercury content of no greater than 10mg per lamp by 1996.

Directing our attention to mercury required for actuating lamps, we studied mercury consumption in fluorescent lamps through their life by determining free-mercury in lamps with a non-destructive method based on cataphoresis phenomenon [1), 2)], of which results are reported below.

Figure 1 shows a schematic diagram of the measurement equipment. 40-W fluorescent lamps in which specific quantities of mercury-dose were operated with DC source,and relationships between the mercury-dose and time(t), which is a time lapse until light intensity at the anode side of a lamp has dropped to 70% of its original value, were studied. Where **t** is the average value of $\mathbf{t}_1$ and $\mathbf{t}_2$ measured in the following way;light intensity at the anode side of a lamp is once droped to70% due to cataphoresis; then, the same lamp is allowed to undergo this sequence two more times with its anode and cathode reversed each time, where time $\mathbf{t}_1$ and $\mathbf{t}_2$ are measured, respectively.

In **Fig.1**, the cooling part provided on the cathode side is to prevent back-diffusion of mercury. Specifications of fluorescent lamps used in our experiment are as follows.

- Types of lamps : FL40W ••• Halo-phosphor (Tc=4,200K), 3-band phosphor (Tc=5,000K)
- Quantities of mercury : 1,2,4 and 8mg (by using glass capsule type)
- Protective film : with and without Alumina protective film.

The mercury working curve obtained from the measurement results is shown in **Fig.2.** Using this curve, free mercury content versus burning time was measured in the above described lamps to determine mercury consumption during life. The results are given in **Table 1** and **Fig.3**.

Regarding mercury consumption of fluorescent lamps through their life, the following was concluded.

1) Mercury consumption of 3-band phosphor is largely reduced by proper use of alumina protective film.
2) With halo-phosphor, reduction of mercury consumption by the use of alumina protective film is insignificant.
3) In the absence of alumina protective film, mercury consumption of 3-band phosphor during life is as large as that of halo-phosphor.
4) Initial quantity of mercury-dose has little influence on mercury consumption during life.

In the experiment using the measurement equipment shown in **Fig.1**, significant emitter spattering from the anode was observed. We are discussing measurements with an external heating cuircuit of the anode side.

References

1) P.W.C.Verheers, Fifth International Symposium on the Science & Technology of Light Souces, p327(1988)
2) H.C.G.Verhaar and B.J.Roelevick, Sixth International Symposium on the Science & Technology of Light Sources, p71(1992)

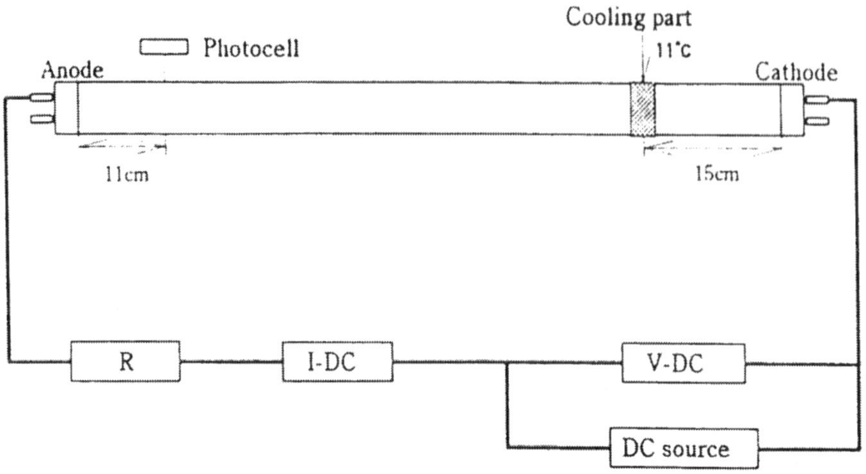

Fig.1 Equipment for Cataphoresis-analysis

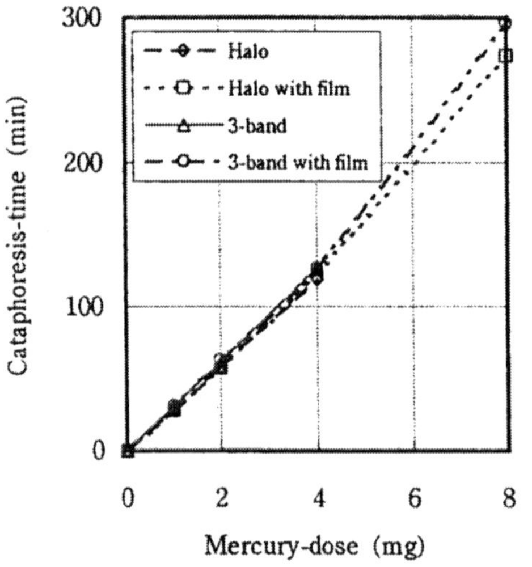

Fig.2 Working curve

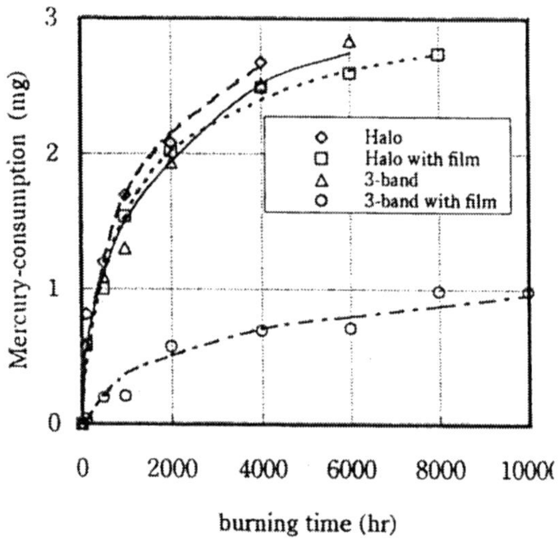

Fig.3 Mercury-consumption during life

Table.1 Mercury-consumption of 40W fluorescent lamps during life

phosphor	protective film	Mercury-consumption (mg)							
		100h	500h	1000h	2000h	4000h	6000h	8000h	10000h
Halo	with	0.58	1.00	1.54	2.02	2.49	2.66	2.74	
3-band		0.03	0.19	0.20	0.57	0.69	0.71	0.98	0.98
Halo	without	0.81	1.20	1.70	2.08	2.68			
3-band		0.59	1.09	1.30	1.94	2.52	2.84		

REUSE OF FLUORESCENT LAMP PHOSPHORS

H.C.G. Verhaar

Philips Lighting B.V., Eindhoven, The Netherlands

In the development of fluorescent lamps, increasingly environmental aspects have to be taken into account. Next to a high luminous efficacy that is required in the user-phase, a reduction of potentially hazardous materials is sought for in the design of new and improvements of present products.
Significant attention of authorities in W-Europe, however is presently concentrating on imposing legislation that stimulates environmentally friendly solutions, i.e. recycling, in the end of life phase of fluorescent lamps.

Objective of this work was to upgrade and reuse /80 phosphors from shrinkage of TL-production as a first step in finding a solution for this recycling problem. Todays modern and highly mechanized mass volume manufacturing of fluorescent lamps always aims to realize a high yield, but inevitably a small amount shrinkage is often generated.
It will be shown that an upgrading process has been developed that allows reuse of the phospors from production reject TL lamps. First the lamps are disassembled by end-cut technology, after which the phosphors are separated with air-pushing. The phosphor upgrading process starts off with the removal of the mainly metallic mercury by distillation. Next a separation process is used to remove the halophosphate phosphors from a mixture with /80 phosphors originating from /841'D' double layer lamps. The halophosphate phosphors with general formula $Ca_5(PO_4)_3(Cl,F):Sb^{3+},Mn^{2+}$ make up for the first layer of about 60 wt.% in these double layer lamps, whereas the second 'triphosphor' layer is made up of a mixture of the blue emitting $BaMgAl_{10}O_{17}:Eu^{2+}$ (BAM), the green-emitting $(Ce,Gd)MgB_5O_{10}:Tb^{3+}$ (CBT), and the red-emitting $Y_2O_3:Eu^{3+}$ (YOX). Various chemical and physical separation techniques can be used. Here a wet-chemical separation technique has been applied, in which the halophosphate phosphors are dissolved either in diluted hydrochloric or in nitric acid. The remaining triphosphor mixture was subsequently rinsed with water and dried, which completes the upgrading process. Advantage of the presently applied washing and dissolution procedure is that none of the phosphors in the triphosphor mixture are dissolved. In previous work a subsequent laborious process is described in which the Europium and Yttrium are regained after dissolution of the red-

emitting YOX followed by precipitation as rare-earth oxalates. The latter can be used in the manufacture of new phosphors [1,2]. The breakthrough in the newly developed process is that the triphosphor mixture from production rejects can directly be reused in new lamps, without having to go through the dissolution process and the required complicated refining of the oxalate precipitate.

The pure /841 mixture, consisting of the phosphors BAM; CBT and YOX, is tested on its reusability in new single layer T8 36W/841 lamps of the New Generation type. For this purpose a small batch of phosphors was made into a new coating suspension, that was used to hand-coat the single layer /841 lamps. Lamp results showed that the light output at 0 and 100 hours of the reference lamp made with the virgin /841 phosphor mixture, and the lamp made with the upgraded phosphor mixture are identical, and the 1000h maintenance is nearly the same. It can thus be concluded that a /841 tri-phosphor mixture from TL production shrinkage can be reused in the manufacture of new /841 lamps. The process implies that the lamps are disassembled by the end-cut/air-push method; followed by mercury distillation and a wet-chemical separation method to remove the halophosphate phosphors.

The subsequent reuse of the glass; the recycling of the metal caps, and the refinement and reuse of the mercury allow virtual complete recycling for production reject T8/800 lamps.
The next step in the process, the reuse of spent phosphors, will close the total recycling chain for fluorescent lamps.

[1] V. Scherer, J. Eger, E. Neukunft, M. Vogel, EU Patent 0 157 249, Osram Gesellschaft (1984)

[2] A. Dutta, L. Dullea, US Patent 5,160,663, GTE Sylvania (1991)

ENVIRONMENTAL ASPECTS OF DISCHARGE LAMPS

H. P. Stormberg
Philips Lighting, Central Development Lighting
PO BOX 80020, 5600 JM Eindhoven, The Netherlands

Introduction

In the early seventies mankind started to realise that our earth needed environmental care and protection against uncontrolled pollution of air, water and soil. It became obvious, that the advantages of fluorescent and high intensity discharge (HID) lamps over incandescent lamps are the lower power consumption per lumen and the resulting lower pollution generated by power plants. The disadvantage of these lamps, however, is that they contain mercury and to a lesser extent radioactive and other toxic materials. When the environmental aspects of discharge lamps are compared to those of incandescent lamps, a number of important factors have to be taken into consideration. This can be done by a life cycle assessment covering the complete 'cradle to grave' product life cycle, determining for each phase what the possible environmental effects are. This product life cycle includes the following three phases:

- The extraction of raw materials, the production of components and the lamp manufacturing.
- The lamp operation.
- Disposal of the used lamp at end of life, eventually re-use or recycling of materials.

In the first section of this paper, the environmental aspects of discharge lamps during the total life cycle are analysed. In the second section, an overview of the legislation and trends in the USA, Europe and Japan is given and in the third section the industrial initiatives to reduce the environmental pollution are summarized.

1. Environmental aspects during life cycle

The lamp life cycle includes the production, operation and disposal of the lamp. The environmental impacts are shown schematically in Fig.1.

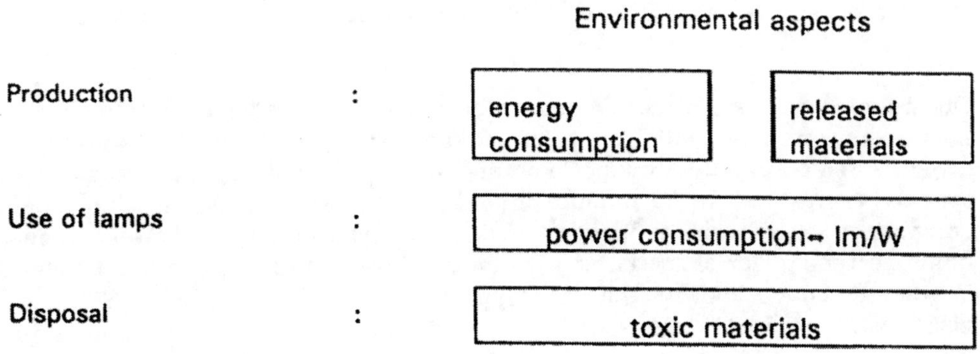

Fig.1: The environmental aspects of discharge lamps in each phase of the life cycle.

The inputs of a lamp factory are energy and raw materials. These two components are converted into lamps, but as a by-product waste materials are generated. Four different flows come out of the factory; one flow of the produced lamps and three different emission flows. This process is depicted in Fig.2. In this analysis of the environmental impact of fluorescent lamps the extraction of raw materials is not considered. The energy used during production will be considered below with the energy needed for operation.

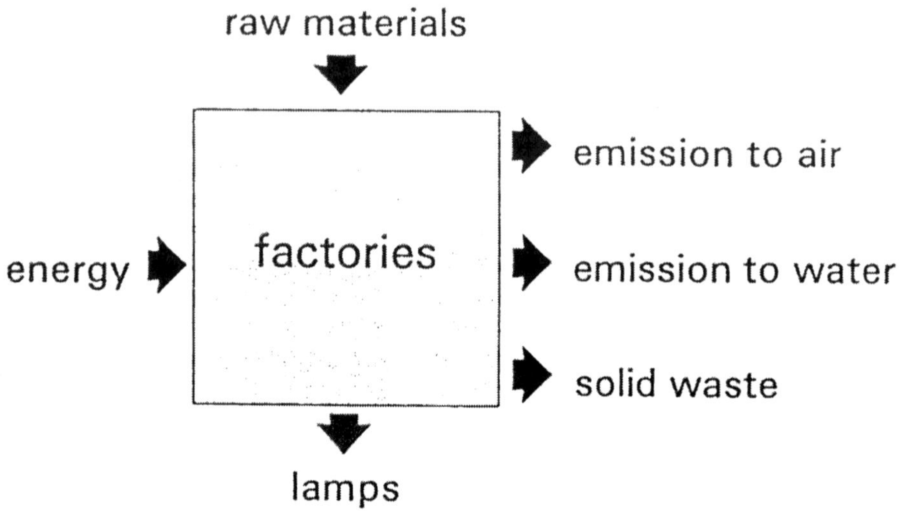

Fig. 2: Material flows at lamp manufacturing

The emission to air can e.g. be caused by organic solvent for the phosphor coating. Water pollution results from cooling and chemical processes in the factory. Last but not least, solid waste originates from the disposal of waste materials and production rejects. Authorities reduce the limits on air, water and soil pollution by legislation and permits.

Mass volume manufacturing always aims to realize a high yield but inevitably a small amount of shrinkage is often generated. Some fraction can be disposed in municipal landfills but mercury containing waste has to be treated as hazardous. This can be vastly reduced by recycling of production rejects which has become possible for fluorescent lamps [1].

The production of a typical fluorescent lamp requires less than 2 kWh electricity. A 36-W fluorescent lamp with a median life-time of 13,000 hours, has a total power consumption of 570 kWh during lamp operation on a choke ballast. It can now be seen that the energy needed for production and disposal of a fluorescent lamp is negligible in comparison to the energy needed during its operation. It is therefore necessary to compare the impact of the latter with those of used materials and lamp manufacturing. Power production impacts the environment by generating e.g. NO_x, SO_2 and CO_2 but also toxic materials like Hg can be released. There are five major sources for electric power: coal, gas, oil, nuclear, hydro. Nuclear power has captured an increasing share of power generation but fossil-fuels are still the dominant source of electricity. The fossil-fuel fraction containing mercury may even increase because new nuclear power plant construction is currently at a standstill in US and some European countries.

H. Muis et al. [2] determined the emission of heavy metals, NO_x and SO_2 per *lumen-hour* for a 36 W fluorescent lamp in the Netherlands, resulting from its energy consumption. The results are compared to those of a 60 W incandescent lamp and shown in table A.

Material (ng/lm-hour)	Fluorescent 36 W			Incandescent 60 W		
	lamp	energy	total	lamp	energy	total
Hg	0.33	0.64	0.97	0	4.45	4.45
Pb	0	19	19	730	136	866
Cu	1326	26	1352	891	185	1076
NO_x	119	21700	21800	1000	152000	153000
SO_2	0	16300	16300	0	114000	114000
radioactivity (Bq/lm-hour)	0	251	251	0	1755	1755

Table A Environmental aspects of a fluorescent lamp compared to those of an incandescent lamp. The data of Muis [2] from 1989 have been adjusted to the present lamp properties.

From table A, it can be concluded that more pollution is put into the environment by the production of the energy needed to operate a fluorescent lamp than by the lamp itself. Furthermore, it can be concluded that less mercury is put into the environment per lumen-hour of a fluorescent lamp than per lumen-hour of an incandescent lamp. The same holds for the other emissions like Pb and NO_x. Only mercury is considered in the remainder of this section, because mercury is the most toxic element put into the environment during the life cycle of discharge lamps and it is furthermore necessary for efficient light generation.

R. Clear and S. Berman [3] determined the average mercury output for electricity production in the US and M. Willemsen [4] did the same for the electricity production in Europe. With these data the total mercury burden of lamps could be calculated. The results for six different lamp types in both the US and Europe are shown in table B. The difference in the luminous efficacy between the incandescent lamp in the US and Europe is due to the lower mains voltage in the US. The higher mercury output in Europe results from the higher mercury content in European coal.

Lamp type (USA)	Hg-dose (mg/lamp)	LO (lm-rated)	Life-time (hrs)	Hg (ng/lm-hr)		
				lamp	energy	total
Fluorescent 32 W	30	2850	20000	0.53	0.59	1.12
CFL 18 W	9	900	10000	0.75	0.95	1.70
Mercury vapour 400 W	75	17500	24000	0.18	1.42	1.60
Metal Halide 400 W	59	40000	20000	0.07	0.62	0.69
HPS 400 W	23	50000	24000	0.02	0.51	0.53
Incandescent 60 W	0	870	1000	0	3.75	3.75
Lamp type (Europe)	**Hg (mg/lamp)**	**LO (lm)**	**Life-time (hrs)**	**Hg (ng/lm-hr)**		
				lamp	**energy**	**total**
Fluorescent 36 W	15	3350	13000	0.33	0.64	0.97
CFL 18 W	5	900	10000	0.56	1.09	1.65
Mercury vapour 400 W	77	20000	15000	0.26	1.23	1.49
Metal Halide 400 W	40	32000	20000	0.06	0.77	0.83
HPS 400 W	20	55000	24000	0.02	0.45	0.47
Incandescent 60 W	0	730	1000	0	4.45	4.45

Table B Mercury released into the environment for different lamps in the United States and in Europe. For US lamps the average mercury dose for that lamp type in the US is given, while for European lamps the most common dose is given. Ballast losses are included.

2. Legislation and trends in the US, Europe and Japan.

Legislation on lamps containing toxic elements differs very much from one country to another. In the US and some European countries, there are strict federal and state laws regulating the disposal of hazardous waste, while in other countries no regulation exists at all.

Radioactivity is a subject of legislation in many countries. The maximum amount of radioactivity which a discharge lamp may contain is limited in most countries, in the US lamps containing Kr^{85} have to be labelled as radioactivity containing products. There are also strict regulations concerning the manufacturing of lamps containing radioactivity.

The difference between legislations on mercury containing discharge lamps starts with the definition of toxic elements. The US government decided that mercury, lead and barium are the most toxic elements used in lamps. Other countries put also antimony on the list of toxic elements. However, they all agree that mercury is the most toxic element on the list, clearly illustrated by the toxicity weight factors as used in the US; 5 for mercury, 0.2 for lead and 0.01 for barium. The legislation on the disposal of lamps containing toxic elements differs also. In many countries lamps are treated as common waste, while in most European countries all discharge lamps containing mercury have to be disposed as hazardous waste. The criteria for the classification 'hazardous waste' are based on the concentration of the toxic substances as e.g. mercury and lead. In several European countries like Austria, Belgium, Germany, Italy, The Netherlands, Switzerland and the Scandinavian countries a lamp collection and recycling infrastructure is operational and disposal is no longer allowed. In the US the mercury that can leach into the ground is measured and only if the level is above a certain threshold has the waste to be treated as hazardous. In the US and Japan, however, small quantities of discharge lamps, like those from households, can be disposed along with the normal trash, while large quantities go to chemical dumps. In the US the lower limit for a generator of hazardous waste is 100 kg per month. That means that the disposal of more than 350 T12-fluorescent lamps per month has to be considered as hazardous waste. But the states with mercury problems in their lakes lowered this limit. The chemical dumps are especially equipped to make sure that no mercury or lead reaches the groundwater. This special equipment and the scarcity of appropriate dumps made disposal quite expensive. The disposal of 1 ton hazardous waste costs 1000 to 2000 USD, while the disposal of common waste only costs 25 to 125 USD. The costs in Europe and the US are roughly comparable. At present a new federal law on fluorescent lamp disposal is under discussion in the US.

On top of the already existing laws, the EPACT-law will be enacted in the US by October 1995. This law regulates the efficiency of energy consuming products. For the lighting industry this means that only lamps with a luminous efficacy above a certain threshold can be sold in the US. The height of this threshold differs for the various lamp types and is determined on basis of the best lamps on the market. The threshold for a 4 ft fluorescent lamp e.g. is 75 lumen per watt. However, it does not require the more efficient, but also more expensive tricolour phosphors.

The political wish to stop the climate change results in recommendations to minimize the emission of green house gases by energy saving. As a consequence the replacement of incandescent lamps by Compact Fluorescent Lamps (CFL) offers an attractive option. Many authorities recommend the use of CFL's and some utility companies in the US and some European countries sometimes subsidise it by attractive payment conditions for the consumer. The total market for CFL's, both integrated and non-integrated, should exceed 200 million lamps in 1995 and will continue to be a strongly expanding market. Another energy saving trend in lighting is the replacement of incandescent and halogen lamps with compact HID lamps. The compact HID lamps are well suited for accent lighting as used in shops. In the first section of this paper it was made clear that energy saving has a large positive environmental impact. The main driving force, however, for replacing a lamp by a more efficient one is still primarily the fact that it saves money and not that it is good for the environment.

Because of the high costs for the disposal of complete lamps as hazardous waste, disassembly of fluorescent lamps has become attractive. When a lamp is disassembled only part of it has to be disposed as hazardous waste, while the other part can be disposed as common waste. For disassembly three different processes are available:

- The lamps are crushed and with a heat treatment the mercury is separated. Then glass and some metal parts can be separated and recycled, the remainder is disposed as hazardous waste.

- The lamp ends are cut off to separate the metal from the glass and phosphors. Then the remainder is crushed. The glass can be re-used for both lighting and non-lighting products, if the mercury is separated.
- As above, but now the phosphors are blown out before crushing. In this manner the glass and the phosphors can be completely separated and both can be re-used or recycled.

The total costs for the collection, transport and disposal of fluorescent lamps lie within the range of 0.8 to 1.0 USD per lamp. In some US-states, recycling of fluorescent lamps is already cheaper than disposal. Therefore, many companies are starting with the recycling of those lamps. In the US more than 20 recycling companies have a total capacity of 100 million lamps (20 % of the total market). In Europe the situation differs widely. The countries in North-West Europe are leading and have the most advanced disassembly technologies. The percentage of recycled lamps is already high in some countries, in leading countries as Germany already more than 70 % of the fluorescent lamps is collected.

3. Industrial initiatives to reduce environmental pollution

3.1 Reduction of the energy consumption

The energy consumption accounts for the largest part of the environmental pollution by discharge lamps. Competition between lighting companies leads to further improvement of the luminous efficacy of all types of lamps. The halogen incandescent lamp with IR-filter and the metal halide lamps show this trend best. The IR halogen incandescent lamp has a luminous efficacy of nearly 30 lm/watt instead of the 13 lm/watt of the 'normal' incandescent lamp and the efficacy of the high pressure mercury lamp improved from 40 lm/watt to about 100 lm/watt by adding metal halides. An overview of the improvement of the luminous efficacy can be found in the 'ALITE 95 Workshop' [5,6,7].

A major breakthrough concerning energy consumption was obtained by the development of integrated compact fluorescent lamps which have a system luminous efficacy between 50 lm/W and 70 lm/W. These lamps were launched in 1980 and have been reduced in size after the development of electronic power supplies. As these lamps can replace incandescent lamps they offer a high potential for energy saving.

Often only the luminous efficacy is discussed, but the luminous maintenance is equally important. The required light level has to be supplied until the end of the service life-time of lamps. This is of major importance because the flux depreciation is strongest at beginning of lamp-life.

The driving force for luminous efficacy improvement is not environmental care, but simply competition. This guarantees that a continuous effort will be made to improve the luminous efficacy even more. Aside from that, an increasing number of authorities will stimulate the replacement of incandescent lamps with compact fluorescent lamps either by legislation or by subsidies.

3.2. Reduction of toxic and non-toxic materials

Often, the amount of material is reduced to make the lamp smaller such that it fits more applications. In other applications like the T8 with colour /800, the reduction in the amount of phosphor is used to lower the cost price. However, reduction of toxic materials used in discharge lamps is mainly done for the environment. Research on the reduction of toxic materials in discharge lamps focuses on the elements lead, radioactive Kr^{85} or Th and mercury.

In the past all stems for discharge lamps were made of lead containing glass. Now newly developed lead free glass can replace the old lead containing glass in most of the lamps. Although no concessions are made to the quality of the lamp, the environmental burden is diminished in lead mines, glass plants and finally on rubbish dumps.

Radioactive Kr^{85} or Th is necessary to ease the ignition of glow switches and high pressure discharge lamps. This radioactivity in the lamp is far below the threshold considered to be dangerous, but it remains a problem during manufacturing and is a point of discussion for ecology groups. Therefore, research projects have been conducted to eliminate or reduce the radioactive elements in lamps. First results suggest that it will be possible to develop such lamps without radioactivity in the long term. Recently, the UV-enhancer has been introduced as a possible solution for this ignition problem [10]. The UV-enhancer is a small additional discharge placed near the discharge envelope. The UV produced by this additional discharge generates electrons at the tungsten electrode and the ignition behaviour of the main discharge is improved.

The reason for using mercury differs significantly for high and low pressure discharge lamps. In HID lamps mercury functions to obtain the appropriate electrical characteristics, while mercury in low pressure fluorescent lamps is essential for the light generation. Therefore, two different solutions were needed to reduce the mercury contents in both lamp types. For high pressure sodium lamps such a solution already exists and the first mercury free high pressure sodium lamp will shortly be introduced. For fluorescent lamps, however, no non-toxic substitute exists with sufficiently high efficiency for UV generation. Research on mercury free dielectric barrier discharges has been done [8]. A compact fluorescent lamp based on this concept resulted in a luminous efficacy of 37 lm/W [9]. But until now the efficient generation of low energetic UV, with wavelengths above 250 nm, causes unresolvable problems. Therefore, research focused on the reduction instead of the replacement of mercury.

A typical 4 ft T8 fluorescent lamp needs only 50 μg mercury in the discharge for optimal performance. However, mercury losses to the glass, the phosphors and the electrodes make it necessary to put much more mercury in the lamp. Therefore, the main objective of the mercury reduction research is to limit the mercury consumption. Some major reductions of the mercury dose have been accomplished in the past and an even further reduction should be possible in the future. In the early days of fluorescent lamps more than 100 mg mercury was dosed in the lamps. At that time dosing of smaller amounts was impossible. A reduction of the mercury dose became possible by new dosing technologies. A further reduction needed measures to reduce the mercury consumption, as new phosphors and a protective layer between glass and phosphor layer. The reduced mercury consumption has also a positive influence on the maintenance of fluorescent lamps. The new generation Philips TLD lamps with a 3 mg mercury dose launched onto the market in 1994 have a nearly perfect maintenance of more than 90% at end of life. Therefore, a lower lamp power can be used in new lighting systems, if light output during the whole life-time has to be guaranteed. One of these new lighting systems will be TL5. This system combines a considerable reduction in total energy saving with a significant reduction of materials.

3.3 Recycling of discharge lamps

Strict legislation on hazardous waste in some countries resulted in new waste management. An infrastructure for collection of discharge lamps is set up and grows with an increasing speed. Most companies focus their activities on fluorescent lamps, because they account for over 90% of the discharge lamps. Now some of the glass is already recycled and re-used in new fluorescent lamps, but in the future more materials from discharge lamps such as the mercury and the phosphors will be recycled. Separation and reutilization processes are known for the metal caps and the mercury and will be applied. The recycling of phosphors is more difficult. Processes to regain the Europium from spent phosphors are known [11, 12], but only recently it was shown that re-use of phosphors from production rejects belongs to the possibilities [1]. The next step in the process, the re-use of spent phosphors, will close the total recycling chain for fluorescent lamps (see Fig.3).

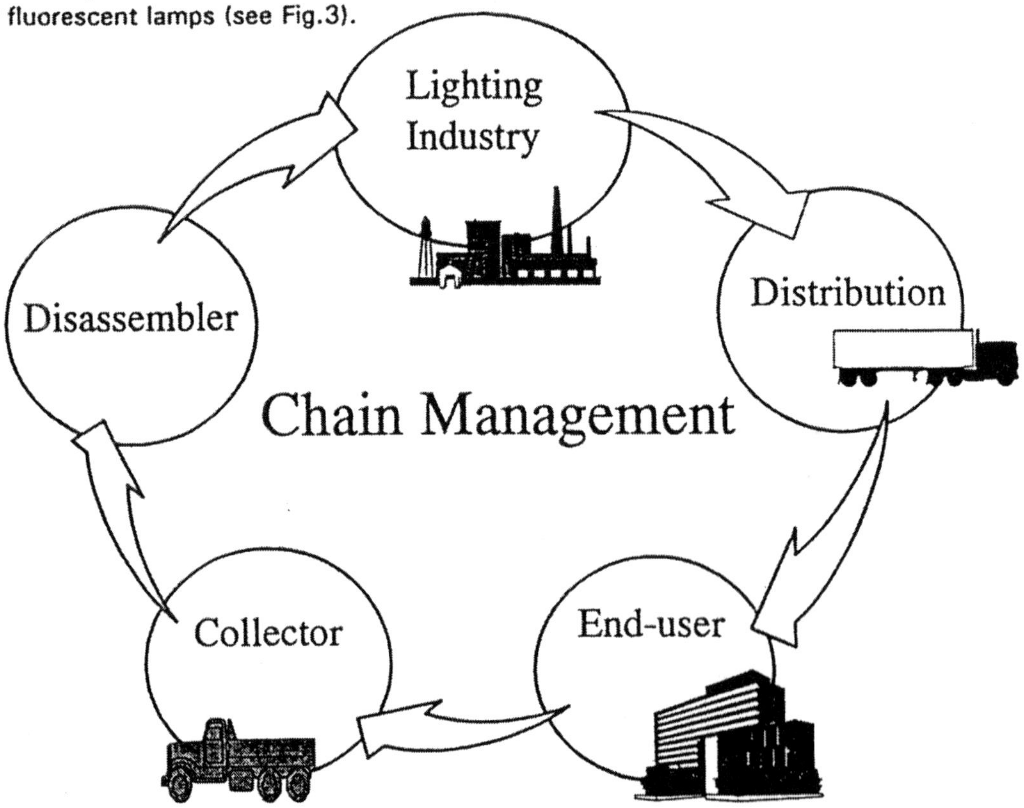

Fig.3: Closing the life-cycle chain in order to use the resources on earth as efficiently as possible.

Unfortunately, the positive prospect of total recycling is slightly spoiled. Even countries, with strict legislation succeed in collecting only approximately 70% of all spent lamps. Of course this percentage will increase if the collection infrastructure improves, but a large part of the lamps will remain disposed on common garbage dumps since extra costs are involved for the customer.

4. Conclusions

- The pollution caused by production of power needed for lamp operation has a larger impact on the environment than the toxic materials used in discharge lamps. Therefore, improvement of the luminous efficacy and the maintenance still benefit the environment more than material reduction.
- Legislation on toxic materials used in lighting products differs significantly from one country to another.
- The lighting industry has reduced the toxic materials in its products. This did not result in a price raise because the average customer is not willing to pay extra for green products.
- Of the materials used in fluorescent lamps, mercury is most hazardous to the environment. Therefore, the lighting industry is continuously decreasing the mercury content. Best on the market is the new generation Philips /800 TLD that contains only 3 mg mercury.
- The recycling of fluorescent lamps is a large step towards the ultimate goal to use the resources on earth as efficiently as possible.

5. References

[1] H. Verhaar: Proc. 7th International Symp. on the Science & Technology of Light Sources, Kyoto, Japan, 1995
[2] H. Muis, A. Posthumus, A. Slob, S. v.d. Sluis: Dutch report about energy and environmental aspects of light sources, Communicatie en Adviesbureau, Rotterdam, 1989
[3] R. Clear and S. Berman: J. of Illum. Eng. Soc., 138-156 (1994)
[4] M. Willemsen: unpublished
[5] J.F. Waymouth, Alite 95 Workshop, Warrenton (USA), 1995 (organised by EPRI, Palo Alto, CA, USA)
[6] J. van Vliet, Alite 95 Workshop, Warrenton (USA), 1995 (organised by EPRI, Palo Alto, CA, USA)
[7] R. Bergman, Alite 95 Workshop, Warrenton (USA), 1995 (organised by EPRI, Palo Alto, CA, USA)
[8] M. Neiger: Dielectric barrier discharges: an unusual light source, 6th International Symp. on the Science & Technology of Light Sources, Budapest, Hungary 1992
[9] F. Vollkommer and L. Hitzschke: patent application PCT, DE94/00380, 1993
[10] G. Zaslavsky, N. Barakitis and W.M.Keeffe: United States Patent 4,818,915, 1989
[11] V. Scherer, J. Eger, E. Neukunft, M. Vogel: EU patent 0 157 249, 1984
[12] A. Dutta, L. Dullea: US patent 5,160,663, 1991

Author Index